Piezoelectric Materials

Piezoelectric Materials

From Fundamentals to Emerging Applications

Volume 2

Edited by Jiagang Wu

WILEY-VCH

The Editor

Prof. Jiagang Wu
Sichuan University
No. 24 South Section 1
Yihuan Road
Chengdu
CH, 610065

Cover Image: © Yuichiro Chino/Getty Images

Library of Congress Card No.: applied for

British Library Cataloguing-in-Publication Data
A catalog record for this book is available from the British Library.

Bibliographic information published by the Deutsche Nationalbibliothek
The Deutsche Nationalbibliothek lists this publication in the Deutsche Nationalbibliografie; detailed bibliographic data are available on the Internet at <http://dnb.d-nb.de>.

Print ISBN: 978-3-527-35471-9
ePDF ISBN: 978-3-527-84121-9
ePub ISBN: 978-3-527-84122-6
oBook ISBN: 978-3-527-84123-3

Typesetting Straive, Chennai, India
Printing and Binding CPI Group (UK) Ltd, Croydon, CR0 4YY

C9783527354719_020524

Contents

Volume 1

Preface

Piezoelectricity, a phenomenon wherein a material generates electrical charges on its surfaces in response to applied mechanical stress or, conversely, yields mechanical strain when subjected to an electric field, serves as a critical bridge between mechanical and electrical energies. This unique functionality has spurred the widespread adoption of piezoelectric materials across a broad spectrum of modern technologies. Examples of these technologies abound and pervade our daily lives. In the realm of consumer electronics, piezoelectric sensors and actuators are integral components of smartphones, enabling precision touch-screen interactions. In the medical field, piezoelectric probes are indispensable in ultrasonic diagnostic equipment, facilitating noninvasive imaging and diagnostic procedures. Additionally, high-precision positioning units in scientific research equipment, ignitors for lighters, and pressure sensors, all leverage the properties of piezoelectric materials to function effectively.

Piezoelectric materials can be categorized based on the form of matter, with classifications including piezoelectric ceramics, single crystals, thin films, and organic materials. Each category boasts specific characteristics, making them ideal for particular applications. Moreover, in alignment with escalating global concerns for environmental protection and sustainable development, piezoelectric materials are often classified as lead-containing or lead-free. This classification underscores the ongoing pursuit of environmentally friendly materials that can deliver the advantages of piezoelectricity without compromising sustainability. The evolving research and development in this field promise exciting advancements and wider adoption of piezoelectric materials in future technologies.

Over the past two decades, the field of piezoelectric materials and associated devices has undergone swift and significant advancements. Both phase-field simulations and neutron scattering experiments have substantiated the exceptional influence of polar nanoregions, which have been instrumental in enhancing the piezoelectric properties of lead-based ceramics and single crystals. Innovative approaches, including new phase boundary engineering and compositional design, have equipped lead-free piezoelectric materials with superior piezoelectricity, which is comparable to some of their lead-free counterparts. The adoption of a multi-scale structural design strategy has consistently shattered previous records in dielectric energy storage, signaling the potential for further advancements in this area. Concurrently, the evolution of microfabrication technology has facilitated the emergence of piezoelectric materials' enticing transduction capabilities in a variety of cutting-edge electronic devices. These include biomedical devices, piezoelectric nanogenerators, and piezoelectric flexible devices, which embody the intersection of technological progress and practical application. Additionally, the utilization of advanced in situ characterization techniques, including transmission electron microscopy, neutron scattering, and high-energy synchrotron radiation, has empowered the researchers to unravel the complex physical mechanisms underlying the observed high performance. These advancements collectively signal a bright future for the continued development and application of piezoelectric materials.

Motivated by the aforementioned advancements, I contend that the present moment provides an opportune juncture for reviewing the research landscape of piezoelectric materials and devices. Such a book proves imperative, not only for early career researchers seeking an informed entry point into the field but also for established scholars aiming to conceptualize and implement novel strategies for enhancing the performance of piezoelectric materials and devices. Although numerous high-quality reviews and books have been previously published, the majority concentrate on specific topics within individual material systems. To obtain a holistic understanding of the field of piezoelectric materials and devices, it is essential to encompass both the fundamental knowledge of piezoelectricity and the state-of-the-art research progress in piezoelectric materials and devices. Bearing this objective in mind, I have undertaken the compilation of this book. My aspiration is that this work will serve as a comprehensive resource, illuminating the path for future research and development in the realm of piezoelectric materials and devices.

This book comprises 20 chapters that delve into various facets of piezoelectricity, a dynamic and evolving field. Chapter 1 provides a solid foundation in the basic principles of piezoelectricity, delineating piezoelectric coefficients, describing piezoelectricity via third-rank tensor, and discussing symmetry considerations. This opening chapter serves as an invaluable resource for readers unfamiliar with piezoelectricity or those seeking to strengthen their foundational understanding. Chapter 2 explores the probing of piezoelectric effects using X-ray diffraction and the observation of the piezoelectric effect in multidomain ferroelectrics. Chapter 3 offers a succinct summary of the evolution and research advancements in lead-based piezoelectric materials, with emphasis on historical progress and modern-day applications. Chapters 4 and 5 deliver a comprehensive overview of lead-free piezoelectric materials, concentrating on four significant material systems: $(K, Na)NbO_3$, $BaTiO_3$, $(Bi, Na)TiO_3$, and $BiFeO_3$. Chapter 6 presents an overview of piezoelectricity in molecular ferroelectrics, focusing on piezoelectricity in different molecular ferroelectrics and piezoresponse force microscopy (PFM)-aided characterization and design. Chapters 7 and 8 review the development and research of $BiScO_3$–$PbTiO_3$ and $Bi_4Ti_3O_{12}$-based high-temperature piezoelectric ceramics, covering aspects like characteristics, fabrication, domain engineering, electrical properties, endurance, and high-temperature applications. Chapter 9 delves into the newly developed texturing method, concentrating on textured piezoceramics, texture–property relationships, and device applications of textured piezoceramics. Chapter 10 reviews the strides made in piezoelectric thin films, focusing particularly on the growth and piezoelectric coefficient measurements of these thin films. Chapter 11 reviews the development of piezoelectric and electro-optic properties in lead-free single crystals, with a focus on (K, Na)NbO_3 and K(Ta, Nb)O_3-based single crystals. Chapter 12 summarizes the electrostrictive effect, touching on its fundamentals and history, reviewing the electrostrictive effect and strain in various ferroelectrics, and providing examples of their applications. Chapters 13 and 14 review ferroelectric materials for dielectric energy storage, including the fundamentals, design, and performance of inorganic, organic, and composite dielectric materials, as well as their applications. Chapter 15 reviews the advances of the electrocaloric effect, concentrating on its occurrence in ceramic films, bulk ceramics, single crystals, and multi-layer ceramics (MLCs), and the designs and demonstrations of ceramics-based cooling devices. Chapter 16 reviews piezo-bioelectronic materials and emerging applications, including materials, fabrication, and emerging biomedical applications of piezo-bioelectronic devices. Chapter 17 reviews the advances of porous piezoelectric materials for energy technologies, concentrating on the types, fabrication techniques, modeling and characterization, and applications of porous piezoelectric materials. Chapter 18 reviews piezoelectric flexible devices, including their working mechanism, materials, structure design, and applications. Chapter 19 reviews the advances of ferroelectrics under compression, including lead-based and lead-free ferroelectrics and their applications. In Chapter 20, the advances of piezoelectricity in biomedical applications are reviewed, concentrating on several representative applications including implantable energy harvesters, ultrasound imaging transducers, bone/tissue regeneration, neural stimulation, and cancer therapy. Each chapter concludes with a section on the future perspective and concluding remarks.

Finally, it is my sincere hope that this book will serve as a valuable resource for those interested in the fascinating world of piezoelectric materials and devices. The broad array of topics covered, from fundamental concepts to cutting-edge research, aims to provide readers with a comprehensive understanding of the field. This book is a testament to the collective efforts of numerous researchers, scientists, and scholars whose contributions have significantly advanced our understanding of piezoelectric materials and their multifaceted applications.

In closing, I would like to emphasize my deep appreciation to all those who have contributed to this endeavor. The work presented herein represents the synthesis of many minds and the culmination of considerable effort. I trust that this book will provide readers with a compelling and enlightening exploration of the world of piezoelectrics.

February 26, 2024

Jiagang Wu
Sichuan University

12

Electrostrictive Effect

Chunlin Zhao[1], Yanli Huang[2], and Jiagang Wu[3]

[1] *Fuzhou University, College of Materials Science and Engineering, No. 2 Xue Yuan Road, University Town, Fuzhou 350108, China*
[2] *Fujian Normal University, College of Photonic and Electronic Engineering, No. 8 Xuefu South Road, Minhou, Fuzhou 350117, China*
[3] *Sichuan University, Department of Materials Science, No. 24 South Section 1, Yihuan Road, Chengdu 610065, China*

12.1 Introduction (Fundamentals and History)

Precision positioning technology, especially micron and nano-positioning technology, is one of the key technologies for modern science and advanced industrial engineering, which has been widely used in advanced fields such as precision machining, nano biomedicine and aerospace. The micro-displacement actuators with high resolution and high positioning accuracy are one of the core components of precision positioning technology. The general micro-displacement actuators usually employ traditional technologies such as mechanical transmission, hydraulic transmission, and magnetostriction. Although micro-displacement can be realized, shortcomings such as a long transmission chain, complex structure, easy heating, and low accuracy are inevitable for traditional micro-displacement actuators. Piezoelectric micro-displacement actuators are widely used because of their high displacement resolution, large load-bearing capacity, small size, and thermal effect [1]. However, the electric field-induced ferroelectric domain switching, domain wall motion, and capacitive loading will lead to inherent strain hysteresis and creep characteristics in piezoelectric materials, which bring great challenges to precise positioning systems. As shown in Figure 12.1a,b, the electrostrictive effect offers unique advantages over the piezoelectric effect, including low strain hysteresis, no remnant strain, weak aging effect, no poling process, temperature stability, faster response time, and a smaller thermal effect. The electrostrictive effect (electrostrictive strain) originates from the displacement of atoms or ions away from the equilibrium position caused by the induced polarization under the external electric field, resulting in remnant polarization and lattice distortion (strain) [4].

For typical piezoelectric materials, such as BTS–0.18BCT piezoceramic with high piezoelectricity, although high electric-induced strain (electrostrain) of >0.1% can be

Piezoelectric Materials: From Fundamentals to Emerging Applications, First Edition. Edited by Jiagang Wu.

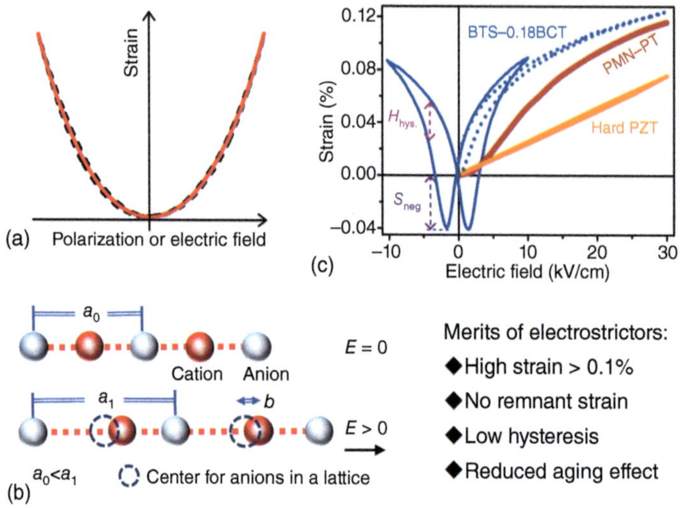

Figure 12.1 (a) Relationship between strain and polarization/electric field in electrostrictive materials. (b) Atomic-scale schematics showing the electrostrictive effect and the advantages of electrostrictive strain. (c) Electric field-induced strain (electrostrain) curves of representative materials with dominant piezoelectric effect (BTS-0.18BCT) and dominant electrostrictive effect (hard PZT and PMN-PT) [2, 3]. Source: (a, b) Adapted from [4]. (c) Adapted from [2, 3].

easily obtained under low electric fields, the strain versus electric field (S–E) curve presents large strain hysteresis (H_{hys}) and high negative strain (S_{neg}) [2]. Although hard lead zirconate titanate (hard PZT) ceramics show a linear S–E curve, their strain value is relatively low [3], as shown in Figure 12.1c, which restricts their application in micro-displacement actuators. In 1980, the electrostrictive strain with a relatively high strain value and no hysteresis was found in $Pb(Mg_{2/3}Nb_{1/3})O_3$ (PMN) ceramic [5, 6], which attracted lots of attention from researchers and industries. Subsequently, researchers have developed a series of electrostrictive materials with high performance, such as PZN, PMN-PT, PZN-PT, Ba, or La-doped PZT ceramics or single crystals [6–9], and carried out a series of studies on chemical modification, performance optimization, mechanism exploration, and new micro-displacement actuators of these electrostrictive materials.

12.2 Characterization of Electrostrictive Coefficient

The electrostrictive effect is fundamental in all dielectric materials. For general piezoelectric materials, the electric field-induced strain (electrostrain) is shown in the following equation [10, 11]:

$$S_{ij} = d_{ijk}E_k + M_{ijkl}E_kE_l + \dots \tag{12.1}$$

$$S_{ij} = g_{ijk}P_k + Q_{ijkl}P_kP_l + \dots \tag{12.2}$$

where d_{ijk} and g_{ijk} are piezoelectric coefficients, P_k and P_l are electric field-induced polarization, and Q_{ijkl} and M_{ijkl} are the electrostrictive coefficients. The first term demonstrates the electrostrain generated by the converse piezoelectric effect. The second term demonstrates the electrostrain generated by the electrostrictive effect, which belongs to the second-order effect. The electrostrictive strain is usually much smaller than the strain induced by the converse piezoelectric effect. However, for piezoelectric and ferroelectric materials with large dielectric constants and electrostrictive coefficients, the electrostrictive strain is observable.

The electrostrictive coefficients Q and M can be determined by measuring the electrostrain versus polarization or electric field curves (S–E or S–P curves), the dielectric constant versus stress curves, the piezoelectric effect of ferroelectrics, lattice parameters, and the dielectric constant under DC bias fields, etc. According to Eqs. (12.1) and (12.2), measuring the S–P and S–E curves is the most direct and accurate method to determine the value of electrostrictive coefficient Q. Figure 12.2a displays the S–P curve of piezoelectric materials with obvious hysteresis behavior induced by domain switching. Under a low driving electric field, this S–P curve shows a large hysteresis due to the orientation of the ferroelectric domain and the movement of the domain wall, so a large deviation occurs between the measuring data and fitting curve, and the obtained Q value is not accurate. Therefore, ferroelectric domain switching and domain wall movement (as well as electric field-induced phase transition) should be avoided when determining the electrostrictive coefficient, as shown in Figure 12.2b. Figure 12.2c,d demonstrate the effect of domain switching on determining the longitudinal electrostrictive coefficient (Q_{33}). If the 180° domain switching occurs, its contribution to strain and polarization is $2d_{33}E$ and $2P_s$, respectively. In general, $2d_{33}E$ is much smaller than $4P_s^2Q_{33}$. Therefore, the obtained Q_{33} value will be smaller when 180° domain switching happens.

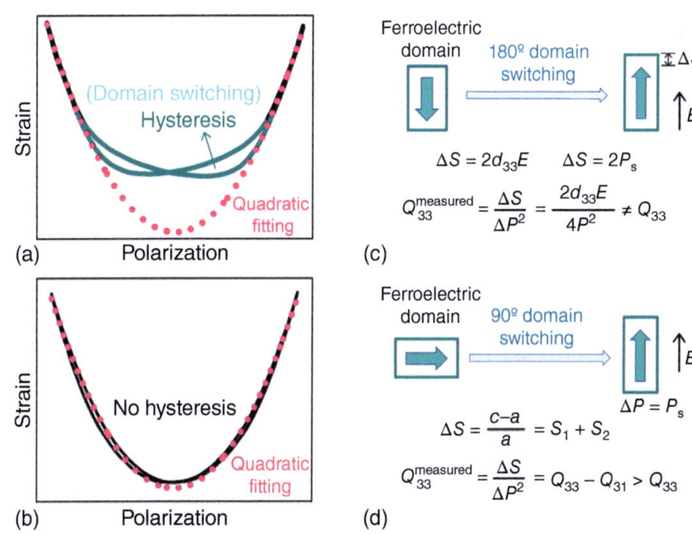

Figure 12.2 (a) S–P curve of piezoelectric materials with obvious hysteresis behavior induced by domain switching. (b) S–P curve of piezoelectric materials with negligible domain switching behavior. Effect of (c) 180° domain switching and (d) 90° domain switching on measuring electrostrictive coefficient Q_{33}. Source: Adapted from [4].

In contrast, 90° domain switching would cause the obtained Q_{33} to be higher due to polarization changes along the vertical direction. In addition, the change in spontaneous polarization direction during phase transition can also affect the measurements [4, 12]. Therefore, to eliminate the influence of domain switching and (electric field-induced) phase transition and obtain an accurate Q value, the measurement of S–P curves should be carried out at high temperature (higher than Curie temperature at paraelectric phase) and high electric field (where the domain motion is limited).

12.3 Electrostrictive Coefficient and Strain in Piezoelectric Materials

Figure 12.3a summarizes the average Q_{33} values of lead-based and lead-free electrostrictive ceramics. Lead-based ceramics, such as PMN-PT [5, 6] and PZT [7, 8], have a moderate Q_{33} of 0.02–0.03 m^4/C^2, which is similar to obtaining Q_{33} in lead-free ST-based [13] and BFO-based ceramics [14, 15], while inferior to lead-free NN- [16–18], BNT- [19–27, 42], KNN- [28–33], and BT-based ceramics [34–40]. The highest Q_{33} in these lead-free systems can reach 0.04–0.07 m^4/C^2. Theoretically, the electrostrictive effect is temperature-independent. As displayed in Figure 12.3b, the Q_{33} of lead-based and lead-free ceramics almost shows similar values with temperature increasing. According to the origin of the electrostrictive effect, the electrostrictive coefficient is related to the characteristic of positive ions in piezoelectric materials. In the centrosymmetric ABO_3-type perovskite ferroelectrics, the equilibrium positions of A- and B-site cations are located at the centers of the O_{12} and O_6 polyhedra, respectively. Therefore, the contribution of cations to the polarizability can be estimated from the average Born effective charge

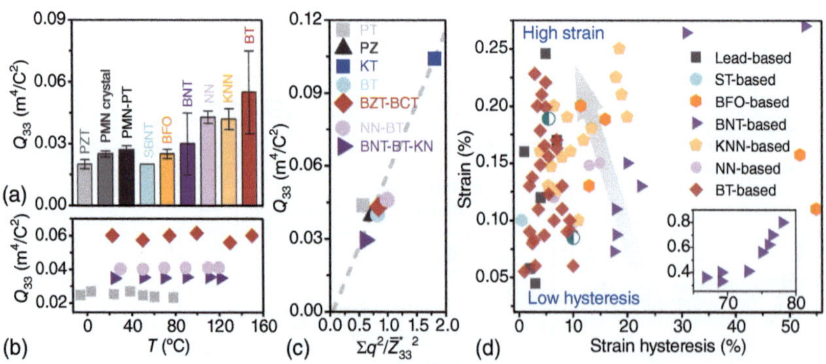

Figure 12.3 (a) Comparison of Q_{33} between lead-based and lead-free electrostrictive ceramics. Source: Adapted from [5–8, 12–40]. (b) Comparison of temperature-dependent Q_{33} for some typical electrostrictive ceramics. Source: Adapted from [8, 18, 27, 35]. (c) Q_{33} and $\sum q^2/\overline{Z}^{*}_{33}{}^2$ of some typical perovskite systems. Source: Adapted from [12, 17, 41]. (d) Comparison of strain and hysteresis in different perovskite electrostrictive ceramics. Source: Adapted from [5–8, 12–40, 42, 43].

$\overline{Z}_{33}^{*}$ [41]. $\overline{Z}_{33}^{*}$ is the derivative of polarization concerning the ion shift and describes the spontaneous polarization in ferroelectrics. When an electric field E_3 is applied, the average S_{33} and P_3 per unit volume can be calculated from the Boltzmann distribution [17]:

$$S_{33} = \left(\frac{3gq^2}{4f^3 r_0}\right) \cdot E_3^2 \tag{12.3}$$

$$P\left(\frac{A}{B}\right) = \overline{Z}_{33}^{*}\left(\frac{A}{B}\right) \cdot D\left(\frac{A}{B}\right) \tag{12.4}$$

where r_0, q, and D are the equilibrium position, the cationic charge, and the relative displacement away from the center of the oxygen octahedron, respectively. The coefficients g and f are related to the crystal structure and properties of the cation, respectively. Thus, Q_{33} is strongly correlated with $\sum q^2/\overline{Z}_{33}^{*2}$ when the differences in lattice parameters and ordering degree between different systems are ignored. Figure 12.3c compares the Q_{33} value of typical perovskite ferroelectrics with corresponding $\sum q^2/\overline{Z}_{33}^{*2}$ [12, 17, 41]. As $\sum q^2/\overline{Z}_{33}^{*2}$ value increases, Q_{33} also increases, indicating that the electrostrictive coefficient is not sensitive to the ferroelectric, antiferroelectric, or paraelectric phases of materials but is sensitive to the cationic properties. This is because the stronger the ionic bond between the cation and oxygen, the larger the value of $\sum q^2/\overline{Z}_{33}^{*2}$, and the larger Q_{33} can be obtained. For example, for the NN-based system, the large Q_{33} is attributed to the synergy effect of the strong A—O ion bond and the large polarization response P_3 caused by the displacement of B-site cation, and then the NN has a large electrostrictive strain [17].

From the perspective of the application, it should not only pay attention to the electrostrictive coefficient but also the electrostrictive strain with low hysteresis under a low electric field and good temperature stability. Figure 12.3d shows the values of electric field-induced strain (electrostrain) and strain hysteresis for different perovskite-type piezoelectric ceramics. The high strain value of ~0.3–0.8% can be obtained in BNT-based ceramics, accompanied by a large hysteresis of ~50–80% [19–27, 42–44]. Other lead-free KNN-, NN-, BFO-, and BT-based ceramics have a similar strain value of ~0.25% when under a low driving electric field, showing small strain hysteresis [14–18, 28–40, 43, 44]. According to the strain hysteresis, it can be divided into three categories: Class 1, hysteresis is between 0% and 10%, including lead-based, BT-based, partly KNN-based, and NN-based ceramics; Class 2, hysteresis is between 10% and 20%, including NN-based, KNN-based, BFO-based, and some BNT-based ceramics; and Class 3, with hysteresis larger than 20%, mainly BNT-based ceramics. To obtain a pure electrostrictive strain of more than 0.1%, different driving electric fields are needed in different systems. The driving electric field of lead-based ceramics (such as PMN–0.1PT) is below 30 kV/cm [6], the driving electric field of BT-based ceramics is about 40–60 kV/cm, and the driving electric field of BNT, NN, and BFO is above 60 kV/cm. Therefore, among lead-free ceramics, the electrostrictive properties of BT-based ceramics are comparable to those of lead-based ceramics, with small strain hysteresis and a low driving electric field.

12.3.1 Optimization of Electrostrictive Strain

The electrostrictive strain (S) of piezoelectric materials can be expressed as $S = QP^2$, and the electric field-induced polarization (P) depends on the dielectric constant (ε), so the electrostrictive strain can be expressed as [4]:

$$S_{33}(E_0) = Q_{33}[P_3(E_0)]^2 = Q_{33}\left[\int_0^{E_0} \varepsilon_{33}(E) \cdot \mathrm{d}E\right]^2 \tag{12.5}$$

According to Eq. (12.5), to obtain a large electrostrictive strain, the ferroelectric material is required to have a large electrostrictive coefficient and a high dielectric constant. The electrostrictive effect originates from ion displacement, so it is generally necessary to regulate the electrostrictive coefficient by regulating the ion properties of the ABO_3 unit cells. At the same time, to obtain a temperature-stable electrostrictive strain, it is necessary that the electrostrictive coefficient and dielectric constant of the material also have good temperature stability. In addition, the electrostrain of piezoelectric materials mainly comes from the inverse piezoelectric effect, the electrostrictive effect, non-180° domain switching, and electric field-induced phase transition, while the strain hysteresis of piezoelectric ceramics is mainly due to non-180° domain switching and electric field-induced phase transition [43, 45]. To obtain electrostrain with low hysteresis, it is necessary to avoid non-180° domain switching and electric field-induced phase transitions. Typical piezoelectric ceramics exhibit electrostrain properties with large hysteresis and poor temperature stability due to their large domain size and the sharp change in dielectric constant at the phase transition.

To obtain low hysteresis and temperature-stable electrostrictive strain in piezoelectric ceramics, the general design strategy is to minimize the domain size of piezoelectric materials through chemical modification and nano-sized and even highly active polar nano-regions (PNRs) are obtained [42, 46]. In this process, the piezoelectric materials gradually show relaxor characteristics with a small change in dielectric constant near the phase transition, so that electrostrictive strains with good temperature stability can be obtained. Through ion doping and composition modification, large electrostrictive strain with low hysteresis and temperature stability can be obtained in piezoelectric materials by inducing relaxor characteristics and making materials that have a high electrostrictive coefficient and permittivity.

12.3.2 Lead-free Piezoelectric Ceramics

12.3.2.1 $BaTiO_3$-based Ceramics

$BaTiO_3$ (BT) ceramics are classical piezoelectric and ferroelectric materials with a high dielectric constant and polarization. At room temperature, BT has a ferroelectric tetragonal phase structure [47]. Due to non-180° ferroelectric domain switching and domain wall movement, its strain curve shows large hysteresis and negative strain, as shown in Figure 12.4a. To obtain a large electrostrain with low hysteresis under a low electric field, strategies include the generation of diffuse phase transition or relaxor characteristics induced by chemical modification (ion doping, oxide addition, or other components) and structure regulation (grain size, domain structure,

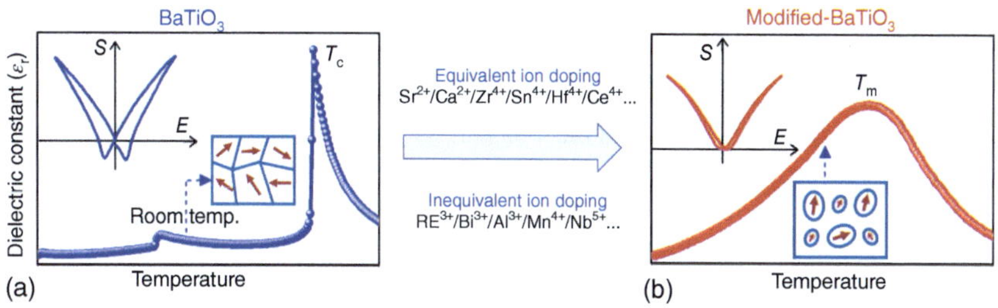

Figure 12.4 Schematics showing temperature dependence of dielectric constant ($\varepsilon_r - T$) curves of (a) pure BT and (b) BT-based relaxor ceramics via chemical modification. The insets show the corresponding strain curves and domain structure.

and non-uniform structure regulation, etc.). Ion doping is the most commonly used chemical modification method. For BT ceramics, A-site doping, B-site doping, or A/B-site co-doping can be carried out [34]. Among them, it can be divided into equivalent doping and aliovalent doping. Ion doping can cause lattice distortion, defect generation, or element enrichment, leading to an inhomogeneous local structure and local stress field, and local inhomogeneity can result in the nanoscale size of ferroelectric domains and even the generation of PNRs [48]. The amounts of doping ions should induce the transition from ferroelectric to relaxor states in BT-based ceramics, as displayed in Figure 12.4b.

The appearance of dielectric dispersion or dielectric relaxor mainly requires two factors: the appearance of random fields and the random interaction of PNRs above the temperature (T_m) at the dielectric maximum [49]. For the equivalent doped BT-based system, the appearance of random fields is related to the enrichment of doped elements at the nanoscale. The destruction of Ti—O bonds in the impurity element-rich region makes PNRs disappear or difficult to nucleate [50]. In addition, the lattice distortion around doping ions allows the charge and local charge centers to be redistributed due to the inconsistent ion radii with Ba^{2+} and Ti^{4+}. All the above results will produce local random fields. The random field produced by equivalent doping is weaker than that produced by aliovalent doping, so the doping content to induce the transition from ferroelectric to relaxor state caused by equivalent doping is higher (usually greater than 20% for BT ceramic) than aliovalent doping.

The commonly used equivalent dopant ions include Ca^{2+}, Sr^{2+}, Sn^{4+}, Hf^{4+}, Zr^{4+}, and Ce^{4+} [34]. For example, for Sn-doped BT ceramic, the dielectric maximum of the $\varepsilon_r - T$ curve is widened at $x = 0.11$, showing dielectric diffusion. The electrostrain with no hysteresis (0.08%, 40 kV/cm) is obtained at room temperature, and Q_{33} is 0.04 m^4/C^2 [51]. The atomic-resolution high-angle annular-dark-field images of $Ba(Ti_{0.8}Sn_{0.2})O_3$ ceramic show that in the region with high Sn content, the displacement of B-site atoms off the center is smaller. It indicates that Sn doping can increase the dielectric properties of PNRs formed in the low Sn content region [52]. After introducing 30% Sr^{2+} into the A site [53] or 10% Ce^{4+} into the B site [54], the ferroelectric-paraelectric phase transition can be shifted to near room temperature or becomes diffusion. Then the low-hysteresis electrostrictive strain of ~0.12–0.15% is achieved under 60–70 kV/cm. It can be seen that the necessary doping content of equivalent ions is usually high. For the aliovalent ion-doped BT-based ceramics, only a low doping content is required to induce a relaxor state. For example, 5% Bi^{3+} doping into A site of $Ba(Ti_{0.93}Sn_{0.07})O_3$ ceramic [55] and a small amount of aliovalent ion Fe^{3+} doped $Ba(Zr_{0.2}Ti_{0.8})O_3$–$x(Ba_{0.7}Ca_{0.3})TiO_3$ (BTZ–xBCT or BCTZ) ceramics [39], their T_C rapidly decreases to near room temperature, and the diffused phase transition occurs. Hysteresis-free electrostrictive strain (0.07–0.09%, 30 kV/cm) and high Q_{33} (0.04 m^4/C^2) are obtained at room temperature.

Bismuth ion (Bi^{3+}) is one of the representative aliovalent doping ions. The hybridization between the lone-pair electrons in the outermost 6p orbital of Bi^{3+} and the 2p orbital of oxygen ions can enhance the polarization strength, and the introduction of various aliovalent ions can enhance the relaxor characteristics in BT. In addition, bismuth-based compounds can reduce sintering temperature, inhibit grain growth, and enhance electrical breakdown strength. For example, the point defect (Bi^{3+} doping) is introduced into the composition near the quadruple convergent point of BTZ–xBCT ceramics, building coexistent glasses with diverse local symmetries near a quadruple point and rendering low energy barriers between different polar states. Then, a large electrostrain performance with low hysteresis (0.21%, 60 kV/cm) and high Q_{33} of ~0.06 m^4/C^2 is obtained [36]. Another representative work is (Li^+, Bi^{3+}) co-doping BT ceramic, in which the long-range ferroelectric ordering can be easily broken by co-doping ion pairs, and polarization nanoclusters and relaxor states are formed. It is found the existence of Li^+ and Bi^{3+} segregations at the nanometer scale and relatively large off-center displacements of Ti^{4+} ions around the Li^+/Bi^{3+}-rich regions. All the results contribute to the large electrostrictive coefficient of $Q_{33} \sim 0.07\ m^4/C^2$ and pure electrostriction without hysteresis behavior [40], as shown in Figure 12.5.

In addition to ion doping, the addition of another compound can also cause the appearance of diffusion or relaxor phase transitions in BT-based ceramics. The adding components are represented by bismuth-based compounds ($BiMO_3$, M is trivalent ions or a combination of multiple ions having a trivalent average valence state) and other perovskite piezoelectric systems. For $Bi(Li_{0.5}Nb_{0.5})O_3$ (BLN) modified BT ceramics (BT–xBLN), Q_{33} of ~0.035–0.045 m^4/C^2 is obtained. A large electrostrain (0.11%, 50 kV/cm) is obtained with relatively poor temperature stability in BT–0.04BLN ceramic because the dielectric constant fluctuates significantly with increasing temperature. The temperature stability of the dielectric constant and electrostrain of BT–0.08BLN ceramic becomes better, while the strain value becomes low (0.02%, 50 kV/cm) due to the low dielectric constant [56]. After adding another perovskite piezoelectric system (e.g. $KNbO_3$), the domain structure can be modulated by introducing defects. The introduction of K^+ and Nb^{5+} can form point defects in the nanoregion of BT ceramic, and induce the formation of acceptor-donor-type defect dipoles, which hinders the growth of ferroelectric domains, thus obtaining a multiphase coexistence of glass ferroelectric phase transition in a local region, so that ceramic can obtain large polarization and a large dielectric constant in a wide temperature range, which contributes to obtaining a large electrostrictive strain (~0.146%) together with ultralow (<3%) hysteresis at 80 kV/cm [57].

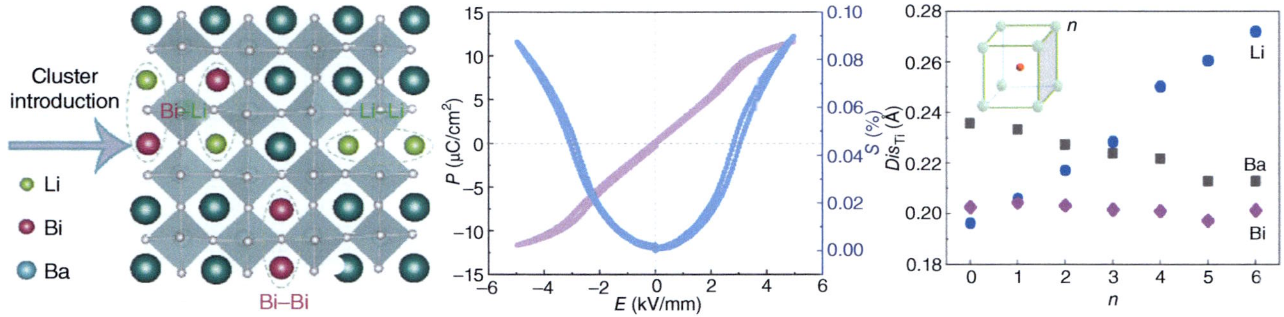

Figure 12.5 Schematic showing structure distortion via cluster introduction design in BT lattice, room-temperature P–E loop, and S–E curve, and the B-site displacement affected by surrounding A-site distribution for (Li^{+}, Bi^{3+}) co-doped ET ceramic. Source: Wang et al. [40]/John Wiley & Sons/CC BY 4.0.

Although pure electrostrictive strain can be obtained by adjusting the relaxor characteristics of BT-based ceramics, it is difficult to simultaneously obtain the relatively high strain and temperature stability. According to the empirical law of the electrostrictive effect, the electrostrictive coefficient Q depends on the order degree of B-site cations in perovskite materials [11]. When B-site cations are arranged in an orderly manner, neighboring ions will regularly fill the surrounding space and form a close packing, so that the activity space of B-site ions is limited, and the deformation per unit polarization is larger, so Q increases, as demonstrated in Figure 12.6a. Based on the above ideas, if the A-site of BT ceramic is doped by small-radius ions, the BT unit cells will be compressed, and the active space of Ti^{4+} is limited, the Q value of BT can be improved. As shown in Figure 12.6b, small-radius ions co-doping BT ceramic at A site [$Ba_{1-x}(Li_{0.5}Ho_{0.5})_xTiO_3$] are obtained by Li^+ and Ho^{3+} ion pairs. It is found that the Li^+–Ho^{3+} ion pairs optimally arrange along [001] direction, minimizing the ferroelectric domain size and inhibiting strain hysteresis. The ion pairs along the [001] direction "compress" the TiO_6 octahedron, so that when Ti^{4+} is shifted under the external electric field, it must cause observable distortion of TiO_6 octahedron, so that the deformation value per unit polarization increases, thus promoting the increase of the electrostrictive coefficient ($Q_{33} \sim 0.06\ m^4/C^2$). In addition, although the dielectric constant of the dielectric peak becomes smaller, the Curie temperature shifts to a higher temperature, and the dielectric peak is still sharp (weak diffusion behavior) after introducing Li^+–Ho^{3+} ion pairs into BT ceramics. Interestingly, the ferro-ferroelectric phase transition of BT ceramics becomes quite diffusion (Figure 12.6c). Finally, high electrostrain (0.12–0.20%, 60 kV/cm) with temperature stability (from room temperature to 150 °C) and low hysteresis (<10%) are obtained in this system, as shown in Figure 12.6d. The superior electrostrictive performance suggests that this BT-based system is a potential lead-free family for applications in the area of actuators [35].

12.3.2.2 ($Bi_{0.5}Na_{0.5}$)TiO_3-based Ceramics

($Bi_{0.5}Na_{0.5}$)TiO_3 (BNT)-based ceramics are typical relaxor ferroelectric materials. Their relaxor phases include the non-ergodic relaxor state and the ergodic relaxor state. The former will irreversibly transform into the long-range ferroelectric phase under an external electric field, while the latter's transformation into the ferroelectric phase is reversible [42, 45]. Pure BNT has two ferroelectric phases below the maximum dielectric temperature (T_m), tetragonal and rhombohedral phases. However, this tetragonal phase is a relaxor phase showing weak polarization, which results in the transition from rhombohedral to tetragonal phase dispersing over a wide temperature range. In this temperature region, the remnant polarization of poled BNT will decrease sharply at a certain temperature, which is called the depolarization temperature (T_d) [42], as shown in Figure 12.7a. Below T_d, the hysteresis loop and strain curve of BNT-based ceramics are similar to those of hard PZT ceramics. Near T_d, the transition from the relaxor phase to the ferroelectric phase occurs under the electric field, showing a double-like hysteresis loop similar to the antiferroelectrics, and a large strain response (0.4–0.8%) with large strain hysteresis (50–80%) is observed. Above T_d, the ferroelectric loop and strain curve

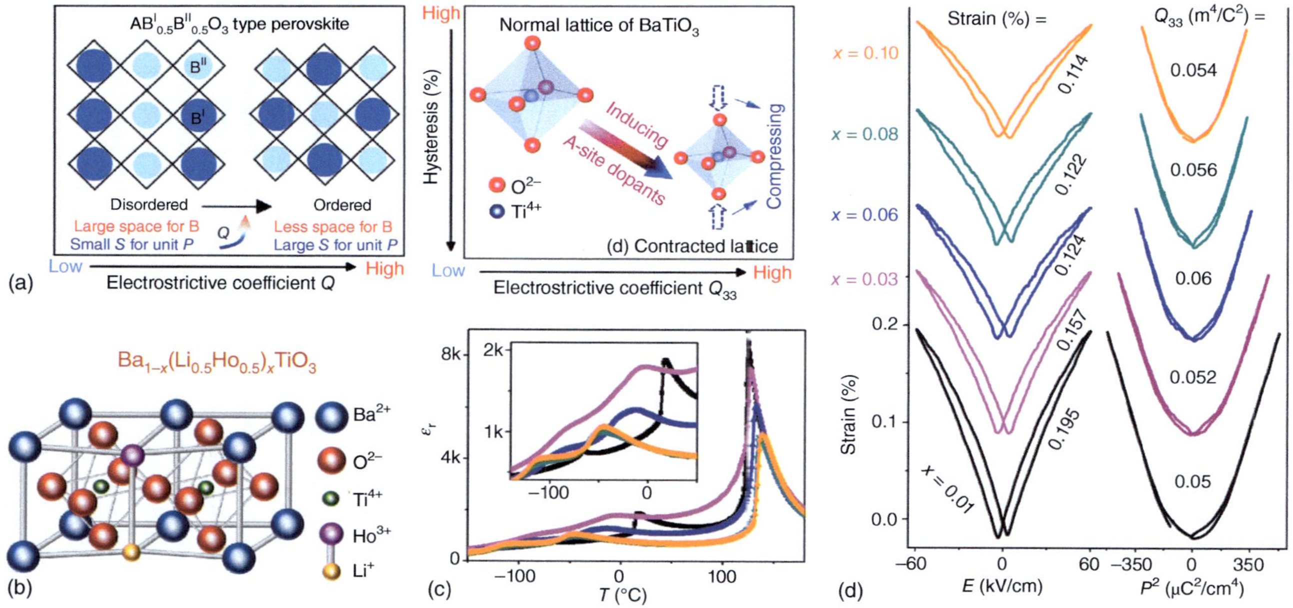

Figure 12.6 (a) Crystallographic model of $AB^{I}_{0.5}B^{II}_{0.5}O_3$ type perovskite and contracted TiO_6 octahedron with A-site doping. (b) Schematic diagram of [001]-oriented Li^{+}-Ho^{3+} doping ion-pairs in BT lattice. (c) $\varepsilon_r - T$ curves and (d) evolution of $S-E$ and $S-P^2$ curves for $Ba_{1-x}(Li_{0.5}Ho_{0.5})_xTiO_3$ ceramics. Source: Reproduced with permission from Ref. [35]; © 2019, Royal Society of Chemistry.

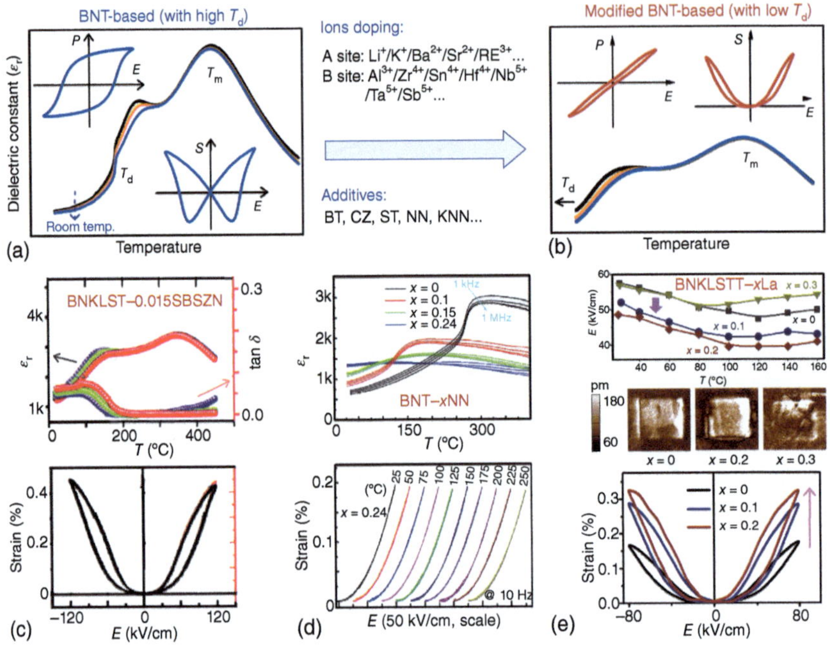

Figure 12.7 Schematics showing the $\varepsilon_r - T$ curves for BNT-based ceramics with (a) high T_d and (b) T_d below room temperature. The insets show the corresponding $P-E$ loops and bipolar $S-E$ curves. $\varepsilon_r - T$ and tan $\delta - T$ curves, and bipolar/unipolar $S-E$ curve for (c) BNKLST-0.15SBSZN ceramic [19] (d) BNT−xNN ceramics [22]. (e) Evolution of critical phase-transition electric field, comparison of the switched domain by PFM, and bipolar $S-E$ curves for BNKLSTT−xLa ceramics [23]. Source: (a–c) Reproduced with permission from Ref. [19]; © 2019, Royal Society of Chemistry. (d) Reproduced with permission from Ref. [22]; © 2020, Royal Society of Chemistry. Source: (e) Reproduced with permission from Ref. [23]; © 2021, Elsevier.

of BNT-based ceramics exhibit extremely small remnant polarization, low residual strain, and low hysteresis due to the characteristics of the ergodic relaxor state [19–27, 42–44].

The electrostrictive effect of pure BNT ceramic is weak, and its Q_{33} is around $0.02\,m^4/C^2$ [58]. To achieve linear and low hysteresis strain, non-180° domain switching and electric field-induced irreversible phase transitions should be avoided. Under the ergodic relaxor state, the electrostrain of BNT-based ceramics is mainly contributed by the electrostrictive effect and the reversible phase transition induced by an external electric field, so they usually present a linear strain response similar to the electrostrictive effect under this state. The electrostrain with electrostrictive characteristics in BNT-based ceramics is also called electrostrictive strain. To obtain the electrostrictive strain, the T_d of BNT-based ceramics should be shifted to below room temperature to obtain the complete ergodic relaxor state. At present, chemical modification is the most effective way to regulate T_d, including ion doping at the A and/or B sites, the addition of other compounds, etc. [42, 44, 45], as shown in Figure 12.7b.

There have been several works to enhance the electrostrictive strain of BNT-based ceramics by chemical modification to regulate the T_d and relaxor states. Based on a complex $[Bi_{0.5}(Na_{0.72}K_{0.18}Li_{0.10})_{0.5}]_{0.95}Sr_{0.05}TiO_3$ (BNKLST) ceramic as the matrix, the $Sr_{0.6}Ba_{0.4}Zr_{0.35}Sb_{0.17}Nb_{0.35}O_3$ (SBSZN) compound is introduced to regulate T_d below room temperature. Then a large electrostrictive strain (>0.2–0.4%) with low hysteresis (~28%) and high Q_{33} (~0.03–0.04 m^4/C^2) is obtained in a wide temperature range (25–125 °C) under a high electric field (120 kV/cm) [19], as displayed in Figure 12.7c. It is also found that high Q_{33} can be obtained when the ergodic relaxor phase is dominant, but a large electric field is needed to reduce Q_{33}. This is because the electrostrictive-like behavior in BNT is derived from the symmetric structure of short-range correlation. In this structure, the polarization vector responds quickly to the external electric field, showing a high Q_{33}. However, the response will be weakened at a high electric field, so the Q_{33} is reduced. In addition, some researchers have optimized the electrostrictive strain by introducing oxides to regulate the appearance of more tetragonal relaxor phases in BNT-based ceramics. By introducing Ta_2O_5 into $Bi_{0.5}(Na_{0.72}K_{0.22}Li_{0.06})_{0.5}TiO_3$ (BNKLT) ceramics, the content of the tetragonal phase is increased. Then, an electrostrictive strain of 0.2% (60 kV/cm) is achieved in this system [25]. In addition, for optimizing the temperature stability of electrostrictive strain, some researchers employ the antiferroelectric system in BNT ceramic to achieve relaxor antiferroelectric behavior. After introducing high-content $NaNbO_3$ (NN), the relaxor phase transition presents in a broad temperature range, among which the highly active antiferroelectric anti-polar nanoregions with an ultrafast response speed are observed. Finally, a high hysteresis-free electrostrictive strain of ~0.20% is achieved from room temperature to more than 250 °C (Figure 12.7d) [22]. Some researchers have used rare-earth ion doping to regulate the local polarization state of the relaxor phase to enhance the electrostrictive strain of BNT-based ceramics. As shown in Figure 12.7e, using Li^+, K^+, Sr^{2+}, and Ta^{5+} co-doped BNT ceramic (BNKLSTT) as the ergodic relaxor-state matrix, the incorporation of La^{3+} at the A site can improve the short-range correlation of polar nanoregions without changing the ergodic relaxor state, reduce the critical electric field of phase transition from relaxor to ferroelectric phase, and thus improve the local polarization. Finally, a high electrostrictive strain of ~0.3% (80 kV/cm) and a low hysteresis of ~8–30% are obtained from room temperature to 100 °C [23].

12.3.2.3 (K,Na)NbO_3-based Ceramics

(K,Na)NbO_3 (KNN) ceramics are binary systems composed of antiferroelectric phase $NaNbO_3$ and ferroelectric phase $KNbO_3$. When the ratio is near 1 : 1, KNN ceramics exhibit enhanced piezoelectric properties. Therefore, most studies are based on $(K_{0.5}Nb_{0.5})NbO_3$ and its adjacent compositions [44, 59, 60]. A relaxor state can be induced in KNN-based ceramics by the addition of ABO_3-type components. KNN–$x$$SrTiO_3$ ceramics are first studied, focusing on their relaxor characteristics and electrostrictive properties. When $x = 0.15$, the ceramics show an obvious relaxor state and pure electrostrictive strain (about 0.006%, 20 kV/cm) [33]. For $BiN_{i2/3}Nb_{1/3}O_3$ (BNN)-doped KNN-based ceramics (KNN–xBNN), as shown in Figure 12.8a, the relaxor phase transition can be obtained when $x > 0.03$. Besides,

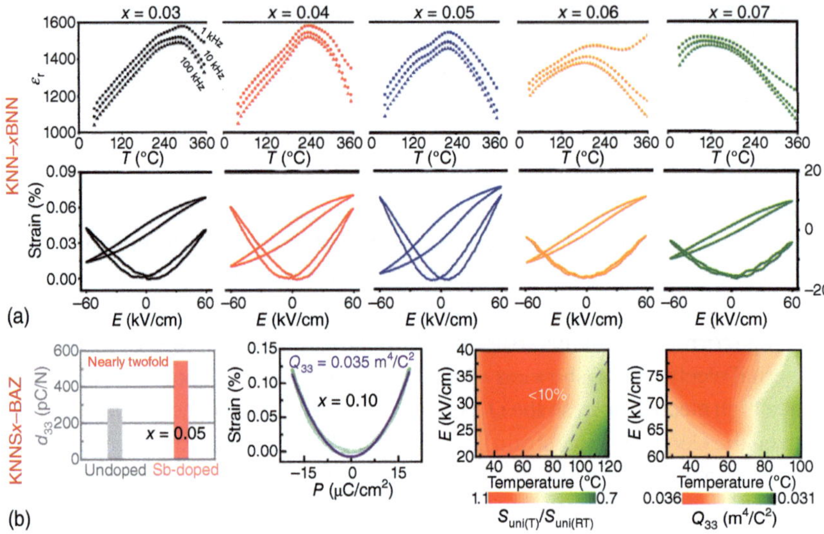

Figure 12.8 (a) ε_r–T curves, P–E loops, and bipolar S–E curves for KNN–xBNZ ceramics [28]. (b) Piezoelectric, strain, and Q_{33} properties for Sb-doped KNNSx–BAZ ceramics [29]. Source: (a) Reproduced with permission from Ref. [28]; © 2020, American Chemical Society. Source: (b) Reproduced with permission from Ref. [29]; © 2021, American Chemical Society.

because A-site small-radius ions distort the NbO_6 octahedron to a small volume, the activity space of B-site ions is restricted, thus the enhanced electrostrictive coefficient (from 0.032 to 0.045 m^4/C^2), and then the low-hysteresis electrostrictive strain (from 0.045% to 0.060%, 60 kV/cm) are achieved [28]. After doping Sb^{5+} and adding $(Bi_{0.5}Ag_{0.5})ZrO_3$ (BAZ), the distorted NbO_6 octahedron and the softened B–O repulsion induce nanoscale domains coexisting with a few PNRs, then causing the relaxor ferroelectric rhombohedral-orthorhombic-tetragonal (R–O–T) multiphase coexistence in a wide temperature range. The temperature-insensitive and low-hysteresis strain (>0.1%, 40 kV/cm) and high Q_{33} of 0.035 m^4/C^2 are obtained at 27–120 °C, along with high piezoelectricity (Figure 12.8b) [29]. After doping Li^+ and adding $SrTiO_3$ (ST), the relaxor ferro-paraelectric phase transition is obtained in KNLN–xST ceramics. The enhanced Q_{33} and high electrostrictive strain (~0.147%) with ultra-low hysteresis are achieved under a high electric field of 300 kV/cm [30]. By doping ions and adding other compounds to tune the phase transition characteristics in KNN-based ceramics, the relaxor ferro-ferroelectric or ferro-paraelectric can be achieved, or the domain structure can be minimized by defect formation to reduce the strain hysteresis. Then it is promising to obtain large electrostrains with low hysteresis in KNN-based ceramics.

12.3.2.4 $NaNbO_3$-based Ceramics

$NaNbO_3$ (NN) belongs to the perovskite structure. Below T_C (about 345 °C), NN only has an antiferroelectric orthogonal phase and a ferroelectric tetragonal phase [61]. NN easily forms a relaxor solid solution with other perovskite compounds, such as NN-BT [17], NN-BCTZ [16], NN-$BaSnO_3$ [62], NN-ST [63], and

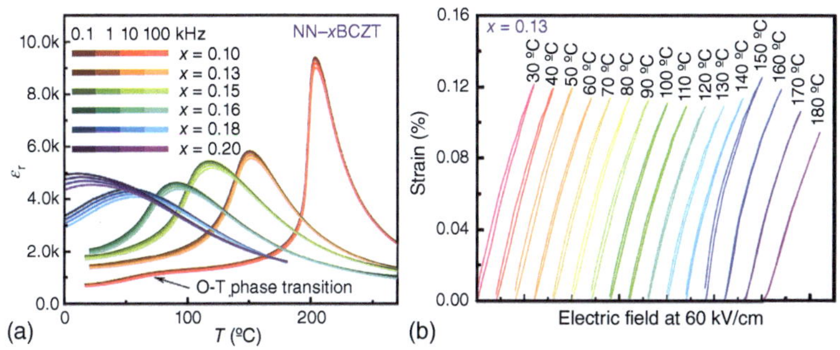

Figure 12.9 (a) ε_r–T curves and (b) unipolar S–E curves of NN–xBCTZ ceramics. Source: Reproduced with permission from [16]; © 2022, Elsevier.

NN-($Bi_{0.5}Li_{0.5}$)TiO_3 [64], and shows good dielectric, piezoelectric, and ferroelectric properties. For example, for the NN-xBT ceramics, the non-ergodic relaxor phase occurs when the BT content is 15%, and a large strain (~0.22%, 100 kV/cm) with low hysteresis (<15%) is obtained. When the BT content is >22%, the coexistence of non-ergodic and ergodic relaxor phases is present in the NN-based ceramics. Then the pure electrostrictive strain (0.15%, 100 kV/cm) and large Q_{33} of 0.046 m^4/C^2 are obtained [17]. It is found that the polar nanoregions of NN-BT ceramics exhibit local tetragonal symmetry. Because of the absence of A lone-pair electrons, the ferroelectric properties of NN-based ceramics are only contributed by B-site ions. Compared with lead- and bismuth-based relaxors with both A- and B-site ion shifts, NN has lower local polarization, a lower driving electric field, a lower strain value, and lower strain hysteresis [65]. In addition, it is found that NN-BT-CZ ceramics exhibit the morphotropic phase boundary (MPB), and high electrostrain (0.15%, 60 kV/cm) with good temperature stability (25–175 °C, ±10%) and low hysteresis (<13%) can be obtained at the phase boundary [66]. After adding a high content of BCTZ, the phase transition of NN-based ceramics can become quite a diffusion. The high electrostrain (0.12% at 60 kV/cm) with excellent temperature stability (variation <10%, 30–160 °C) and ultra-low hysteresis (<6%) can be obtained. It is found that the high strain property is dominated by the intrinsic electrostrictive effect of more than 80% [16], as shown in Figure 12.9. However, it is a key problem that NN-based ceramics usually need a high-driving electric field that can obtain relatively high electrostrain. Thus, it is necessary to regulate the activity of local polar nanoregions by composition design to achieve high strain under a low electric field.

12.3.2.5 $BiFeO_3$-based Ceramics

$BiFeO_3$ (BFO) is the only single-phase multiferroic material with both magnetic and ferroelectric properties at room temperature. Because the sintering temperature window of BFO is too narrow for easy ion volatilization, it is difficult to prepare BFO-based ceramics without a few defects and secondary phases. Bi^{3+} volatilization and Fe^{3+} valency change will lead to a large number of cation defects and oxygen vacancies, which ultimately makes the leakage current quite large, so it is difficult to obtain the saturated hysteresis loops in BFO-based ceramics. At present, researchers mainly focus on improving the sintering characteristics and electrical properties of BFO-based ceramics by ion doping ($Mn^{2+/3+}$, La^{3+}, Dy^{3+}, Sm^{3+}, Ti^{4+}, etc.) and adding other compounds (BT, $PbTiO_3$, $PbZrO_3$, etc.) [44, 67]. Among them, for BFO–*x*BT ceramics, it is found that composition modification can make the transition from the non-ergodic relaxor state to the ergodic relaxor state, thereby reducing strain hysteresis [68]. For example, in $(Ba_{0.8}Ca_{0.2})ZrO_3$ (BCZ) modified BFO–0.33BT ceramic, the non-ergodic relaxor state of ceramics can be transformed to the ergodic relaxor state, thus the enhanced Q_{33} of 0.029 m^4/C^2 and a high electrostrain (0.4%, 60 kV/cm) is obtained while accompanying a high strain hysteresis of 33% [15]. To decrease the strain hysteresis, researchers employ Sc^{3+} doping and add 10%$Bi_{1.05}(Zn_{0.5}Hf_{0.5})O_3$ (BZH) to tune the phase transition and domain structure. As shown in Figure 12.10a, the relaxor phase transition can be obtained in BFS–*x*BT–BZH ceramics ($x \geq 40\%$). Then a moderate electrostrain value (~0.1%, 60 kV/cm) with reduced hysteresis (~20%) and Q_{33} of 0.022 m^4/C^2 are obtained (Figure 12.10b). It is found that this relaxor BFO-based ceramics show a special microstructure, that is, the coexistence of grains with strong piezoelectric response polar clusters (to ensure high polarization) and grains with weak piezoelectric response polar nanoregions (to ensure rapid polarization response) (Figure 12.10c) [14]. However, the modified BFO-based ceramics still have a large strain hysteresis. At high temperatures, BFO-based ceramics can obtain almost hysteresis-free strain, and the maximum strain increases with the increase in temperature, which is

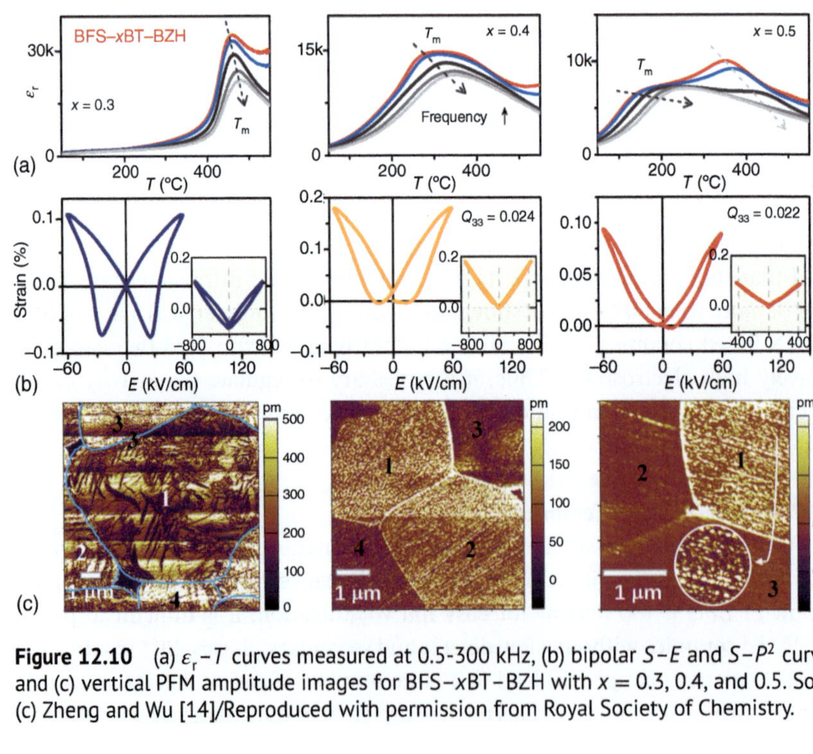

Figure 12.10 (a) ε_r–T curves measured at 0.5-300 kHz, (b) bipolar S–E and S–P^2 curves, and (c) vertical PFM amplitude images for BFS–*x*BT–BZH with $x = 0.3$, 0.4, and 0.5. Source: (c) Zheng and Wu [14]/Reproduced with permission from Royal Society of Chemistry.

particularly special in lead-free ceramic systems. While it is difficult to obtain high-temperature electrostrain due to their large high-temperature conductivity. Therefore, the structure-property modification and related physical mechanism of BFO-based ceramics need to be further explored.

12.3.3 Lead-based Ferroelectrics

12.3.3.1 Lead-based Single Crystals

In 1980, the high electrostrictive effect and pure electrostrictive strain were found in $Pb(Mg_{2/3}Nb_{1/3})O_3$ (PMN) ceramics and single crystals [5, 6]. The hydrostatic electrostrictive coefficient of PMN single crystal is ~0.006 m^4/C^2, and the electrostrictive strain is about ~0.02–0.04% (~25 kV/cm) with no strain hysteresis at −6.5 to 23.8 °C, as shown in Figure 12.11. These works attract extensive attention from fundamental research and industry applications. PMN ceramic and single crystal are the typical relaxor ferroelectrics, and the remarkable electrostrictive effect of PMN mainly results from the relaxor structure and high dielectric property of PMN ferroelectric. Other lead-based perovskite relaxor ferroelectrics also exhibit high electrostrictive strain, such as $Pb(Zn_{1/3}Nb_{2/3})O_3$ (PZN) [69–71], $Pb(Mg_{1/3}Nb_{2/3})O_3$–$PbTiO_3$ (PMN-PT) ferroelectrics [6, 72].

PMN-PT ferroelectric is the most well-known electrostrictive ferroelectric because of its higher dielectric, ferroelectric, piezoelectric, and electrostrictive properties [46]. For example, although there is an observable domain switching process, PMN–0.28PT single crystal has a large and low-hysteresis electrostrain of ~0.38% under 80 kV/cm, which is similar to PZN single crystal (~0.43%) and much higher than PMN single crystal (~0.17%). The domain switching process is observed under a low electric field (<10 kV/cm), while the pure and high electrostrictive effect can be found under a high electric field (>10 kV/cm). It is found that PMN–0.28PT single crystal exhibits electric field-dependent and anisotropic electrostrictive coefficients. The structure origin is assigned to the different transformation abilities of TiO_6 octahedra along different directions or under different driving electric fields. Finally, the maximum Q_{33} is found to be ~0.055 m^4/C^2 along the (001) direction [73].

12.3.3.2 Lead-based Ceramics

Figure 12.12 gives the phase diagram of $(1-x)$PMN–xPT relaxor ferroelectrics [74]. Pure PMN ferroelectric exhibits the paraelectric phase near room temperature.

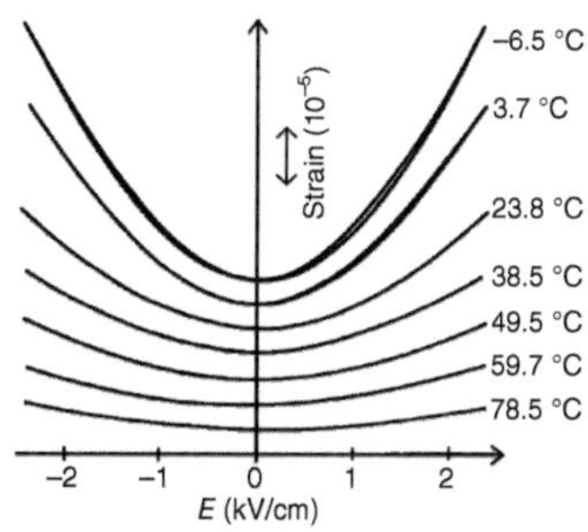

Figure 12.11 Bipolar S–E curves at different temperatures for PMN single crystal. Source: Reproduced with permission from [5]; © 1980, American Institute of Physics.

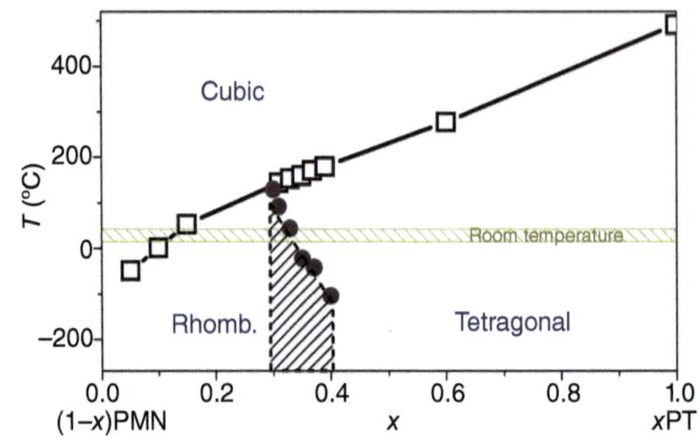

Figure 12.12 Phase diagram of (1 − x)PMN−xPT ferroelectrics. Source: Reproduced with permission from [74]; © 2009, John Wiley and Sons.

Thus, the electrostrictive strain of PMN ceramic is relatively small, about ~0.01% under 20 kV/cm. After introducing large-polarization $PbTiO_3$ (PT), the Curie temperature of PMN–xPT ceramics rapidly increases to near room temperature at $x = 0.13$ and above room temperature at $x > 0.13$. In this process, the electrostrictive effect still shows a dominant role in the strain response of PMN–xPT ceramics when $x \leq 0.13$. The electrostrictive strain rapidly increases to ~0.043 (~23 kV/cm) for PMN–0.13PT ceramic with no strain hysteresis. After further increasing the content of PT, the room-temperature phase structure becomes the rhombohedral phase, then the rhombohedral-tetragonal phase boundary and tetragonal phase. The ferroelectric polarization becomes higher, and PMN–xPT ceramics gradually change from relaxor ferroelectric to typical ferroelectric in this process, and the ferroelectric domain switching behavior and strain hysteresis also become observable [6].

For PMN–xPT ceramics at $x > 0.13$, if the dominant electrostrictive effect and electrostrictive strain are desired, it is necessary to measure the electrostrain at a high temperature near Curie temperature. For example, in PMN–0.325PT ceramic, the Curie temperature is around ~160–170 °C. Therefore, the obvious strain hysteresis and domain switching behavior present in this ceramic is below Curie temperature. And the electrostrictive strain of ~0.133% (~25 kV/cm) with quite small strain hysteresis is obtained in non-textured PMN–0.325PT ceramic (R-PMN-PT), with the Q_{33} of ~0.021 m^4/C^2. In addition, when this ceramic is textured, the electrostrain and electrostrictive properties can be further improved. The textured PMN–0.325PT ceramic (T-PMN-PT) exhibits a more stable electrostrain property than that of non-textured ceramic in a wide temperature range. The electrostrain is elevated to ~0.16%, and Q_{33} improves to ~0.052 m^4/C^2 for textured ceramic. It is found that the electrostrictive effect mainly contributes to the enhancement of electrostrain for textured PMN–0.325PT ceramic [75].

For lead-based ceramics, making the phase transition relaxor is also an effective way to decrease strain hysteresis and improve the electrostrictive effect. For example, the ε_r–T curves of $Pb(Ni_{1/3}Nb_{2/3})O_3$–$0.1Pb(In_{0.5}Nb_{0.5})O_3$–$0.3PbZr_{0.1}Ti_{0.9}O_3$ (PNZT) lead-based relaxor ferroelectric ceramic become more relaxor and diffusion after doping 1.5% Sm^{3+}, as presented in Figure 12.13a. The hysteresis-free electrostrain of ~0.12% is obtained under 30 kV/cm (Figure 12.13b), and the electrostrain can be further elevated to >0.18% under 80 kV/cm. Besides, the electrostrictive coefficient

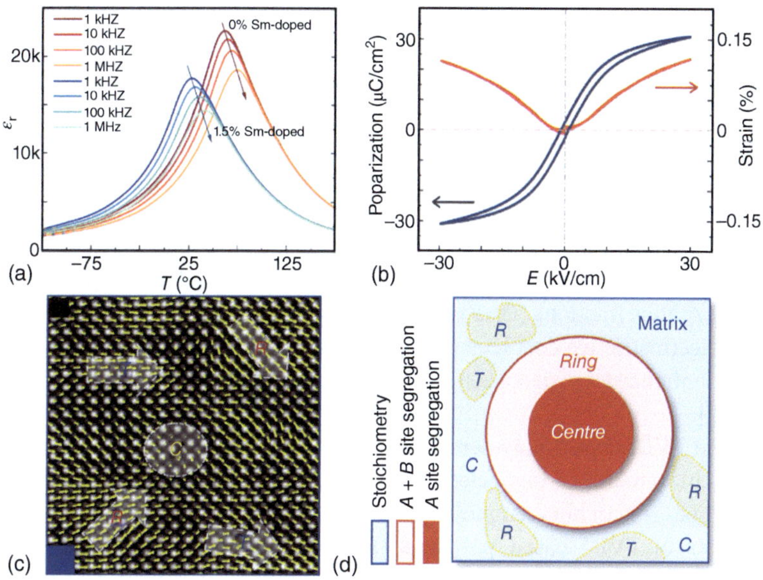

Figure 12.13 (a) $\varepsilon_r - T$ curves of undoped and 1.5% Sm-doped PNZT ceramics. (b) $P-E$ and bipolar $S-E$ curves of Sm-PNZT ceramic. (c) Atomically resolved HAADF-STEM image showing the coexistence of R, T, and C phase nanoregions. (d) Schematic of the endotaxial nanoprecipitation structure. Source: Reproduced with permission from [76]; © 2022, John Wiley and Sons.

M_{33} can be elevated to $\sim 5\text{–}8 \times 10^{-16}$ m^2/V^2. It is observed that the more relaxor characteristic and such high electrostrictive properties result from the special domain structure with endotaxial nanoprecipitation embedding in the polar matrix of coexisting rhombohedral-tetragonal-cubic (R–T–C) nanoregions (Figure 12.13c). The ferroelectric matrix contributes to the high strain response. The segregation of doped Sm^{3+} and light B-site cations brings about the randomly distributed endotaxial nanoprecipitation with a nonpolar central region, then facilitating the strong restoring force to diminish the strain hysteresis (Figure 12.13d) [76].

12.3.4 Other Piezoelectric Materials

12.3.4.1 Ferroelectric Polymers

The inorganic ferroelectric ceramics and crystals have high piezoelectric and strain performances, while their electrostrictive coefficients are relatively small. Besides, the high processing temperature and poor mechanical flexibility make them difficult for applications on flexible or film-integrated devices. Different from the inorganic, ferroelectric polymers possess large strain and electrostrictive properties and show soft mechanical capacity; thus, these materials are desirable for soft piezoelectric actuators, sensors, energy harvesting devices, etc. The representative ferroelectric polymer is the poly(vinylidene fluoride-trifluoroethylene) (PVDF-TrFE)-based copolymer, which has achieved a series of progresses on piezoelectric and (electrostriction) electrostrain properties [77, 78]. For example, the effect of thermal annealing on the structure and electrostrictive properties of P(VDF-TrFE-CTFE) (CTFE is chlorotrifluoroethylene) random terpolymer are studied, as shown in Figure 12.14. It is found that this terpolymer embraces three phases, including crystals, oriented, and isotropic amorphous fractions. A giant electrostrictive property with a transverse strain of ~5% (2000 kV/cm) is obtained in the paraelectric terpolymer above the Curie temperature. This superior electrostrictive performance is attributed to mechano-electrostriction due to conformational transformation at low fields (<30 MV/m) and electrostatic repulsion among field-induced ferroelectric nanodomains in bulky CTFE units that generate taut-tie molecules (TTM) and oriented amorphous fractions (OAF) at high fields (>50 MV/m). It is observed that the highest electrostrictive coefficient Q_{31} is ~20.5 m^4/C^2 above Curie temperature and can be enhanced by decreasing the dielectric nonlinearity of ferroelectric polymers [79].

Most of the PVDF-based polymers have shown low dielectric constant, which limits the achievement of high electrostrictive and electrostrain properties. To this end, some researchers employ conductive fillers to enhance the dielectric properties of PVDF-based polymers. For example, the P(VDF-TrFE-CFE) (CFE is chlorofluoro-ethylene) (PVTC) is selected as the matrix, and the poly(3-hexylthiophene)-*b*-poly(methyl methacrylate) (P3HT-*b*-PMMA, PTM) is selected as the conductive filler. It is found that the dielectric constant of the PVTC-PTM blend increases from ~54 to ~90 (100 Hz) when 1.5 wt% PTM is introduced, and the transverse electrostrain increases from ~0.6% (PVTC) to ~1.15% (PVTC–1.5%PTM), showing near two-time enhancement, which is consistent with the improvement of dielectric constant. The enhanced electrostrain property mainly originates from the electrostrictive effect, which is the result of the rearrangement of polar groups (the crystalline phase transition from paraelectric to ferroelectric phase) affected by the homogenously dispersed PTM conducting fillers in the PVTC matrix [80].

12.3.4.2 Molecular Ferroelectrics

Ferroelectric polymers (e.g. PVDF) have a large electrostrictive coefficient, while their poor crystallinity and large coercive field (>50 MV/m) restrict their applications. In recent years, molecular ferroelectrics have drawn lots of attention due to their merits of low acoustic impedance, structural tunability, and mechanical flexibility. Previous studies have shown that the high Curie temperature, spontaneous polarization, piezoelectric coefficient, and electrocaloric property were achieved in molecular ferroelectrics [81, 82]. Besides, a large electrostrictive coefficient was also observed in some molecular ferroelectrics [83, 84]. For example, a large Q_{33} of 0.80 m^4/C^2 was achieved in an organic supramolecular ferroelectric of [(NH_3-TEMPO)([18]crown-6)](ReO_4). In addition, a lead halide perovskite ferroelectric of (EATMA)$PbBr_4$ (EATMA = (2-aminoethyl)-trimethylammonium) single crystal, its EATMA cation is surrounded by corner-sharing $PbBr_6$ octahedra by hydrogen bondings. When the EATMA cation is replaced by EATMP ((2-aminoethyl)-trimethylphosphanium), the EATMP cation is well ordered to give an ordered alignment of dipoles because the heavier P atom and longer P—C bond

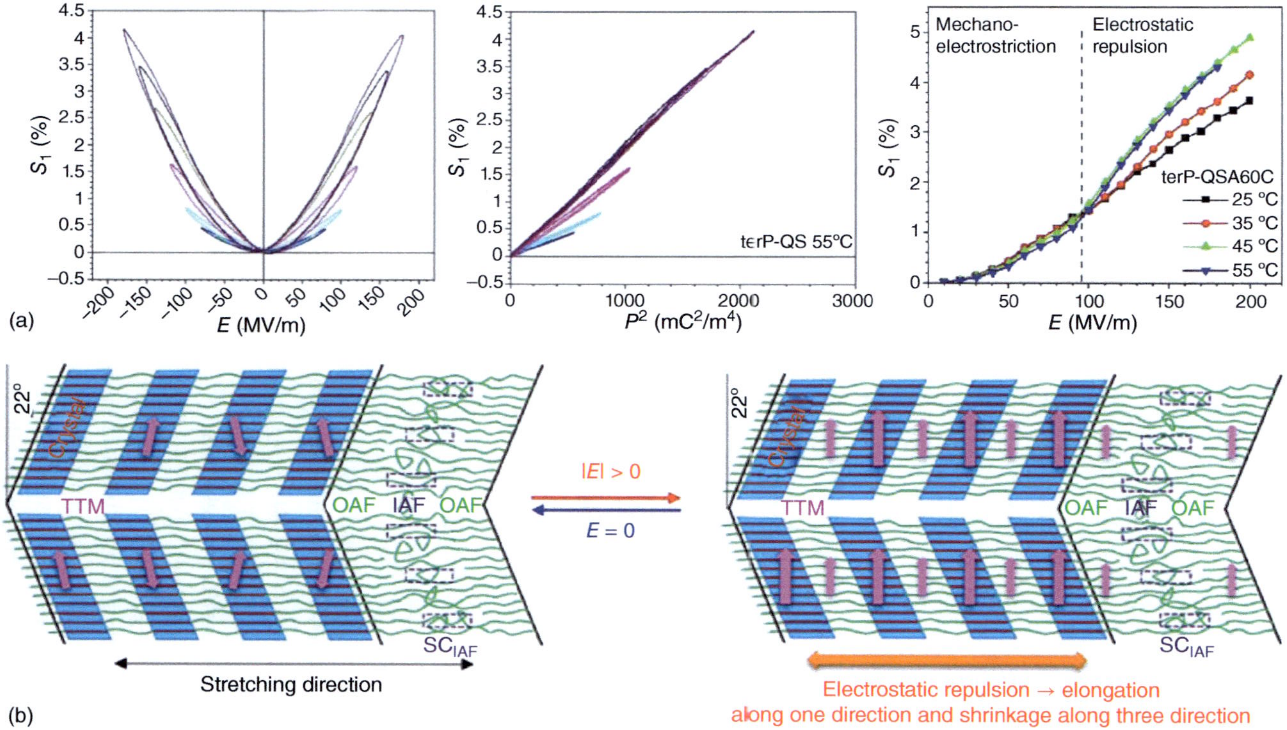

Figure 12.14 (a) Bipolar S–E curves and unipolar S–P^2 curves at 55 °C, and electric field-dependent strain values for P(VDF-TrFE-CTFE) terpolymers. (b) Electrostatic repulsion among polarized nanodomains in oriented and isotropic amorphous crystals (OAF and IOF) under >50 MV/m. Source: Reproduced with permission from [79]; © 2021, American Chemical Society.

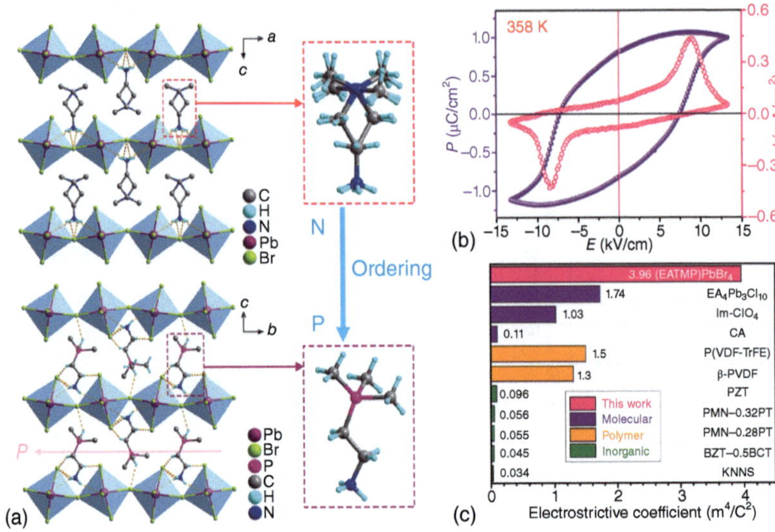

Figure 12.15 (a) Schematics along the *b*-axis of (EATMA)$PbBr_4$ with the orientationally disordered EATMA cations and (EATMP)$PbBr_4$ with ordered EATMP cations. (b) *P*–*E* loops of (EATMP)$PbBr_4$. (c) Comparison of electrostrictive coefficients (Q_{33} or Q_{22}) among (EATMP)$PbBr_4$, some inorganic, polymer, and molecular materials. Source: Reproduced with permission from [83]; © 2021, American Chemical Society.

length give the potential energy barrier of tumbling motion of EATMP cation. As shown in Figure 12.15a, the pink arrow indicates the direction of spontaneous polarization. This ferroelectric undergoes a ferroelectric phase transition with a high T_C of 534 K and shows the typical ferroelectric hysteresis loop (Figure 12.15b). More importantly, the giant Q_{22} of ~3.96 m^4/C^2 is achieved in this hybrid perovskite ferroelectric, which far exceeds the electrostrictive coefficient of PVDF and inorganic ferroelectrics [83], as shown in Figure 12.15c. However, the electric field-induced strain value seems to be very small in molecular ferroelectrics, even though they possess a giant electrostrictive coefficient, which is hardly employed in actual application. Therefore, further exploration of the electrostriction strain in molecular ferroelectrics is necessary.

12.3.4.3 Hybrid Perovskite Ferroelectrics

Organic-inorganic hybrid perovskites have attracted extensive interest for potential applications in next-generation photovoltaic devices. Organic-inorganic hybrid perovskites show a similar structure to normal inorganic ferroelectric perovskites. These types of materials have been found to show good electromechanical strain and electrostrictive properties [85, 86]. For example, methylammonium lead triiodide perovskites ($MAPbI_3$) single crystals are found to show a large compressive electrostrictive strain of >1%, while this high strain can only be observed at a low frequency [87, 88]. In addition, this high electric field-induced strain does not derive from the intrinsic piezoelectric, ferroelectric, or electrostrictive effects, thermal expansion, Maxwell stress, local polar fluctuation, or methylammonium cation ordering. The structure's origin is found to be the lattice deformation induced by additional defects when applying bias voltages. Under a high frequency, the electrostrictive coefficient is found to be −45.7 to −730 nm^2/V^2 after low-frequency AC poling, which is correlated with the rise of defects from the ion migration under reverse bias. This high electrostrictive coefficient is the largest value among all reported electrostrictive materials (e.g. 120 nm^2/V^2 for $Pb(Zn_{1/3}Nb_{2/3})O_3$–$0.08PbTiO_3$ ceramic, −113 nm^2/V^2 for P(VDF-TrFE) polymer) [87]. The large electrostrictive strain of $MAPbI_3$ may give extensive potential applications in sonar transducers, actuators, etc.

12.4 Applications

The electrostrictive effect can achieve a low-hysteresis, fast, and nano- or micro-sized strain response under an external electric field. Therefore, electrostrictive materials show great potential in new electromechanical actuators to replace conventional piezoelectric actuators [89]. Nowadays, the worldwide market for piezoelectric actuators is projected to be near $10 billion per year [90], so the development of electrostrictive actuators draws lots of attention. For example, the electrostrictive/piezoelectric multimaterial monomorphs actuator is fabricated by layered manufacturing and co-firing techniques, in which the PMN–0.35PT ceramic acts as the piezoelectric element to offer high strain response and the PMN–0.1PT ceramic acts as the electrostrictive element to offer low strain hysteresis. As a result, the multimaterial monomorph actuator exhibits a moderate dielectric constant, ferroelectric polarization, and deflection displacement (~11 μm, 3 kV/cm) with quite low hysteresis compared to monolithic monomorph actuators [91].

Electrostrictive polymers have been the subject of much interest and research in flexible actuator applications [92]. The flexible polymer actuators greatly extend the application possibility compared to conventional hard ceramic actuators, such as flexible MEMS resonators, transducers, acoustical devices/loudspeakers, and adaptive thin-shell reflectors. For example, researchers use the alternating-current electric field to enhance the electrostrictive amplitude based on the P(VDF-TrFE) polymers. Although the electrostrictive response is non-linear, it can provide a straightforward approach to generate improved strain amplitude of micromachined cantilevers with integrated P(VDF-TrFE) polymers for electro-mechanical actuation. It is noted that this electrostrictive actuator can be used as a resonant microsystem with a resonance frequency of f_0, which needs to be excited at half the resonance frequency at $1/2f_0$ [93].

Since the development of low-power electronics, the possibility of using electrostrictive materials for energy harvesting has been investigated in recent years, especially using electrostrictive polymers [94]. Nowadays, several studies have shown that electrostrictive polymers exhibit suitable applications for microgenerators because of their flexibility, processability, and possible property optimization with the incorporation of dielectric or conductive fillers. For example, the polyurethane polymer filled with 1% carbon black nano-powder exhibits a large electrostrictive strain (~3%, 100 kV/cm). The output current of energy harvesting based on this polymer generator can be ~4 μA, and the harvesting power can be ~90 μW under the direct-current exciting signal. The output current and harvesting power can be improved by ~27% and ~75% under a trapezoidal exciting signal, respectively [95]. This harvesting energy can be sufficient to power a microsystem.

12.5 Conclusions

Compared with the piezoelectric effect, the electrostrictive effect belongs to the second-order effect and is often ignored. With the innovation of material design and preparation technology, materials with large electrostrictive effects have been reported, and the importance of electrostrictive effects has been further paid attention to. The development of large electrostrictive materials also provides an opportunity for the development of new micro-displacement actuators. Given the electrostrictive effect and electrostrictive strain, the research progress of lead-free electrostrictive ceramics in recent years and the methods of regulating the electrostrictive characteristics are overviewed, and the advantages and disadvantages of the electrostrictive properties of lead-based and lead-free materials are compared. At present, some progress has been made in the study of the electrostrictive effect of lead-free bulk ceramics, and the electrostrictive performance is comparable to that of lead-based electrostrictive ceramics. However, there is still a certain distance to be left before a comprehensive understanding of the electrostrictive effect and the effective research and development of devices, mainly including:

(1) Most studies are based on longitudinal electrostrictive coefficient Q_{33} and longitudinal strain S_{33}, and the electrostrictive coefficient of a material is a fourth-order tensor. For bulk ceramics, the transverse electrostrictive coefficient Q_{11} and transverse electrostrictive strain S_{11} are also important, and therefore, further research is needed.
(2) Developing relaxor characteristics is the main way to enhance the electrostrictive properties and reduce strain hysteresis. However, severe relaxor characteristics will greatly reduce the electrostrictive strain, which makes it difficult to meet the needs of practical applications. The construction of highly active polar nanoregions, or nanodomains, in relaxor ferroelectrics plays an important role in improving the electrostrictive properties. The effective regulation of polar nanoregions to obtain the optimal electrostrictive strain properties needs more in-depth and systematic studies.
(3) Ferroelectric polymers usually need quite a high driving electric field (>1000 kV/cm) to obtain the observable electrostrain property. To realize large electrostrictive and strain properties under a relatively low electric field, the fundamental electrostriction mechanism of ferroelectric polymers needs to be better understood.
(4) Clarifying the relationship between the electrostrictive effect and other electrical properties can provide a new design idea for optimizing the electrostrictive

performance of electrostrictive materials. However, there is a lack of deep and overall understanding of the relationships between the electrostrictive effect and other electrical properties.

(5) To improve the reliability and durability of the electrostrictive actuator, reduce the driving voltage, and increase the displacement, it is necessary to optimize and innovate the material preparation process.

(6) Lead-free electrostrictive materials are still in the stage of material research and development, and there are few reports on the design and preparation of relevant electrostrictive devices. Therefore, it is necessary to strengthen the research on lead-free electrostrictive devices and promote their practical applications.

References

1 Uchino, K. (1996). *Piezoelectric Actuators and Ultrasonic Motors*. Springer Science & Business Media.

2 Zhao, C., Wu, H., Li, F. et al. (2018). Practical high piezoelectricity in barium titanate ceramics utilizing multiphase convergence with broad structural flexibility. *JACS* 140 (45): 15252–15260.

3 Park, S.E. and Shrout, T.R. (1997). Ultrahigh strain and piezoelectric behavior in relaxor based ferroelectric single crystals. *J. Appl. Phys.* 82 (4): 1804–1811.

4 Li, F., Jin, L., Xu, Z., and Zhang, S. (2014). Electrostrictive effect in ferroelectrics: an alternative approach to improve piezoelectricity. *Appl. Phys. Rev.* 1 (1): 011103.

5 Uchino, K., Nomura, S., Cross, L.E. et al. (1980). Electrostrictive effect in lead magnesium niobate single crystals. *J. Appl. Phys.* 51 (2): 1142–1145.

6 Cross, L.E., Jang, S.J., Newnham, R.E., and Uchino, S. (1980). Large electrostrictive effects in relaxor ferroelectrics. *Ferroelectrics* 23 (1): 187–191.

7 Leung, K.M., Liu, S.T., and Kyonka, J. (1980). Large electrostrictive effect in Ba:PZT and its application. *Ferroelectrics* 27 (1): 41–43.

8 Liu, Y., Wen, B., and Ma, J. (1997). Large electrostriction and microstructure of La:PZT ceramics. *Ferroelectrics* 196 (1): 77–80.

9 Uchino, K. (2022). Electrostrictive and piezoelectric effects in relaxor ferroelectrics-historical background. *IEEE Trans. Ultrason. Ferroelectr. Freq. Control* 69 (11): 3013–3036.

10 Newnham, R.E. (2005). *Properties of Materials: Anisotropy, Symmetry, Structure*. Oxford University Press on Demand.

11 Uchino, K., Nomura, S., Cross, L.E. et al. (1981). Electrostrictive effect in perovskites and its transducer applications. *J. Mater. Sci.* 16 (3): 569–578.

12 Huang, Y., Zhao, C., and Wu, J. (2022). Electrostrictive effect of lead-free perovskite ceramics. *J. Chin. Ceram. Soc.* 50 (3): 575–586.

13 Ang, C. and Yu, Z. (2006). High, purely electrostrictive strain in lead-free dielectrics. *Adv. Mater.* 18 (1): 103–106.

14 Zheng, T. and Wu, J. (2020). Mesoscale origin of dielectric relaxation with superior electrostrictive strain in bismuth ferrite-based ceramics. *Mater. Horiz.* 7 (11): 3011–3020.

15 Liu, X., Zhai, J., and Shen, B. (2019). Novel bismuth ferrite-based lead-free incipient piezoceramics with high electromechanical response. *J. Mater. Chem. C* 7 (17): 5122–5130.

16 Jing, R., Hu, Q., Zhang, L. et al. (2022). Ultra-slim electrostrains with superior temperature-stability in lead-free sodium niobate-based ferroelectric perovskite. *J. Materiomics* 8 (6): 1230–1238.

17 Zuo, R., Qi, H., Fu, J. et al. (2016). Giant electrostrictive effects of $NaNbO_3$-$BaTiO_3$ lead-free relaxor ferroelectrics. *Appl. Phys. Lett.* 108 (23): 232904.

18 Lu, X., Hou, L., Jin, L. et al. (2018). Origin of composition-insensitive electrostrictive coefficient and continuous decrease of domain wall density in $(1-x)NaNbO_3-xBaTiO_3$ lead-free ferroelectrics. *J. Eur. Ceram. Soc.* 38 (9): 3127–3135.

19 Yin, J., Liu, G., Zhao, C. et al. (2019). Perovskite $Na_{0.5}Bi_{0.5}TiO_3$: a potential family of peculiar lead-free electrostrictors. *J. Mater. Chem. A* 7 (22): 13658–13670.

20 Li, T., Liu, C., Ke, X. et al. (2020). High electrostrictive strain in lead-free relaxors near the morphotropic phase boundary. *Acta Mater.* 182: 39–46.

21 Li, G., Shi, C., Zhu, K. et al. (2022). Achieving synergistic electromechanical and electrocaloric responses by local structural evolution in lead-free BNT-based relaxor ferroelectrics. *Chem. Eng. J.* 431: 133386.

22 Qi, H. and Zuo, R. (2020). Giant electrostrictive strain in $(Bi_{0.5}Na_{0.5})TiO_3$-$NaNbO_3$ lead-free relaxor antiferroelectrics featuring temperature and frequency stability. *J. Mater. Chem. A* 8 (5): 2369–2375.

23 Deng, A. and Wu, J. (2021). Optimized strain properties with small hysteresis in BNT-based ceramics with ergodic relaxor state. *J. Eur. Ceram. Soc.* 41 (10): 5147–5154.

24 Liu, X., Han, J., Huang, Y. et al. (2021). Large electrostrictive coefficient with high electro-strain dominated by modified ergodic state in BNT-based solid solutions. *J. Am. Ceram. Soc.* 104 (3): 1391–1401.

25 Han, J., Yin, J., and Wu, J. (2020). BNT-based ferroelectric ceramics: electrical properties modification by Ta_2O_5 oxide addition. *J. Am. Ceram. Soc.* 103 (1): 412–422.

26 Hao, J., Xu, Z., Chu, R. et al. (2016). Structure evolution and electrostrictive properties in $(Bi_{0.5}Na_{0.5})_{0.94}Ba_{0.06}TiO_3$-$M_2O_5$ (M = Nb, Ta, Sb) lead-free piezoceramics. *J. Eur. Ceram. Soc.* 36 (16): 4003–4014.

27 Liu, X., Xue, S., Ma, J. et al. (2018). Electric-field-induced local distortion and large electrostrictive effects in lead-free NBT-based relaxor ferroelectrics. *J. Eur. Ceram. Soc.* 38 (14): 4631–4639.

28 Wu, B., Zhao, C., Huang, Y. et al. (2020). Superior electrostrictive effect in relaxor potassium sodium niobate based ferroelectrics. *ACS Appl. Mater. Interfaces* 12 (22): 25050–25057.

29 Zhang, N., Lv, X., Zhang, X.-x. et al. (2021). Feasible way to achieve multifunctional (K, Na)NbO_3-based ceramics: controlling long-range ferroelectric ordering. *ACS Appl. Mater. Interfaces* 13 (50): 60227–60240.

30 Zhang, L., Jing, R., Huang, Y. et al. (2022). Ultrahigh electrostrictive effect in potassium sodium niobate-based lead-free ceramics. *J. Eur. Ceram. Soc.* 42 (3): 944–953.

31 Quan, Q., Fan, H., Shen, Q. et al. (2022). Large electrostrictive effect and dielectric properties of $(K_{0.5}Na_{0.5})NbO_3$–$BaZrO_3$ ceramics. *J. Eur. Ceram. Soc.* 42 (5): 2195–2203.

32 Zhao, L., Zheng, H., Ma, Z. et al. (2022). Insights into the correlation between strain and electrostrictive coefficient of potassium sodium niobate based ceramics from relaxor structure. *Ceram. Int.* 49 (3): 4614–4621.

33 Bobnar, V., Malič, B., Holc, J. et al. (2005). Electrostrictive effect in lead-free relaxor $K_{0.5}Na_{0.5}NbO_3$–$SrTiO_3$ ceramic system. *J. Appl. Phys.* 98 (2): 024113.

34 Zhao, C., Huang, Y., and Wu, J. (2020). Multifunctional barium titanate ceramics via chemical modification tuning phase structure. *InfoMat* 2 (6): 1163–1190.

35 Huang, Y., Zhao, C., Yin, J. et al. (2019). Giant electrostrictive effect in lead-free barium titanate-based ceramics via A-site ion-pairs engineering. *J. Mater. Chem. A* 7 (29): 17366–17375.

36 He, L., Wang, D., Xu, M. et al. (2021). Large electrostrain with nearly-vanished hysteresis in eco-friendly perovskites by building coexistent glasses near quadruple point. *Nano Energy* 90: 106519.

37 Huang, Y., Zhao, C., and Wu, J. (2018). Giant electrostrictive responses and temperature insensitive strain in barium titanate-based ceramics. *Adv. Electron. Mater.* 4 (7): 1800075.

38 Jin, L., Huo, R., Guo, R. et al. (2016). Diffuse phase transitions and giant electrostrictive coefficients in lead-free Fe^{3+}-doped $_{0.5}Ba(Zr_{0.2}Ti_{0.8})O_3$–$0.5(Ba_{0.7}Ca_{0.3})TiO_3$ ferroelectric ceramics. *ACS Appl. Mater. Interfaces* 8 (45): 31109–31119.

39 Li, F., Jin, L., and Guo, R. (2014). High electrostrictive coefficient Q_{33} in lead-free $Ba(Zr_{0.2}Ti_{0.8})O_3$–$x(Ba_{0.7}Ca_{0.3})TiO_3$ piezoelectric ceramics. *Appl. Phys. Lett.* 105 (23): 232903.

40 Wang, L., Qi, H., Deng, S. et al. (2023). Design of superior electrostriction in $BaTiO_3$-based lead-free relaxors via the formation of polarization nanoclusters. *InfoMat* 5 (1): e12362.

41 Zhong, W., King-Smith, R., and Vanderbilt, D. (1994). Giant LO-TO splittings in perovskite ferroelectrics. *Phys. Rev. Lett.* 72 (22): 3618.

42 Zhou, X., Xue, G., Luo, H. et al. (2021). Phase structure and properties of sodium bismuth titanate lead-free piezoelectric ceramics. *Prog. Mater Sci.* 122: 100836.

43 Hao, J., Li, W., Zhai, J., and Chen, H. (2019). Progress in high-strain perovskite piezoelectric ceramics. *Mater. Sci. Eng., R* 135: 1–57.

44 Wu, J. (2018). *Advances in Lead-free Piezoelectric Materials*. Springer.

45 Viola, G., Tian, Y., Yu, C. et al. (2021). Electric field-induced transformations in bismuth sodium titanate-based materials. *Prog. Mater Sci.* 122: 100837.

46 Li, F., Zhang, S., Damjanovic, D. et al. (2018). Local structural heterogeneity and electromechanical responses of ferroelectrics: learning from relaxor ferroelectrics. *Adv. Funct. Mater.* 28 (37): 1801504.

47 Acosta, M., Novak, N., Rojas, V. et al. (2017). $BaTiO_3$-based piezoelectrics: fundamentals, current status, and perspectives. *Appl. Phys. Rev.* 4 (4): 041305.

48 Shvartsman, V.V. and Lupascu, D.C. (2012). Lead-free relaxor ferroelectrics. *J. Am. Ceram. Soc.* 95 (1): 1–26.

49 Kleemann, W. (2006). *Frontiers of Ferroelectricity*. Springer.

50 Yin, J., Shi, X., Tao, H. et al. (2022). Deciphering the atomic-scale structural origin for large dynamic electromechanical response in lead-free $Bi_{0.5}Na_{0.5}TiO_3$-based relaxor ferroelectrics. *Nat. Commun.* 13: 6333.

51 Jin, L., Qiao, J., Hou, L. et al. (2018). A strategy for obtaining high electrostrictive properties and its application in barium stannate titanate lead-free ferroelectrics. *Ceram. Int.* 44 (17): 21816–21824.

52 Shi, T., Xie, L., Gu, L., and Zhu, J. (2015). Why Sn doping significantly enhances the dielectric properties of $Ba(Ti_{1-x}Sn_x)O_3$. *Sci. Rep.* 5 (1): 1–4.

53 Jin, L., Luo, W., Hou, L. et al. (2019). High electric field-induced strain with ultra-low hysteresis and giant electrostrictive coefficient in barium strontium titanate lead-free ferroelectrics. *J. Eur. Ceram. Soc.* 39 (2–3): 295–304.

54 Ang, C., Yu, Z., Jing, Z. et al. (2002). Piezoelectric and electrostrictive strain behavior of Ce-doped $BaTiO_3$ ceramics. *Appl. Phys. Lett.* 80 (18): 3424–3426.

55 Fang, M., Ji, Y., Zhang, Z. et al. (2018). Re-entrant relaxor-ferroelectric composite showing exceptional electromechanical properties. *NPG Asia Mater.* 10 (11): 1029–1036.

56 Lu, X., Wang, L., Jin, L. et al. (2018). Ultra-low hysteresis electrostrictive strain with high thermal stability in $Bi(Li_{0.5}Nb_{0.5})O_3$-modified $BaTiO_3$ lead-free ferroelectrics. *J. Alloys Compd.* 753: 558–565.

57 Huang, Y., Zhao, C., Wu, B., and Wu, J. (2020). Multifunctional $BaTiO_3$-based relaxor ferroelectrics toward excellent energy storage performance and electrostrictive strain benefiting from crossover region. *ACS Appl. Mater. Interfaces* 12 (21): 23885–23895.

58 Zhang, S.T., Kounga, A.B., Jo, W. et al. (2009). High-strain lead-free antiferroelectric electrostrictors. *Adv. Mater.* 21 (46): 4716–4720.

59 Wu, J., Xiao, D., and Zhu, J. (2015). Potassium-sodium niobate lead-free piezoelectric materials: past, present, and future of phase boundaries. *Chem. Rev.* 115 (7): 2559–2595.

60 Li, J.F., Wang, K., Zhu, F.Y. et al. (2013). $(K, Na)NbO_3$-based lead-free piezoceramics: fundamental aspects, processing technologies, and remaining challenges. *J. Am. Ceram. Soc.* 96 (12): 3677–3696.

61 Mishra, S., Choudhury, N., Chaplot, S. et al. (2007). Competing antiferroelectric and ferroelectric interactions in $NaNbO_3$: neutron diffraction and theoretical studies. *Phys. Rev. B* 76 (2): 024110.

62 Aydi, A., Khemakhem, H., Boudaya, C. et al. (2004). New ferroelectric and relaxor ceramics in the mixed oxide system $NaNbO_3$–$BaSnO_3$. *Solid State Sci.* 6 (4): 333–337.

63 Xu, H., Navrotsky, A., Su, Y., and Balmer, M.L. (2005). Perovskite solid solutions along the $NaNbO_3$-$SrTiO_3$ join: phase transitions, formation enthalpies, and implications for general perovskite energetics. *Chem. Mater.* 17 (7): 1880–1886.

64 Xie, A., Zuo, R., Qiao, Z. et al. (2021). $NaNbO_3$–$(Bi_{0.5}Li_{0.5})TiO_3$ lead-free relaxor ferroelectric capacitors with superior energy-storage performances via multiple synergistic design. *Adv. Energy Mater.* 11 (28): 2101378.

65 Zuo, R., Qi, H., Fu, J. et al. (2017). Multiscale identification of local tetragonal distortion in $NaNbO_3$–$BaTiO_3$ weak relaxor ferroelectrics by Raman, synchrotron X-ray diffraction, and absorption spectra. *Appl. Phys. Lett.* 111 (13): 132901.

66 Zuo, R., Qi, H., and Fu, J. (2016). Morphotropic $NaNbO_3$–$BaTiO_3$–$CaZrO_3$ lead-free ceramics with temperature-insensitive piezoelectric properties. *Appl. Phys. Lett.* 109 (2): 022902.

67 Wu, J., Fan, Z., Xiao, D. et al. (2016). Multiferroic bismuth ferrite-based materials for multifunctional applications: ceramic bulks, thin films and nanostructures. *Prog. Mater Sci.* 84: 335–402.

68 Wang, T.-H., Tu, C.-S., Ding, Y. et al. (2011). Phase transition and ferroelectric properties of $x BiFeO_3$–$(1-x)BaTiO_3$ ceramics. *Curr. Appl Phys.* 11 (3): S240–S243.

69 Kumar, U., Cross, L.E., and Halliyal, A. (1992). Pyroelectric and electrostrictive properties of $(1-x-y)$PZN–xBT–yPT ceramic solid solutions. *J. Am. Ceram. Soc.* 75 (8): 2155–2164.

70 Halliyal, A. and Safari, A. (1994). Synthesis and properties of lead zinc niobate: $Pb(Zn_{1/3}Nb_{2/3})O_3$ based relaxor ferroelectrics. *Ferroelectrics* 158 (1): 295–300.

71 Liu, S.-F., Park, S.-E., Cross, L.E., and Shrout, T.R. (2002). Temperature dependence of electrostriction in rhombohedral $Pb(Zn_{1/3}Nb_{2/3})O_3$–$PbTiO_3$ single crystals. *J. Appl. Phys.* 92 (1): 461–467.

72 Li, J., Li, J., Wu, H.-H. et al. (2021). Influence of phase transitions on electrostrictive and piezoelectric characteristics in PMN–30PT single crystals. *ACS Appl. Mater. Interfaces* 13 (32): 38467–38476.

73 Li, F., Jin, L., Xu, Z. et al. (2013). Electrostrictive effect in $Pb(Mg_{1/3}Nb_{2/3})O_3$–$x PbTiO_3$ crystals. *Appl. Phys. Lett.* 102 (15): 152910.

74 Lima, J., Paraguassu, W., Freire, P. et al. (2009). Lattice dynamics and low-temperature Raman spectroscopy studies of PMN-PT relaxors. *J. Raman Spectrosc.* 40 (9): 1144–1149.

75 Zeng, J., Zhao, K., Ruan, W. et al. (2016). Contribution to the large and stable electric field induced strain for textured $Pb(Mg_{1/3}Nb_{2/3})_{0.675}Ti_{0.325}O_3$ ceramics. *Appl. Phys. Lett.* 109 (5): 052905.

76 Liu, Y., Deng, S., Li, J. et al. (2022). High-performance electrostrictive relaxors with dispersive endotaxial nanoprecipitations. *Adv. Mater.* 34 (36): 2204743.

77 Liu, Y., Aziguli, H., Zhang, B. et al. (2018). Ferroelectric polymers exhibiting behaviour reminiscent of a morphotropic phase boundary. *Nature* 562 (7725): 96–100.

78 Kang, G.-D. and Cao, Y.-M. (2014). Application and modification of poly(vinylidene fluoride) (PVDF) membranes – a review. *J. Membr. Sci.* 463: 145–165.

79 Zhu, Z., Rui, G., Li, R. et al. (2021). Enhancing electrostrictive actuation via strong electrostatic repulsion among field-induced nanodomains in a relaxor ferroelectric poly(vinylidene fluoride-*co*-trifluoroethylene-*co*-chlorotrifluoroethylene) random terpolymer. *ACS Appl. Mater. Interfaces* 13 (35): 42063–42073.

80 Kim, I.J., Cho, K.Y., Kim, E. et al. (2020). Development of high dielectric electrostrictive PVDF terpolymer blends for enhanced electromechanical properties. *Nanomaterials* 11 (1): 6.

81 Shi, P.-P., Tang, Y.-Y., Li, P.-F. et al. (2016). Symmetry breaking in molecular ferroelectrics. *Chem. Soc. Rev.* 45 (14): 3811–3827.

82 Pan, Q., Xiong, Y.-A., Sha, T.-T., and You, Y.-M. (2021). Recent progress in the piezoelectricity of molecular ferroelectrics. *Mater. Chem. Front.* 5 (1): 44–59.

83 Zhang, H.-Y., Zhang, Z.-X., Chen, X.-G. et al. (2021). Large electrostrictive coefficient in a two-dimensional hybrid perovskite ferroelectric. *JACS* 143 (3): 1664–1672.

84 Huang, C.R., Li, Y., Xie, Y. et al. (2021). The first high-temperature supramolecular radical ferroics. *Angew. Chem. Int. Ed.* 60 (30): 16668–16673.

85 Long, G., Sabatini, R., Saidaminov, M.I. et al. (2020). Chiral-perovskite optoelectronics. *Nat. Rev. Mater.* 5 (6): 423–439.

86 Zhang, T., Xu, K., Li, J. et al. (2022). Ferroelectric hybrid organic-inorganic perovskites and their structural and functional diversity. *Natl. Sci. Rev.* 10 (2): nwac240.

87 Chen, B., Li, T., Dong, Q. et al. (2018). Large electrostrictive response in lead halide perovskites. *Nat. Mater.* 17 (11): 1020–1026.

88 Li, W., Man, Z., Zeng, J. et al. (2021). Poling effect on the electrostrictive and piezoelectric response in $CH_3NH_3PbI_3$ single crystals. *Appl. Phys. Lett.* 118 (15): 151905.

89 Gao, X., Yang, J., Wu, J. et al. (2020). Piezoelectric actuators and motors: materials, designs, and applications. *Adv. Mater. Technol.* 5 (1): 1900716.

90 Hong, C.-H., Kim, H.-P., Choi, B.-Y. et al. (2016). Lead-free piezoceramics – where to move on? *J. Materiomics* 2 (1): 1–24.

91 Hall, A., Allahverdi, M., Akdogan, E., and Safari, A. (2005). Piezoelectric/electrostrictive multimaterial PMN-PT monomorph actuators. *J. Eur. Ceram. Soc.* 25 (12): 2991–2997.

92 Cottinet, P.-J., Guyomar, D., Guiffard, B. et al. (2011). Electrostrictive polymer composite for energy harvesters and actuators. *J. Polym. Eng.* 31: 133–140.

93 Hafner, J., Teuschel, M., Disnan, D. et al. (2021). Large bias-induced piezoelectric response in the ferroelectric polymer P(VDF-TrFE) for MEMS resonators. *Mater. Res. Lett.* 9 (4): 195–203.

94 Lallart, M., Cottinet, P.J., Guyomar, D., and Lebrun, L. (2012). Electrostrictive polymers for mechanical energy harvesting. *J. Polym. Sci., Part B: Polym. Phys.* 50 (8): 523–535.

95 Benbella, A., Matrane, I., Eddiai, A. et al. (2022). Role of polarization signals effects of electrostrictive polymers in the energy harvesting. *Mater. Today Proc.* 66: 359–364.

13

Ferroelectric Materials for Dielectric Energy Storage: Fundamentals and Applications

Haibo Zhang, Hua Tan, Shuaikang Huang, and Mohsin A. Marwat

Huazhong University of Science and Technology, School of Materials Science and Engineering, State Key Laboratory of Material Processing and Die and Mould Technology, Luoyu Road 1037, Wuhan 430074, PR China

13.1 Introduction

In recent years, with the significant increase in energy consumption, the demand for the development of environmentally friendly and low-cost renewable energy sources has been increasing, and the worldwide consumption of energy has increased significantly [1]. Electricity, as a secondary energy source, plays a leading role in human daily life, industrial manufacturing, and scientific research because of its cost-effectiveness, versatility, and ease of transportation. Compared with traditional fossil fuels, electric energy generated from renewable resources can effectively cope with resource depletion and reduce environmental pollution [2]. Electric energy storage is an important topic in the field of energy storage, in addition to the fossil energy, solar energy, and wind energy. Currently, the mainstream energy storage devices include batteries, dielectric capacitors, electrochemical capacitors, and fuel cells. Especially, dielectric capacitors are widely used in wearable energy harvesting, inverters, high-voltage power grids, aerospace, oil drilling, and electromagnetic pulse weapons due to their ultra-high charge and discharge rates and extremely high power density [3, 4].

Dielectric capacitors can generally be divided into two main categories (ceramic capacitors and film capacitors) according to the materials used. Ceramic capacitors, also known as multilayer ceramic capacitors, consist of ceramic dielectric foils with printed electrodes (inner electrodes) laminated in a staggered manner, sintered at high temperature to form a ceramic chip, and then sealed with metal layers (outer electrodes) at both ends of the chip to form a monolithic-like structure. With the rapid development of the world electronics industry, multilayer chip ceramic capacitors, a basic component of the electronics industry, are also moving forward at an incredible rate, increasing by 10–15% every year [5, 6]. With the improvement in the reliability and integration of multilayer ceramic capacitors, the scope of their use is getting wider, and they are used in various electronic machines and electronic devices for military and civilian use, such as computers,

Piezoelectric Materials: From Fundamentals to Emerging Applications, First Edition. Edited by Jiagang Wu.

telephones, program-controlled switches, precision test instruments, and radar communications. Film capacitor is constructed by using a metal foil as an electrode and winding it with plastic film, such as polyethylene, polypropylene, polystyrene, or polycarbonate, from both ends to form a cylinder. Film capacitors are mainly used in electronics, home appliances, communication, electric power, electrified railroad, hybrid cars, wind power, and solar power. The steady development of these industries has driven the growth of the film capacitor market. With the development of technology level, the renewal cycle of many industries such as electronics, home appliances, and communication are becoming shorter. Film capacitors, with their good electrical performance and high reliability, have become indispensable electronic components to promote the renewal of the abovementioned industries. In the coming years, with the further development of digitalization, informatization, and network construction and the increased investment of the country in power grid construction, electrified railroad construction, energy-saving lighting, and hybrid cars, as well as the upgrading of consumer electronic products, the market demand of film capacitors will further show a rapid growth trend [7, 8].

The core of these capacitors are dielectric ceramics and dielectric organic materials. This chapter first introduces the fundamentals of energy storage dielectric materials, mainly including key parameters, polarization mechanisms, and breakdown mechanisms. Then, the industrial preparation of two main types of dielectric capacitors, multilayer ceramic capacitors and organic film capacitors, is introduced. Finally, the practical applications of dielectric capacitors are summarized with the national development requirements.

13.2 Fundamentals of Dielectric Energy Storage Materials

13.2.1 Key Parameters of Dielectric Energy Storage Materials

Generally, energy storage density is defined as energy in per unit volume (J/cm^3), which is calculated by

$$W = \int_0^{D_{max}} E dD \tag{13.1}$$

where W is the total energy density; E is the applied electric field; D_{max} is the maximum electric displacement at E; dD is the increment of electric displacement per unit of the electric field [9].

For dielectric materials, D is an unmeasurable microscopic physical quantity and is usually expressed by polarization as follows:

$$D = \varepsilon_0 E + P \tag{13.2}$$

where ε_0 is the vacuum dielectric constant of 8.854×10^{-12} F/m; P is the polarization.

Meanwhile, P is dependent on E as follows:

$$P = \varepsilon_0 \chi E \tag{13.3}$$

where χ is the dielectric polarization coefficient.

So that Eq. (13.2) can be expressed as

$$D = \varepsilon_0(\chi + 1)E = \varepsilon_0\varepsilon_r(E)E \tag{13.4}$$

where $\varepsilon_r(E)$ is the relative dielectric constant at E [10].

Therefore, Eq. (13.1) after the change is as follows:

$$W = \int_0^{E_{max}} \varepsilon_0\varepsilon_r(E)EdE \tag{13.5}$$

In particular, for linear dielectrics (LDs) (sometimes also for paraelectrics (PEs)), the dielectric constant is electric field independent, so the energy density can be written as:

$$W = \frac{1}{2}\varepsilon_0\varepsilon_r E^2 \tag{13.6}$$

It should be noted that W is the sum of effective energy density (W_{rec}) and energy loss (W_{loss}). In practice, W_{rec} is more important than W in evaluating the energy storage performances of dielectric materials (for LDs, $W_{rec} = W$). W_{rec} is determined by the area enclosed by the discharge curve of its P–E loop and the polarization axis, as shown in Figure 13.1. The equation is given as follows:

$$W_{rec} = \int_{P_r}^{P_{max}} EdP \tag{13.7}$$

Energy storage efficiency (η) is another important parameter to evaluate energy storage performances of dielectric materials, which is expressed as [10]:

$$\eta = \frac{W_{rec}}{W} \times 100\% = \frac{W_{rec}}{W_{rec} + W_{loss}} \times 100\% \tag{13.8}$$

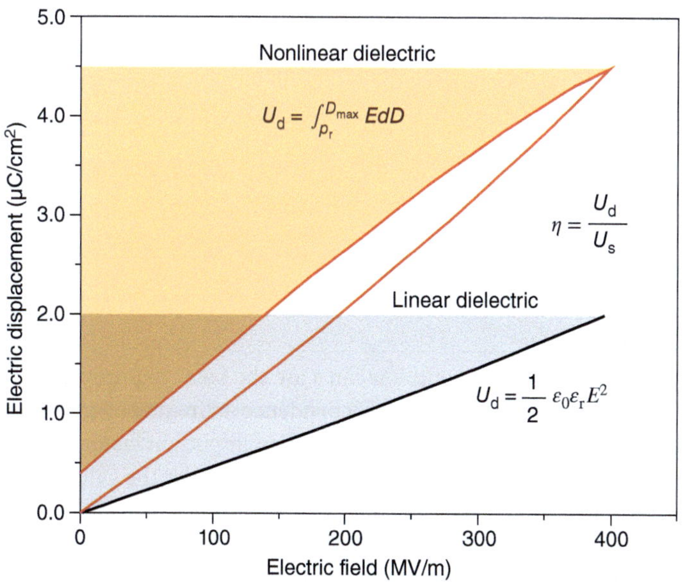

Figure 13.1 Characteristic D–E hysteresis loops of linear and nonlinear dielectrics under the same electric field. The shaded area is the effective energy density of the dielectric. Source: Huang et al. [11]. Reproduced with permission of Taylor & Francis.

where W_{loss} is the energy loss during the discharge process, which equals the area enclosed by the P–E loop in number.

W_{loss} is mainly dissipated as heat, and a higher value means a stronger negative effect on the service life of ceramic capacitors. Therefore, it is vital that P–E loops gradually go slim to enhance η and then improve its practical application.

According to Eqs. (13.7) and (13.8), P–E loops go slim, accompanied by high P_{max}, low P_r (i.e. high $\triangle P = P_{max} - P_r$), and high breakdown strength (E_b), which become the key issues in optimizing the energy storage characteristics of dielectric materials.

Energy storage performances of dielectric materials can be reflected by P–E loops measured by a modified Sawyer–Tower circuit. Meanwhile, the energy storage characteristics of capacitors, including effective discharging time ($t_{0.9}$) and power density (P), are more accurately reflected by the charging–discharging curve recorded at a specific RLC circuit (circuit consisting of resistance, inductance, capacitance). In the charging process, the potential difference (Φ) between two surfaces of the capacitor equals the applied voltage (V), representing the charging process is finished. Then, the vacuum switches automatically and quickly rotate to achieve the discharging process [12]. Current flows to the oscilloscope via a load (R_0), which is recorded as wave function as a function of time. The discharging density is given by

$$J(t) = \frac{\int_0^t V(t)I(t)dt}{\varphi} = \frac{\tau V_0^2}{2\varphi R_0}\left(1 - e^{-\frac{2t}{\tau}}\right) \tag{13.9}$$

where φ is the effective volume; τ is the relaxation time; $V(t)$ is the voltage as a function of time; $I(t)$ is the current as a function of time.

It is reported that the value of J calculated by Eq. (13.9) is generally smaller than that of W_{rec}. For the reasons, it may be closely related to two factors. First, the domains cannot switch and orientate promptly due to fast discharge speed, and thus, the energy is not completely released. Second, the circuit has an equivalent series resistance that generates Joule energy. In order to acquire higher value of J and W_{rec}, reducing domain size (relaxor ferroelectrics [RFEs]) may be an effective method.

Power density (P) is also an important parameter of dielectric ceramic capacitors, which is determined as follows [13]:

$$P = \frac{\pi f \varepsilon_0 \varepsilon_r E_b^2}{2 \tan \delta} \tag{13.10}$$

where f is the testing frequency.

It can be seen that high ε_r, high E_b, and low $\tan \delta$ are the basic requirements for achieving high P. Meanwhile, P has an obvious dependence on frequency (f), which means high f corresponds to higher P. In addition, $t_{0.9}$ is an important factor in influencing the practical application of ceramic capacitors. The smaller the value of $t_{0.9}$, the stronger the pulse current can be generated in a short time. It is worth mentioning that the value of $t_{0.9}$ is dependent on not only the intrinsic properties of the dielectric materials but also some external factors [14].

Whether considering dielectric materials or pulse capacitors, the reliability of work is a crucial factor influencing their application scenes. Generally, the reliability of work requires keeping the stability of electric properties of material and device under thermal, electric, mechanical, magnetic, and external field stimulus.

Temperature stability usually requires that dielectric properties and polarization of the material have a gentle fluctuation as a function of temperature. For example, the dielectric constant needs to maintain a ±15% variation over a temperature range from −55 to 125 °C (X7R), to 150 °C (X8R), and even to 300 °C, especially in high-temperature working environments. Therefore, one should know the Joule heat categories: dielectric loss or leakage current, and to solve corresponding issues. Generally, different electric field conditions such as cycle number, frequency, voltage category, and magnitude, all require keeping good energy storage stability. In particular, fatigue endurance is a very important issue to influence capacitor working capability. In addition, mechanics and magnetic fields both influence the polarization behavior of materials, and their importance may play a crucial role in future multifunctional coupling requirements. Therefore, the evaluation criterion of reliability of working should be a complicated project.

13.2.2 Polarization Mechanisms

The key parameters in energy storage materials were introduced in the Section 13.2.1, and it can be seen that the polarization strength and breakdown strength play a decisive role in the energy storage density of a material, and the mechanism of polarization and breakdown will be briefly introduced below.

Polarization is an important indicator to evaluate the dielectric and energy storage properties of dielectric materials. Polarization is defined as total dipole moments in a dielectric per unit volume [15–18]. It is related to the dielectric constant under the homogeneous applied field and is calculated as follows [19]:

$$P = (\varepsilon_r - 1)\varepsilon_0 E \tag{13.11}$$

where P is the dielectric polarization; ε_0 is the permittivity of vacuum; E is the applied electric field. Based on the relaxation and resonance regimes, the polarization can be divided into four different types: electronic polarization (P_e), ionic polarization (P_i), dipolar polarization (P_d), and interfacial polarization (P_{int}), as shown in Figure 13.2 [16, 20].

The mechanism of electronic polarization is shown in Figure 13.3a. Atoms have a symmetrical structure, which exhibits electrical neutrality in the absence of an electric field. When an electric field is applied, the lighter electrons move more readily than the nucleus. The center of the negative charge deviates from the center of the positive charge, which creates an induced dipole moment called electronic polarization [21, 22]. Electronic polarization exists in all dielectrics, but the dipole moment generated by electronic polarization is very small and will disappear after removing the external electric field.

The principle of ionic polarization is shown in Figure 13.3b. In the initial state, the dipole moments created by anions and cations are neutralized by each other, and the whole does not exhibit polarization characteristics. When an electric field is applied in a certain direction, cations move in the positive direction of the electric field, while

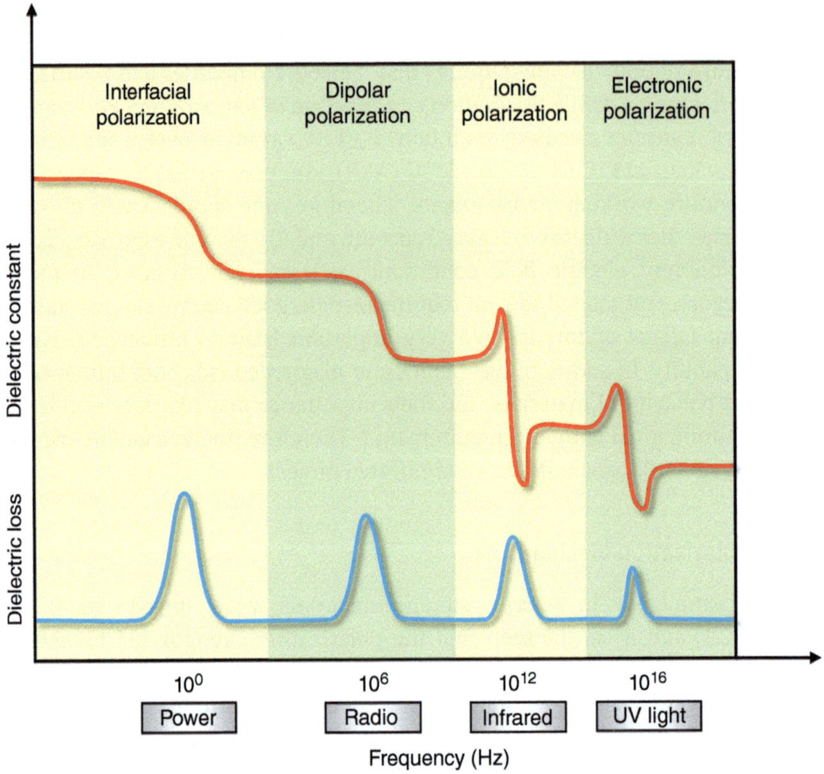

Figure 13.2 Different types of polarizations and their frequency dependencies. Dielectric constant and corresponding losses depicted with red and blue lines, respectively.

anions move in the opposite direction. The positive dipole moment increases, and the negative dipole moment decreases in the dielectric, resulting in ionic polarization [15, 17]. The enhancement of ionic polarization can lead to a large improvement in polarization properties, which is one of the sources of the high ε_r of ceramics such as $BaTiO_3$. Electronic polarization and ionic polarization are intrinsic to the molecule, which can follow the electric field rapidly and are independent of temperature [23].

Dipolar polarization, also known as orientation polarization, is the orientation of the intrinsic dipoles of molecules or molecular chains in the direction of the electric field. As shown in Figure 13.3c, under an electric field, the internal dipole is subjected to electric field forces and has a tendency to form an ordered arrangement in the electric field direction, thus exhibiting orientation polarization [24]. Temperature and frequency affect the dipolar polarization of dielectrics. Too high or too low temperature will hinder the orderly arrangement of dipoles and thus reduce the dipole polarization of the dielectric. As the frequency of the electric field increases, the change in dipole orientation cannot follow the change in the electric field, which results in a weak dipole polarization [25].

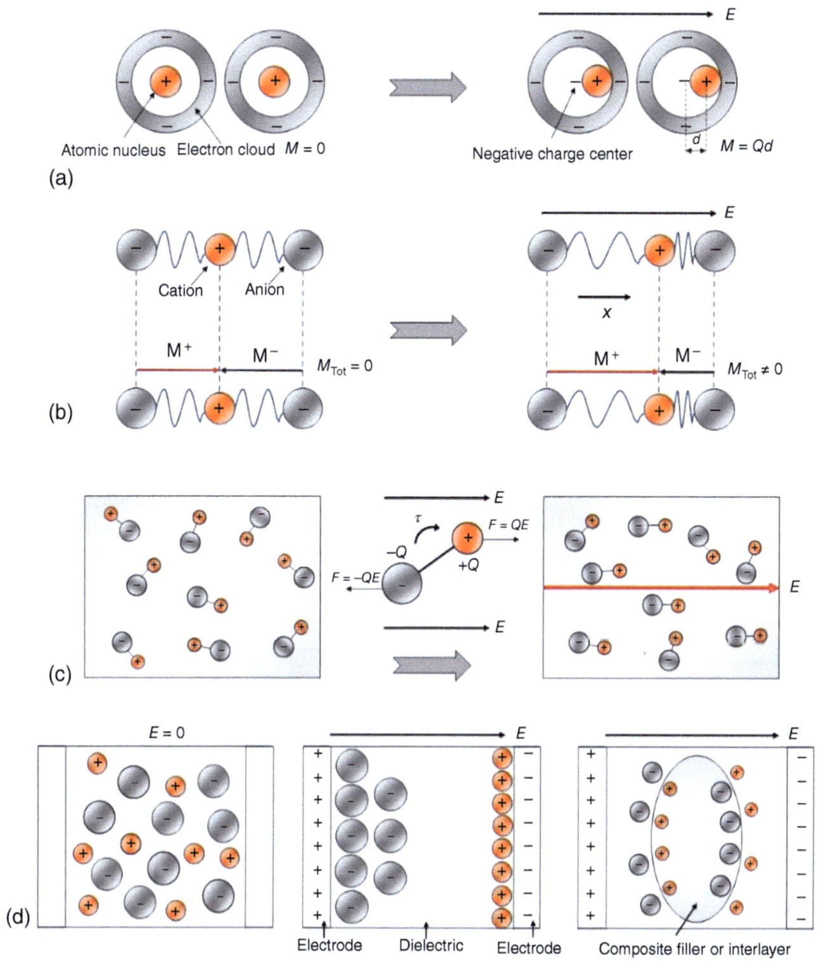

Figure 13.3 Schematic diagrams of the four polarizations: (a) electronic polarization, (b) ionic polarization, (c) dipolar polarization, (d) interfacial polarization. Source: Huang et al. [11]. Reproduced with permission of Taylor & Francis.

Interfacial polarization, also called Maxwell–Wagner–Sillars polarization, occurs mainly at the two-phase interface of dielectrics. As shown in Figure 13.3d, interfacial polarization is generated by the accumulation of interfacial charges, which is related to the difference in the dielectric properties of the materials on both sides of the interface [26]. The interfacial polarization is the main reason for the high ε_r of polymer–inorganic composites. Dipolar polarization and interfacial polarization are greatly affected by temperature, and the ε_r will change accordingly. During the discharge process, the discharge of dipolar polarization and interfacial polarization is slow. It takes a certain time for the dipole to return to the equilibrium state, which can only occur in the frequency range shown in the relaxation region of Figure 13.2 [18].

13.2.3 Breakdown Mechanisms

The breakdown of solid dielectrics is a transformation process in which the dielectric changes from an insulating state to a good conductive state under high electric fields. The lowest critical voltage that causes breakdown is called the breakdown voltage or breakdown strength [27–29]. In order to determine the dielectric breakdown strength, the two-parameter Weibull distribution function is widely utilized for the analysis. The cumulative function of Weibull distribution is expressed as follows:

$$P(E) = 1 - \exp\left[-\left(\frac{E_{\mathrm{i}}}{E_{\mathrm{w}}}\right)^{\beta}\right] \tag{13.12}$$

where $P(E)$ is the cumulative probability of electrical failure under a certain electric field; E_{i} is the experimental breakdown strength; E_0 is the Weibull characteristic breakdown strength, which represents the cumulative failure probability of 63.2%; β is the shape factor related to the scatter of the breakdown data; and a higher β value means a higher reliability of the data.

The breakdown strength of polymer dielectrics is affected by its own characteristics, such as the thickness, band gap, grain size, microstructure, and geometry of the tested sample. Due to the randomness of the breakdown experiment, the same material can exhibit a wide range of breakdown strengths. Except for the material sample itself, the measured breakdown strength is influenced by the applied voltage type and frequency, the electrode shape and contact with the tested sample, the ambient medium, the environmental conditions (temperature, humidity, and pressure), and so on [29–31]. Although the sophisticated mechanisms of dielectric breakdown behaviors are not yet fully understood, some conventional breakdown theories are used to explain the breakdown process of dielectric materials. Breakdown can be divided into intrinsic breakdown, electromechanical breakdown, thermal breakdown, and partial discharge breakdown according to different principles. In the following, we will introduce several breakdown mechanisms in detail.

13.2.3.1 Intrinsic Breakdown

There are two main types of intrinsic breakdown mechanism: electronic breakdown and avalanche breakdown [14]. Electronic breakdown is assumed to occur when electrons in dielectrics gain sufficient energy from the electric field to cross the forbidden energy gap between the valence band and conduction band, as shown in Figure 13.4a. More electrons are accumulated in the conduction band when this process is repeated, eventually leading to breakdown [33]. In ideal dielectrics, the conduction band and the valence band are separated by a large energy gap. In practice, however, dielectrics usually contain some imperfections in their structure, such as missing atoms (vacancies) and foreign atoms (impurities). These imperfections may trap free electrons and form defect energy levels separated by the energy ΔV from the bottom of the conduction band. The electronic breakdown strength (E_{ele}) can be expressed by [14]:

$$E_{\mathrm{ele}} = \frac{A_{\mathrm{ele}}}{\sqrt{\Delta V}} \exp\left(\frac{\Delta V}{2k_{\mathrm{B}}T}\right) \tag{13.13}$$

where A_{ele} is a constant; k_{B} is the Boltzmann constant; T is the temperature in K.

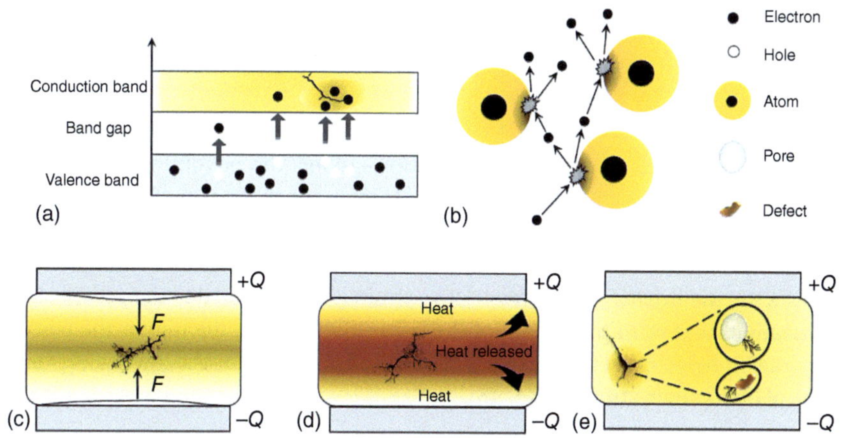

Figure 13.4 Schematic diagram of different types of breakdown mechanisms: (a) electronic breakdown, (b) electron avalanche breakdown, (c) electromechanical breakdown, (d) thermal breakdown, and (e) partial discharge breakdown. Source: Liu et al. [32]. Reproduced with permission of The Royal Society of Chemistry.

The other type of intrinsic breakdown mechanism is avalanche breakdown. An electron within a dielectric drifts from the cathode toward the anode under the applied electric field [34, 35]. As shown in Figure 13.4b, the electron can gain energy from the field and lose energy by colliding with lattice atoms. If the gained energy exceeds the ionization potential, an additional electron will be liberated due to the collision with the first electron. This process is repeated like a chain reaction, leading to an electron avalanche [36, 37]. Breakdown will occur when the avalanche exceeds a critical size. The critical avalanche size requires about 10^{12} electrons, corresponding to 40 ionizing collisions by the initiating electron. Thus, the critical avalanche breakdown strength in a dielectric with thickness d can be expressed by:

$$E_{ava} = \frac{k}{\ln\left(\frac{a_0 d}{40}\right)} \tag{13.14}$$

where k and a_0 are constants depending on the dielectrics. It can be seen that the thicker the dielectric is, the lower the critical breakdown strength will be, and this is one of the reasons why thin films can survive higher electric fields compared to bulk ceramics. The collision ionization process proceeds as a chain reaction causing the number of electrons to increase sharply, resulting in an electron avalanche. The avalanche breakdown shows a high-temperature dependence [38–40].

13.2.3.2 Electromechanical Breakdown

High voltage applied to a dielectric induces charges on its surface. The electrostatic attraction of the surface charges results in compressive forces, as shown in Figure 13.4c. When the compressive forces on the dielectric exceed its mechanical compressive strength, the dielectric collapses [41–43]. If the substance deforms without plastic deformation setting in, the highest electromechanical breakdown strength (E_{em}) is:

$$E_{em} = \frac{d}{d_0} E_c \approx 0.6\sqrt{\frac{Y}{\varepsilon_r \varepsilon_0}} \tag{13.15}$$

where the factor 0.6 corresponds to the point where mechanical instability occurs and Y is the Young's modulus. This equation ignores the possibility of instability occurring at a lower electric field because of stress concentration at imperfections, the dependence of Y on time, or stress and plastic flow [44]. In addition, electromechanical breakdown is also associated with electrostriction, which is ubiquitous in all dielectric materials. The electrostrictive strain is proportional to the square of the polarization, that is, to the square of the permittivity. Thus, the higher the permittivity, the higher the electrostrictive strain, which accounts for the lower electromechanical breakdown strength [45, 46].

13.2.3.3 Thermal Breakdown

Thermal breakdown originates from the internal thermal instability of dielectrics. Under the action of a high electric field, dielectric loss can cause energy dissipation to be consumed as heat, which increases the temperature of the dielectrics. The enhanced temperature increases the conductivity of the dielectrics exponentially, and the resulting higher conductivity loss will bring more heat. Thermal breakdown may occur when the rate of heat generation exceeds the rate of heat dissipation, as shown in Figure 13.4d [47–49]. Thermal breakdown is related to the characteristics of dielectrics and is affected by the cooling conditions and ambient temperature during operation. Thermal breakdown is the main cause of the high-temperature breakdown of many dielectric materials, and the existence of thermal breakdown greatly reduces the safety and cycle life of capacitors at high temperatures [50–53].

13.2.3.4 Partial Discharge Breakdown

Under the working voltage, long-term breakdown is related to partial discharge, which includes external (e.g. at particle/matrix interfaces), internal (e.g. in voids and pores), as well as corona discharges [54–56]. Normally, the form of partial discharge is shown as a pulse with a duration of less than 1 μs. Among these, internal discharges are more common and serious in partial discharges. It is inevitable that the dielectric materials contain cavities and pores. These cavities and pores are filled with some impurities with a lower dielectric constant than that of the matrix material, which causes the electric field near cavities and pores to be higher than that at other positions. Moreover, cavities and pores can withstand lower breakdown strength due to structural defects. Therefore, partial discharge and local breakdown are more prone to occur in the cavities and pores [57–59]. Short-term partial discharge has little effect on the insulation performance of the dielectric material. However, long-term partial discharge can cause electrical treeing, erosion, and degradation of dielectric materials, which will lead to a decrease in the overall insulation properties of dielectrics. Under the long-term influence of the electric field, temperature, and other factors, the solid dielectrics undergo a slow chemical change, gradually deteriorating their insulating ability, and finally changes from an insulating state to a good conductive state under the normal working voltage [57].

13.3 Design and Performance of Dielectric Energy Storage Devices

13.3.1 Multilayer Ceramics Capacitors

Multilayer ceramic capacitors (MLCCs) are an important class of chip components due to their small size, low dielectric loss, high specific capacitance, and low price. MLCCs are used in a large number of electronic products, such as cell phones, scanners, and computers, due to their small size, low dielectric loss, high specific capacitance, and low price. The rapid development of MLCC and its good application prospects have led to the development of related dielectric materials, whose application areas require high dielectric constants ε, good dielectric properties $\Delta C/C$, high insulation resistivity ρ, and low dielectric loss, tan δ [60, 61].

From the present reports, the main dielectric systems used for MLCC are $BaTiO_3$-based, lead-based, $BaTiO_3$, and lead-based composite and tungsten bronzite dielectric ceramic materials. Although most of the dielectric materials used in MLCCs, whether at the experimental stage or commercially available, are $BaTiO_3$-based materials, this is related to the characteristics of $BaTiO_3$ itself, which is a typical perovskite structure with an internal O-atom forming an oxygen octahedral structure and a Ti atom located at the center of the oxygen octahedron, which has a large displacement space for Ti ions due to the large space of the oxygen octahedron. In particular, the temperature dependence of the dielectric constant near the Curie temperature T_C (about 130 °C) has a strong influence on the dielectric constant, and it is found that when the temperature is higher than the Curie temperature T_C, the dielectric constant of $BaTiO_3$ can be drastically reduced. The pure $BaTiO_3$ has the disadvantages of a large temperature change rate, easy aging of the dielectric, high dielectric loss, and does not meet the application requirements of the product, so it can be used in the actual product by means of doping modification, microstructure adjustment, and process optimization to obtain a more desirable effect. Here, we first introduce the preparation process of MLCCs; the schematic is shown in Figure 13.5 [62].

13.3.1.1 Preparation of Casting Slurry

The casting slurry with good fluidity is prepared by mixing the calcined ceramic powder with a solvent, suspension dispersant, binder, and plasticizer at certain ratios. The volume fraction of the ceramic powder in the slurry is often greater than 50% to ensure the qualified strength and relative density of the ceramic layer.

13.3.1.2 Tape-casting Process

The mixed slurry is then transferred to the tape-casting machine, where the slurry is scraped on the tape by a doctor blade. The thickness of the casted slurry and the resultant ceramic green body is controlled by adjusting the height of the blade. After volatilization of the excessive solvent in the slurry, a ceramic tape green body with good elasticity and flexibility can be obtained.

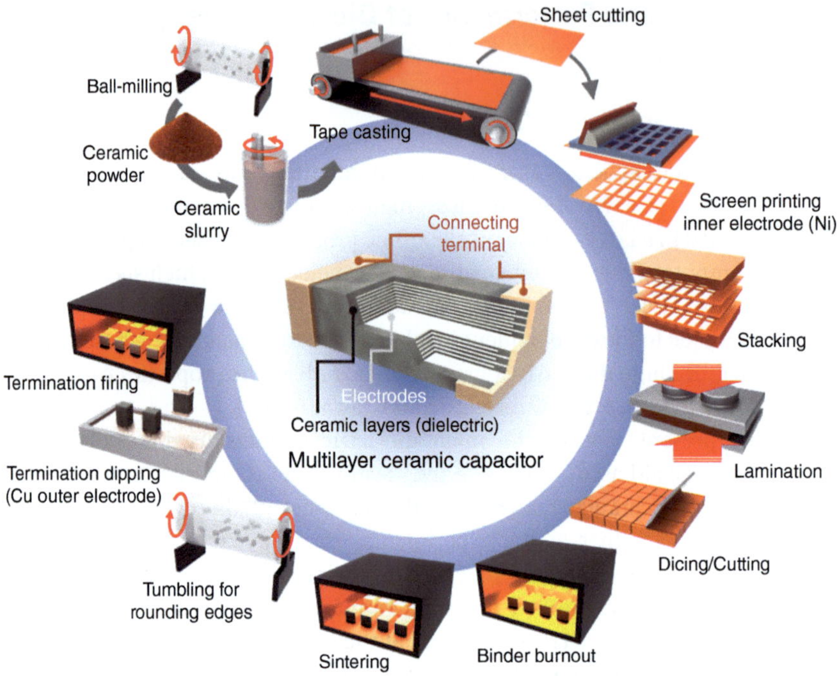

Figure 13.5 Schematics of the MLCC architecture and its fabrication process. Source: Hong et al. [62]. Reproduced with permission of The Royal Society of Chemistry.

13.3.1.3 Preparation of Inner Electrodes

The dried green body is cut into a designed shape, and a thin electrode layer is printed on one side of the tape by screen printing.

13.3.1.4 Lamination

Tens of layers of green bodies printed with internal electrodes are stacked. The electrodes of adjacent green bodies are stacked in a misaligned manner to form positive and negative electrodes. Then, the stacked green bodies are subjected to warm isostatic pressing and cut into individual actuator green bodies.

13.3.1.5 Degreasing and Co-firing

The actuator green body is first heated to 200–300 °C at a slow rate to remove the remnant organics. Then, the heating temperature is raised to the ceramic sintering temperature for the co-firing of the ceramic layer and the inner electrodes.

13.3.1.6 External Electrode Printing, Packaging, and Testing

The sintered ceramic is chamfered to fully expose the inner electrode. Then, the electrode slurry was applied to the exposed internal electrodes so that the internal electrode on the same side is connected to form the external electrode. A low-temperature sintering for the ceramic is performed again to ensure the connection of the inner and outer electrodes. Finally, an electrical performance test is performed to eliminate defective products.

The ceramic layers prepared by tape casting often show high density, low porosity, and small grain size. These characteristics are all thought to be conducive to the capacitors to achieve a higher breakdown of the electric field. In theory, the breakdown strength of MLCC increases exponentially with the decrease in the thickness of a single ceramic layer. For example, for the Al_2O_3-modified $BaTiO_3$ dielectric layer-based capacitor, its breakdown field strength is significantly increased from 184 to 66 kV/cm when the thickness of the single dielectric layer reduces from 63 to 12 μm. The energy storage density of MLCC also increased from 1.7 to 4 J/cm^3 [63].

The selection of ceramic materials in MLCCs is one of the most critical factors to determine energy storage performances. With the development of dielectric materials and phase engineering, high energy storage density and efficiency have been achieved for dielectric energy storage applications. For example, the MLCCs based on $0.85BaTiO_3$–$0.15Bi(Zn_{0.5}Ti_{0.5})O_3$ were reported to have an energy storage density of 2.8 J/cm^3 at 330 kV/cm [64]. By optimizing the composition, a reduction in the thickness of each layer to approximately 5 μm, and using the two-step sintering (TSS) method to avoid the grain growth, Wang et al. successfully prepared MLCCs based on $0.87BaTiO_3$–$0.13Bi(Zn_{2/3}(Nb_{0.85}Ta_{0.15})_{1/3})O_3$ (BT–BZNT) with high energy storage density (10.5 J/cm^3) at 1047 kV/cm [65]. Furthermore, BT–BZNT MLCC also shows good thermal stability, with only a relatively low variation (i.e. $<\pm5\%$) can be observed over a temperature range of −75 to 150 °C at 400 kV/cm.

In MLCCs, the electric breakdown of ferroelectric ceramics is often caused by strain-induced mechanical breakdown under the electric field. This strain effect has been found to be strongly linked with the grain orientation of dielectric ceramics in MLCC. For example, by analyzing the ⟨100⟩, ⟨110⟩, and ⟨111⟩-textured single-layer 0.65BNT–0.35SBT capacitors using finite element simulation, Li et al. found that ⟨111⟩-oriented ceramic has the smallest local strain and stress concentration at the electrode end. Its elastic energy density is also much lower than ⟨100⟩ and ⟨110⟩ textured samples. As a result, the breakdown field strength of ⟨111⟩-textured 0.65BNT–0.35SBT MLCC can reach up to 1030 kV/cm, accompanied by a large energy storage density of 21.5 J/cm^3 [66]. This is the highest value in the reported lead-free MLCCs so far. Compared with non-textured MLCC (with a breakdown strength of 64 kV/cm), the breakdown strength of the ⟨111⟩ textured sample increased by 16 times (1030 kV/cm). This is primarily attributed to the reduced strains in the textured sample, which has much more pores than the non-textured ceramics. These holes can reduce the possibility of microcrack initiation and crack-driven dielectric breakdown. Thus, the local electric field around the pores is reduced, and the growth rate of the partial discharge tree is also slowed. In addition, the difference between the maximum polarization of non-textured MLCC and ⟨111⟩-textured MLCC is small, which helps to increase the energy storage density.

MLCC is a key electronic component used in almost all areas of electronics, and the number and variety of applications are steadily growing. Over the past few decades, MLCC manufacturing has focused on accelerating miniaturization, improving volumetric efficiency, enhancing environmental stability and reliability, and reducing costs, all of which have been made possible by improvements in material formulation and processing techniques. Most ceramic capacitors are based on $BaTiO_3$, which can be chemically or physically modified to exhibit the desired temperature-stable dielectric properties. Therefore, future research on MLCC dielectric ceramics should focus on the following aspects: improving the dielectric constant of the material and improving the dielectric temperature characteristics. How to find suitable additive components and develop new material systems to increase the upper limit of the working temperature as much as possible while maintaining a relatively high dielectric constant; the performance of $BaTiO_3$-based dielectric materials changes greatly after the introduction of other components, but the microscopic mechanisms and theoretical studies around $BaTiO_3$ materials are not enough, so the basic theoretical studies are also an important research direction for this material.

13.3.2 Organic Thin Film Capacitors

The successful large-scale preparation of dielectric polymer films with excellent macroscopic properties and reasonable structure using mature film preparation processes is an important prerequisite for industrial applications of thin film capacitors. At present, large-scale industrial production of dielectric films mainly includes two processes: (i) polymer solution casting process based on the solvent method; (ii) polymer cast plate biaxial stretching process based on the melt method. As the large-scale film preparation by solution casting is still in its infancy, there are many problems with the quality of the prepared film, while taking into account factors such as environmental protection, the biaxial stretching process is mainly used for large-scale production of industrial films at present [67].

The biaxial orientation process is an important process in the manufacturing of industrial dielectric film capacitors, which enables highly ordered alignment of polymer chains and is suitable for large-scale, large-area, continuous production of industrial capacitor films with uniform thickness, high structural quality isotropy, and superior capacitive performance. As shown in Figure 13.6, the biaxial orientation system includes an extrusion section, an orientation section, rollers and

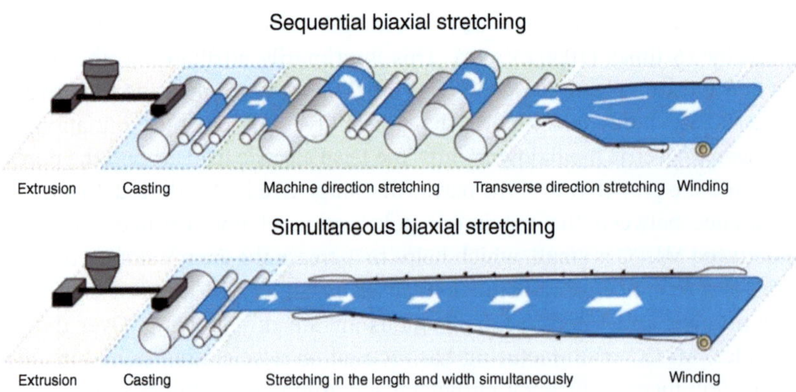

Figure 13.6 Typical tenter-frame biaxial stretching processes for the fabrication of polymer films. Source: Fan et al. [22]. Reproduced with permission of Elsevier Ltd.

clamps, and a winding section, which can be divided into two categories: sequential biaxial orientation and simultaneous biaxial orientation. Conventional biaxially oriented equipment is a double-blown film equipment with a tensioning frame. The film thickness of the double-blown bubble equipment is usually around 15–250 μm, which is too thick for a capacitor, while the tension frame can produce films less than 15 μm thick and is widely used for mass production [68]. The first step of the biaxial orientation process is the extrusion section, where the polymer material is heated to its melting point and extruded through a die to produce flat sheets. The second step is the continuous or simultaneous biaxial stretching of the extruded flat sheets. Two directions are used for stretching: one is machine direction (MD), which is the direction in which the flat sheet moves along the machine (length direction); the other is transverse direction (TD), which is perpendicular to MD (width direction). While simultaneous biaxial orientation is more desirable for improving the quality of stretch films, continuous biaxially oriented films are more popular in large-scale industrial production due to their relatively simple design and lower equipment costs [68].

The two main advantages of biaxial stretching includes : (i) the polymer film after biaxial stretching has a more uniform thickness distribution, which can prevent the electric field distortion due to uneven thickness, thus meeting the demand of high working electric fields. The biaxially stretched polymer film has fewer defects and significantly improved breakdown performance. In addition, compared with other film preparation processes, biaxial stretching can obtain a thinner film thickness. Since the energy density per unit volume is inversely proportional to the film thickness, simultaneous improvement of dielectric permittivity and breakdown strength is a difficult task in the short term, and therefore, it is a feasible and effective method to achieve high energy density film capacitors by reducing the dielectric film thickness. The film thickness obtained by stretching is only 2–3 μm, and the thickness uniformity is good, with a variation of no more than 5% over a large area of film [21]; (ii) biaxial stretching facilitates the complete stretching and orientation of polymer molecular chains, which is very beneficial for improving the crystallinity of films and obtaining high breakdown strength.

The biaxially stretched dielectric polymer film still needs to go through a complex process as shown in Figure 13.7, to finally obtain the finished film capacitor, whose subsequent steps mainly include surface metallization, film slitting, winding, end metallization, healing, end soldering, painting, and end testing [69, 70]. First, hundreds of nanometers of zinc–aluminum alloy need to be deposited on the stretched film surface by physical vapor deposition (PVD) method as the surface electrode. Then the metallized film is slit to the target size by film slitting according to the actual demand, and next, the initial film capacitor is prepared by a high-speed winding process. Then the end face of the original film capacitor is capped and metallized with zinc–aluminum alloy or gold. A healing process is then performed to eliminate film defects with surge voltages. The leads are then soldered to the terminal surfaces of the original film capacitors, followed by a lacquer treatment to protect the film capacitors from damage by external environments. Finally, in order

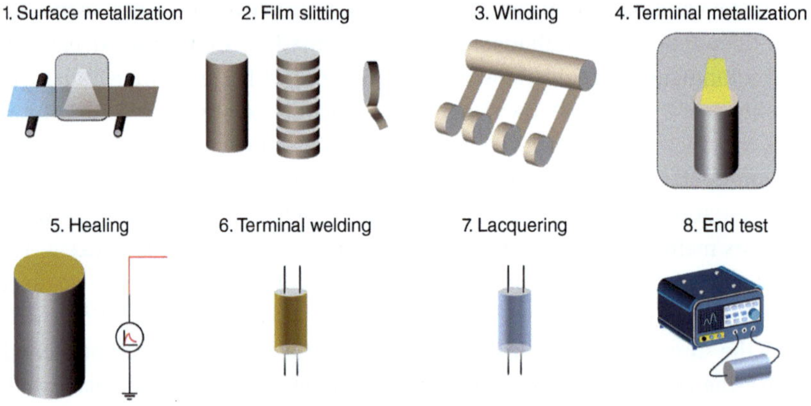

Figure 13.7 Typical manufacturing process flow for the fabrication of polymer film capacitors. Source: Feng et al. [21]. Reproduced with permission of American Chemical Society.

to test the capacitors to meet qualification requirements, end test of the capacitors is required [71].

Despite the excellent properties exhibited by the biaxially oriented film preparation process, which meets the industrial demand for large-scale film preparation, the complex process of preparing capacitors from dielectric polymers still leaves many problems for the conversion of laboratory results to industry. Compared with the industrial requirements for film size (over 100 m in length and 150 mm in width), the area of laboratory-prepared films is much smaller, and it is difficult for the laboratory-obtained films to meet the needs of large-area film preparation processes [72]. In addition, with the gradual increase in energy storage requirements, higher energy density per unit volume of capacitors is required, and the preparation of ultra-thin films (less than 5 μm) is essential. However, as the thickness decreases, the problem of thickness uniformity arises, and small deviations in film thickness can lead to severe electric field concentration and, thus, deterioration of capacitive performance. Therefore, the problem of performance degradation due to large-area film preparation from laboratory results to industrial scale requires the joint efforts of scientists and engineers from all related fields to obtain high-energy density film capacitors suitable for industrial production by rational optimization of the process.

13.4 Applications

In recent years, with the rapid development of defense military technology in the field of electromagnetic energy equipment, such as electromagnetic pulse simulation, directed energy weapon (DEW), and directed energy high technology (DEAT), the requirements of pulse power system for high power output are getting higher. As the core part of the whole large system, the pulse power system, in practical applications, is required to release all the stored energy to the load in the shortest time and to release a large current in an instant, which is the key to directly affect the output of electromagnetic energy equipment. The energy storage module in pulse power can be divided into inductive energy storage, capacitive energy storage, inertial energy storage, and chemical energy storage according to the principles of energy storage. Compared to several commonly used pulse power energy storage components, pulse energy storage capacitors, which can realize instantaneous high power conversion between electromagnetic energy and kinetic energy, have the highest releasable power density (MW level), the highest operating voltage (kV level), fast charging and discharging rate (μs–ns level), and long cycle life (more than 10^4 times), and have an irreplaceable role in the field of pulse power systems. For defense, it can be used as a power source for fast start-up devices of military vehicles such as armored vehicles and tanks and as a pulse power source for new concept laser weapons and electromagnetic rail guns; for civil applications, it can be used in artificial intelligence, offshore oil and gas and geological exploration, medical technology, excimer lasers, and other fields. In addition, as a kind of capacitor with a special purpose, a pulse energy storage capacitor is also a key component planned for development by all countries of the world. Therefore, high-energy density pulse dielectric energy storage devices have received wide attention from academia and industry in recent years. With the urgent need for miniaturization and lightweight development of pulsed power devices, the development of high-performance dielectric energy storage materials with high energy storage density, high energy storage efficiency, and can work under very high electric field is more and more urgent.

References

1 Panwar, N.L., Kaushik, S.C., and Kothari, S. (2011). Role of renewable energy sources in environmental protection: a review. *Renewable Sustainable Energy Rev.* 15: 1513–1524. https://doi.org/10.1016/j.rser.2010.11.037.

2 Sinsel, S.R., Riemke, R.L., and Hoffmann, V.H. (2020). Challenges and solution technologies for the integration of variable renewable energy sources—a review. *Renewable Energy* 145: 2271–2285. https://doi.org/10.1016/j.renene.2019.06.147.

3 Castro, M., Kumar, B., Feller, J.F. et al. (2011). Novel e-nose for the discrimination of volatile organic biomarkers with an array of carbon nanotubes (CNT) conductive polymer nanocomposites (CPC) sensors. *Sens. Actuators, B* 159: 213–219. https://doi.org/10.1016/j.snb.2011.06.073.

4 Hao, X. (2013). A review on the dielectric materials for high energy-storage application. *J. Adv. Dielectr.* 03: 1330001. https://doi.org/10.1142/s2010135x13300016.

5 Wang, G., Lu, Z., Li, Y. et al. (2021). Electroceramics for high-energy density capacitors: current status and future perspectives. *Chem. Rev.* 121: 6124–6172. https://doi.org/10.1021/acs.chemrev.0c01264.

6 Palneedi, H., Peddigari, M., Hwang, G.-T.T. et al. (2018). High-performance dielectric ceramic films for energy storage capacitors: progress and outlook. *Adv. Funct. Mater.* 28: 1803665. https://doi.org/10.1002/adfm.201803665.

7 Nasreen, S., Treich, G.M., Baczkowski, M.L. et al. (2017). Polymer dielectrics for capacitor application. In: *Kirk-Othmer Encyclopedia of Chemical Technology* (ed. O. Kirk Othmer), 1–29. Hoboken, NJ: Wiley. https://doi.org/10.1002/0471238961.koe00036.

8 Burke, A. (2021). Prospects for the development of high energy density dielectric capacitors. *Appl. Sci.* 11: 8063. https://doi.org/10.3390/app11178063.

9 Dang, Z.M., Yuan, J.K., Yao, S.H., and Liao, R.J. (2013). Flexible nanodielectric materials with high permittivity for power energy storage. *Adv. Mater.* 25: 6334–6365. https://doi.org/10.1002/adma.201301752.

10 Cheng, L., Gao, H., Liu, K. et al. (2022). Research progress on multilayer-structured polymer-based dielectric nanocomposites for energy storage. *Macromol. Mater. Eng.* 307: 2100822. https://doi.org/10.1002/mame.202100822.

11 Huang, S., Liu, K., Zhang, W. et al. (2023). All-organic polymer dielectric materials for advanced dielectric capacitors: theory, property, modified design and future prospects. *Polym. Rev.* 63: 515–573. https://doi.org/10.1080/15583724.2022.2129680.

12 Chu, B., Zhou, X., Neese, B. et al. (2006). Relaxor ferroelectric poly(vinylidene fluoride-trifluoroethylene-chlorofluoroethylene) terpolymer for high energy density storage capacitors. *IEEE Trans. Dielectr. Electr. Insul.* 13: 1162–1169. https://doi.org/10.1109/TDEI.2006.1714943.

13 Choi, D.-H., Randall, C., Furman, E., et al. Energy and power densities of capacitors and dielectrics. *2015 IEEE International Workshop on Integrated Power Packaging (IWIPP)*, Chicago, IL, USA (3–6 May 2015). IEEE. pp. 52–55. https://doi.org/10.1109/IWIPP.2015.7295976.

14 Yang, L., Kong, X., Li, F. et al. (2019). Perovskite lead-free dielectrics for energy storage applications. *Prog. Mater. Sci.* 102: 72–108. https://doi.org/10.1016/j.pmatsci.2018.12.005.

15 Dakin, T.W. (2006). Conduction and polarization mechanisms and trends in dielectric. *IEEE Electr. Insul. Mag.* 22: 11–28. https://doi.org/10.1109/MEI.2006.1705854.

16 Huan, T.D., Boggs, S., Teyssedre, G. et al. (2016). Advanced polymeric dielectrics for high energy density applications. *Prog. Mater. Sci.* 83: 236–269. https://doi.org/10.1016/j.pmatsci.2016.05.001.

17 O'Hara, A., Balke, N., and Pantelides, S.T. (2022). Unique features of polarization in ferroelectric ionic conductors. *Adv. Electron. Mater.* 8: 2100810. https://doi.org/10.1002/aelm.202100810.

18 Qiu, J., Gu, Q., Sha, Y. et al. (2022). Preparation and application of dielectric polymers with high permittivity and low energy loss: a mini review. *J. Appl. Polym. Sci.* 139: 52367. https://doi.org/10.1002/app.52367.

19 Kim, M.P., Um, D.S., Shin, Y.E., and Ko, H. (2021). High-performance triboelectric devices via dielectric polarization: a review. *Nanoscale Res. Lett.* 16: 35. https://doi.org/10.1186/s11671-021-03492-4.

20 Henini, M. (2002). Principles of electronic materials and devices (second edition). *Microelectron. J.* 33: 681. https://doi.org/10.1016/s0026-2692(02)00042-3.

21 Feng, Q.K., Zhong, S.L., Pei, J.Y. et al. (2022). Recent progress and future prospects on all-organic polymer dielectrics for energy storage capacitors. *Chem. Rev.* 122: 3820–3878. https://doi.org/10.1021/acs.chemrev.1c00793.

22 Fan, B., Zhou, M., Zhang, C. et al. (2019). Polymer-based materials for achieving high energy density film capacitors. *Prog. Polym. Sci.* 97: 101143. https://doi.org/10.1016/j.progpolymsci.2019.06.003.

23 Thakur, V.K. and Gupta, R.K. (2016). Recent progress on ferroelectric polymer-based nanocomposites for high energy density capacitors: synthesis, dielectric properties, and future aspects. *Chem. Rev.* 116: 4260–4317. https://doi.org/10.1021/acs.chemrev.5b00495.

24 Baldwin, A.F., Ma, R., Huan, T.D. et al. (2014). Effect of incorporating aromatic and chiral groups on the dielectric properties of poly(dimethyltin esters). *Macromol. Rapid Commun.* 35: 2082–2088. https://doi.org/10.1002/marc.201400507.

25 Chen, Y., Du, Y.K., Yue, Y.C. et al. (2020). Correlation between energy storage density and differential dielectric constant in ferroelectrics. *J. Electron. Mater.* 49: 659–667. https://doi.org/10.1007/s11664-019-07786-3.

26 Wei, J. and Zhu, L. (2020). Intrinsic polymer dielectrics for high energy density and low loss electric energy storage. *Prog. Polym. Sci.* 106: 101254. https://doi.org/10.1016/j.progpolymsci.2020.101254.

27 Laghari, J.R. and Sarjeant, W.J. (1992). Energy-storage pulsed-power capacitor technology. *IEEE Trans. Power Electron.* 7: 251–257. https://doi.org/10.1109/63.124597.

28 Nehm, F., Klumbies, H., Richter, C. et al. (2015). Breakdown and protection of ALD moisture barrier thin films. *ACS Appl. Mater. Interfaces* 7: 22121–22127. https://doi.org/10.1021/acsami.5b06891.

29 Shen, Z.H., Wang, J.J., Jiang, J.Y. et al. (2019). Phase-field modeling and machine learning of electric-thermal-mechanical breakdown of polymer-based dielectrics. *Nat. Commun.* 10: 1843. https://doi.org/10.1038/s41467-019-09874-8.

30 Rodríguez, A.E., Morgan, W.L., Touryan, K.J. et al. (1991). An air breakdown kinetic model. *J. Appl. Phys.* 70: 2015–2022. https://doi.org/10.1063/1.349487.

31 Tan, D.Q. (2020). The search for enhanced dielectric strength of polymer-based dielectrics: a focused review on polymer nanocomposites. *J. Appl. Polym. Sci.* 137: 49379. https://doi.org/10.1002/app.49379.

32 Liu, X.J., Zheng, M.S., Chen, G. et al. (2022). High-temperature polyimide dielectric materials for energy storage: theory, design, preparation and properties. *Energy Environ. Sci.* 15: 56–81. https://doi.org/10.1039/d1ee03186d.

33 Gerson, R. and Marshall, T.C. (1959). Dielectric breakdown of porous ceramics. *J. Appl. Phys.* 30: 1650–1653. https://doi.org/10.1063/1.1735030.

34 Owate, I.O. and Freer, R. (1992). Ac breakdown characteristics of ceramic materials. *J. Appl. Phys.* 72: 2418–2422. https://doi.org/10.1063/1.351586.

35 Seitz, F. (1949). On the theory of electron multiplication in crystals. *Phys. Rev.* 76: 1376–1393. https://doi.org/10.1103/PhysRev.76.1376.

36 Guo, H., Duan, B., Wu, H., and Yang, Y. (2019). Breakdown mechanisms of power semiconductor devices. *IETE Tech. Rev.* 36: 243–252. https://doi.org/10.1080/02564602.2018.1450652.

37 Ling, Y.M. (2007). Study on breakdown characteristics of a low-pressure dielectric barrier discharge. *J. Plasma Phys.* 73: 417–426. https://doi.org/10.1017/S0022377806004715.

38 Shen, Z.H., Wang, J.J., Lin, Y. et al. (2018). High-throughput phase-field design of high-energy-density polymer nanocomposites. *Adv. Mater.* 30: 1704380. https://doi.org/10.1002/adma.201704380.

39 Sun, Y., Bealing, C., Boggs, S., and Ramprasad, R. (2013). 50+ Years of intrinsic breakdown. *IEEE Electr. Insul. Mag.* 29: 8–15. https://doi.org/10.1109/MEI.2013.6457595.

40 Sun, Y., Boggs, S.A., and Ramprasad, R. (2012). The intrinsic electrical breakdown strength of insulators from first principles. *Appl. Phys. Lett.* 101: 132906. https://doi.org/10.1063/1.4755841.

41 Cai, Z., Wang, X., and Li, L. (2019). Phase-field modeling of electromechanical breakdown in multilayer ceramic capacitors. *Adv. Theory Simul.* 2: 1800179. https://doi.org/10.1002/adts.201800179.

42 Li, J., Zhang, Y., Xia, Z. et al. (2001). Action of space charge on aging and breakdown of polymers. *Chin. Sci. Bull.* 46: 796–800. https://doi.org/10.1007/BF02900426.

43 Liu, L., Chen, H., Li, B. et al. (2015). Thermal and strain-stiffening effects on the electromechanical breakdown strength of dielectric elastomers. *Appl. Phys. Lett.* 107: 62906. https://doi.org/10.1063/1.4928712.

44 Zhang, L., Wang, Q., and Zhao, X. (2011). Mechanical constraints enhance electrical energy densities of soft dielectrics. *Appl. Phys. Lett.* 99: 171906. https://doi.org/10.1063/1.3655910.

45 Rahimabady, M., Chen, S., Yao, K. et al. (2011). High electric breakdown strength and energy density in vinylidene fluoride oligomer/poly(vinylidene fluoride) blend thin films. *Appl. Phys. Lett.* 99: 142901. https://doi.org/10.1063/1.3645619.

46 Nomura, S., Uchino, K., and Newnham, R.E. (1980). Large electrostrictive effects in relaxor ferroelectrics. *Ferroelectrics* 23: 187–191. https://doi.org/10.1080/00150198008018801.

47 Huang, X., Xie, L., Yang, K. et al. (2014). Role of interface in highly filled epoxy/$BaTiO_3$ nanocomposites. Part II-effect of nanoparticle surface chemistry on processing, thermal expansion, energy storage and breakdown strength of the nanocomposites. *IEEE Trans. Dielectr. Electr. Insul.* 21: 480–487. https://doi.org/10.1109/TDEI.2013.004166.

48 Pukhov, A.A. (1999). Thermal mechanism of microwave breakdown in HTSC films: theoretical arguments and experimental evidence. *Supercond. Sci. Technol.* 12: 102–104. https://doi.org/10.1088/0953-2048/12/2/008.

49 Shen, Z.H., Wang, J.J., Jiang, J.Y. et al. (2018). Phase-field model of electrothermal breakdown in flexible high-temperature nanocomposites under extreme conditions. *Adv. Energy Mater.* 8: 1800509. https://doi.org/10.1002/aenm.201800509.

50 Somogyi, K. (1998). Some critical conditions of the thermal breakdown in semiconductors. *Sens. Actuators, A* 71: 58–62. https://doi.org/10.1016/S0924-4247(98)00176-9.

51 Xu, K.Q., Zhao, K.Y., Gu, Y. et al. (2015). Local elastic and thermal behaviors of dielectric breakdown regions in multilayer ceramic capacitors. *Phys. Status Solidi RRL* 9: 745–748. https://doi.org/10.1002/pssr.201510343.

52 Li, Q., Chen, L.L.Q., Gadinski, M.R. et al. (2015). Flexible higherature dielectric materials from polymer nanocomposites. *Nature* 523: 576–579. https://doi.org/10.1038/nature14647.

53 Claude, J., Lu, Y., and Wang, Q. (2007). Effect of molecular weight on the dielectric breakdown strength of ferroelectric poly(vinylidene fluoride-chlorotrifluoroethylene)s. *Appl. Phys. Lett.* 91: 212904. https://doi.org/10.1063/1.2816327.

54 Feng, X., Xiong, Q., Gattozzi, A.L., and Hebner, R.E. (2021). Partial discharge experimental study for medium voltage DC cables. *IEEE Trans. Power Deliv.* 36: 1128–1136. https://doi.org/10.1109/TPWRD.2020.3002508.

55 Garner, A.L., Loveless, A.M., Dahal, J.N., and Venkattraman, A. (2020). A tutorial on theoretical and computational techniques for gas breakdown in microscale gaps. *IEEE Trans. Plasma Sci.* 48: 808–824. https://doi.org/10.1109/TPS.2020.2979707.

56 Huang, X., Li, Y., Liu, F. et al. (2014). Electrical properties of epoxy/POSS composites with homogeneous nanostructure. *IEEE Trans. Dielectr. Electr. Insul.* 21: 1516–1528. https://doi.org/10.1109/TDEI.2014.004314.

57 Li, S., Yin, G., Chen, G. et al. (2010). Short-term breakdown and long-term failure in nanodielectrics: a review. *IEEE Trans. Dielectr. Electr. Insul.* 17: 1523–1535. https://doi.org/10.1109/TDEI.2010.5595554.

58 Li, Z., Sheng, G., Jiang, X., and Tanaka, T. (2017). Effects of inorganic fillers on withstanding short-time breakdown and long-time electrical aging of epoxy composites. *IEEJ Trans. Electr. Electron. Eng.* 12: S10–S15. https://doi.org/10.1002/tee.22548.

59 Lu, B., Feng, Q., and Sun, H. (2019). The effect of environmental temperature on negative corona discharge under the action of photoionization. *IEEE Trans. Plasma Sci.* 47: 149–154. https://doi.org/10.1109/TPS.2018.2872504.

60 Sakabe, Y. (1997). Multilayer ceramic capacitors. *Curr. Opin. Solid State Mater. Sci.* 2: 584–587. https://doi.org/10.1016/S1359-0286(97)80049-6.

61 Covaci, C. and Gontean, A. (2022). "Singing" multilayer ceramic capacitors and mitigation methods—a review. *Sensors* 22: 3869. https://doi.org/10.3390/s22103869.

62 Hong, K., Lee, T.H., Suh, J.M. et al. (2019). Perspectives and challenges in multilayer ceramic capacitors for next generation electronics. *J. Mater. Chem. C* 7: 9782–9802. https://doi.org/10.1039/c9tc02921d.

63 Wang, H., Zeng, C., Liu, B., and Wang, X. (2019). Energy storage properties of surface-modified $BaTiO_3$ ceramic films for multilayer capacitors applications. *J. Adv. Dielectr.* 09: 1950027. https://doi.org/10.1142/S2010135X19500279.

64 Kumar, N., Ionin, A., Ansell, T. et al. (2015). Multilayer ceramic capacitors based on relaxor $BaTiO_3$-$Bi(Zn_{1/2}Ti_{1/2})O_3$ for temperature stable and high energy density capacitor applications. *Appl. Phys. Lett.* 106: 252901. https://doi.org/10.1063/1.4922947.

65 Zhao, P., Wang, H., Wu, L. et al. (2019). High-performance relaxor ferroelectric materials for energy storage applications. *Adv. Energy Mater.* 9: 1803048. https://doi.org/10.1002/aenm.201803048.

66 Li, J., Shen, Z., Chen, X. et al. (2020). Grain-orientation-engineered multilayer ceramic capacitors for energy storage applications. *Nat. Mater.* 19: 999–1005. https://doi.org/10.1038/s41563-020-0704-x.

67 Chen, Q., Shen, Y., Zhang, S., and Zhang, Q.M. (2015). Polymer-based dielectrics with high energy storage density. *Annu. Rev. Mater. Res.* 45: 433–458. https://doi.org/10.1146/annurev-matsci-070214-021017.

68 Wang, D. and Dang, Z.M. (2018). Processing of polymeric dielectrics for high energy density capacitors. In: *Dielectric Polymer Materials for High-Density Energy Storage* (ed. Z.M. Dang), 429–446. Elsevier https://doi.org/10.1016/B978-0-12-813215-9.00012-9.

69 Breil, J. (2011). Future trends for biaxially oriented films and orienting lines. In: *Biaxial Stretching of Film: Principles and Applications* (ed. M.T. DeMeuse), 240–273. Elsevier https://doi.org/10.1533/9780857092953.2.240.

70 Ding, L., Shao, L., and Bai, Y. (2014). Deciphering the mechanism of corona discharge treatment of BOPET film. *RSC Adv.* 4: 21782–21787. https://doi.org/10.1039/c4ra02289k.

71 Barbaro, G., Galdi, M.R., Di Maio, L., and Incarnato, L. (2015). Effect of BOPET film surface treatments on adhesion performance of biodegradable coatings for packaging applications. *Eur. Polym. J.* 68: 80–89. https://doi.org/10.1016/j.eurpolymj.2015.04.027.

72 Tan, D.Q. (2020). Review of polymer-based nanodielectric exploration and film scale-up for advanced capacitors. *Adv. Funct. Mater.* 30: 1808567. https://doi.org/10.1002/adfm.201808567.

14

Ferroelectric Materials for Dielectric Energy Storage: Progress and Material Design

Haibo Zhang, Hua Tan, Shuaikang Huang, and Mohsin A. Marwat

Huazhong University of Science and Technology, School of Materials Science and Engineering, State Key Laboratory of Material Processing and Die and Mould Technology, Luoyu Road 1037, Wuhan 430074, PR China

14.1 Introduction

In this chapter, we had a detailed understanding of the fundamentals and applications of dielectric materials, and the progress of energy storage and dielectric properties of dielectric materials determined the development of energy storage capacitors. Dielectric energy storage materials can be generally classified into five categories according to the *P–E* loops, which are (i) linear dielectrics (LDs), (ii) paraelectrics (PEs), (iii) ferroelectrics (FEs), (iv) relaxor ferroelectrics (RFEs), and (v) antiferroelectrics (AFEs). The *P–E* schematic for different kinds of dielectric materials is well represented in the work of Palneedi et al. [1], as shown in Figure 14.1. For LD ceramics, such as TiO_2 ceramics, linear *P–E* loops are obtained with a stable dielectric constant under electric-field excitation. They usually possess high breakdown field (E_b) values and ultralow dielectric loss (tan δ). However, a low dielectric constant (ε_r) and energy storage density (W) are obstacles to realizing practical applications [2]. For PEs, they have dipoles instead of FE domains and can quickly return to their initial nonpolar state after the electric field is removed. Therefore, their electric hysteresis lines are extremely fine in shape, have weak nonlinear characteristics, and do not have hysteresis phenomena under ideal conditions. However, although the relative permittivity of PEs is higher than that of LDs, it is significantly lower than that of several other types of nonlinear dielectrics. For PEs, the Curie temperature is generally lower than room temperature, so they exhibit almost the same properties as LDs at room temperature and above. By now, there is still some controversy in the classification, so they are not classified separately here and are classified as LDs. The typical PE material strontium titanate (ST), for example, has a relative permittivity of about 300 [3]. Normal FE ceramics, such as undoped $BaTiO_3$ and $PbTiO_3$ ceramics, usually possess large maximum polarization (P_{max}) remnant polarization (P_r), and moderate E_b. Energy storage efficiency is extremely low due to the switch of irreversible electric domains under the electric field, and an increase in recoverable energy storage density (W_{rec}) cannot be achieved. In contrast, the

Piezoelectric Materials: From Fundamentals to Emerging Applications, First Edition. Edited by Jiagang Wu.

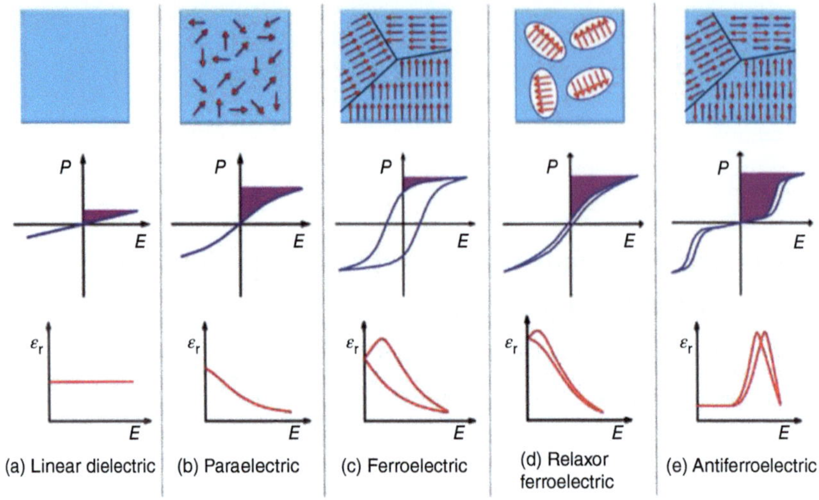

Figure 14.1 A schematic graph of the *P*–*E* loops for (a) linear dielectrics (LDs); (b) paraelectrics, (c) ferroelectrics (FE); (d) relaxor ferroelectrics (RFEs); and (e) antiferroelectrics (AFE). Source: Palneedi et al. [1]. Reproduced with permission of Wiley-VCH.

RFE and AFE dielectrics are very suitable for energy storage applications. They both have a large P_{max} and a low P_r and thus a large energy storage density is obtained. In particular, for RFE ceramics ($Bi_{0.5}Na_{0.5}TiO_3$-based ergodic RFE ceramics), FE order is induced by the electric field, and when the field is switched off, the FE domains relax back to the ergodic relaxor state with near-zero polarization, thus ensuring a large P_{max}, low P_r, and high energy storage density. For the AFEs, double-hysteresis is obtained when the electric field surpasses the critical field, thus guaranteeing higher energy density. Notably, the appearance of double-hysteresis contradicts the expected high efficiency [4]. In addition, there are classifications of relaxor antiferroelectrics (RAFEs) in literature. There have been many studies on RAFEs in recent years, they present the same *P*–*E* loops as RFEs, such as NN-SBT [5], NN-BNT [6], and so on. Since RAFEs are developed based on AFE materials, they are classified as AFEs in this work.

This chapter reviews the milestones of dielectric materials and the research progress in recent years. In Sections 14.2 and 14.3, the current problems of dielectric material systems and their modification strategies are presented from two aspects: inorganic and organic materials, respectively. Meanwhile, in Section 14.4, the improvement strategies of different composites to achieve high energy storage density are presented.

14.2 Inorganic Dielectric Materials

As described above, inorganic dielectric energy storage materials can also be classified into LD, PE, FE, RFE, and AFE by the *P*–*E* curves. Among them, the PEs are usually compared with LDs due to their similar *P–E* loops above room temperature. In addition, the *P–E* curves of AFEs are essentially similar to those of RFEs. Among them, FEs are not suitable for energy storage applications because of their large residual polarization, which leads to very low energy storage density and efficiency. However, the FEs can be transformed into RFEs by appropriate doping modifications, thus improving the energy storage performance. In the following, the latest research progress on inorganic dielectric materials will be detailed, starting with three kinds of different dielectric ceramics: LD ceramics, RFE ceramics, and AFE ceramics. Note that, dielectric glass ceramics and ceramic films are also discussed and included in the following subsections.

14.2.1 Liner Dielectric Ceramics

LD ceramics usually possess characteristics of low ε_r and tan δ, as well as moderate E_b. It is thereby hard to obtain high W_{rec} under high electric field. In this regard, the research of LDs is mainly concentrated on increasing ε_r or improving polarization behavior based on maintaining high E_b. In terms of PEs (there are also reports about superparaelectric RFEs [7]), they have Curie temperatures below room temperature and weakly nonlinear *P–E* hysteresis. Therefore, at room temperature and above, PEs have practically the same properties as LDs. And there is still some controversy in many studies about the classification of these two, such as ST, which is classified as an LD in some literature and as PE in others. Therefore, in this subsection, the PEs are also combined into the LD and presented together. It should be noted that in the study of LDs, most researchers calculate the theoretical energy storage density only by measuring the breakdown field strength and dielectric constant. The authors feel that such a method is not comparable to the energy storage density calculated by *P–E* curve measurements because the dielectric constant cannot remain constant when chemical composition changes.

14.2.1.1 TiO_2-Based Ceramics

TiO_2 is a typical LD, with characteristics of low E_b (~4 kV/cm), low tan δ (<0.1%), dielectric constant (~110), and wide band gap (~3.2 eV), which limits its applications [8, 9]. Generally, TiO_2 has three crystal structures including orthorhombic brookite, tetragonal rutile, and anatase. The rutile phase is more stable and easier to synthesize than others, and hence gains more attention in ceramic bulks and films [10, 11].

In order to enhance the energy storage density, one approach is to increase the ε_r. In 2013, Hu et al. [12] first reported that (In, Nb) co-doped TiO_2 ceramics displayed a giant dielectric constant (>10^4) as well as low tan δ (<5%) and possessed good temperature and frequency stabilities over a wide temperature range (80–450 K). They explained that "triangular" $In_2^{3+}V_0^{\bullet\bullet}Ti^{3+}$ and "diamond"-shaped $Nb_2^{5+}Ti^{3+}A_{Ti}$ (A = $Ti^{3+}/In^{3+}/Ti^{4+}$) defect complexes are strongly correlated, giving rise to large defect-dipole clusters containing highly localized electrons that are together responsible for the excellent properties observed in co-doped TiO_2. The temperature dependences of the dielectric properties and the structure of the In and Nb defect complexes are shown in Figure 14.2. Since then, a large number

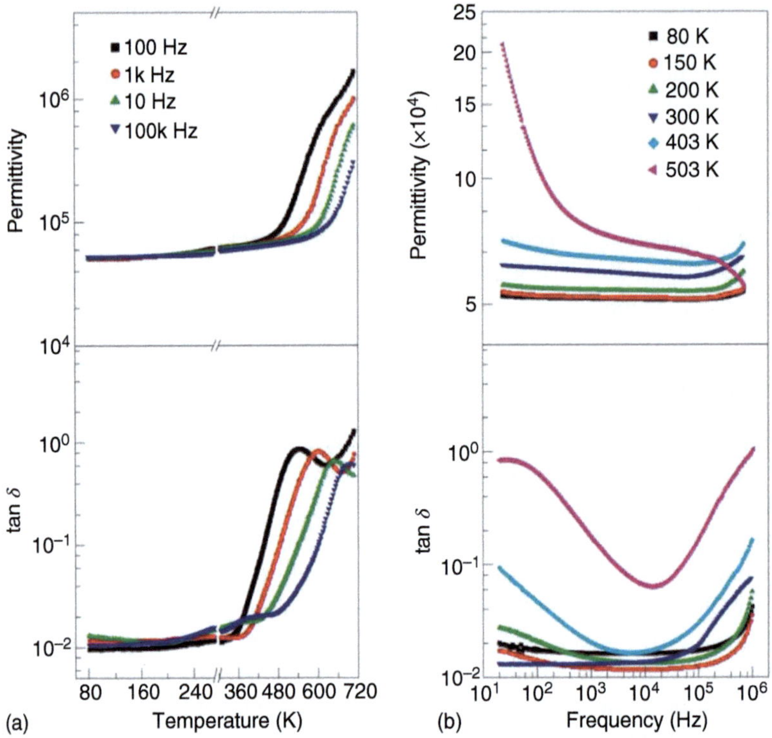

Figure 14.2 (a) Dielectric permittivity of 10% (Nb + In) co-doped TiO_2 measured from 80 to 700 K at given frequencies. A narrow data break occurs near the room temperature region (approximately 270–300 K), as two experimental set-ups were used for the low-temperature and high-temperature measurements, respectively. (b) Frequency dependences of dielectric permittivity and loss of 10% (Nb + In) co-doped TiO_2 measured at the given temperatures. Source: Hu et al. [12]. Reproduced with permission of Springer Nature.

of co-doped TiO_2-based energy storage ceramics have been investigated, such as: (Al, La) co-doped TiO_2 [13], (Ta, Al) co-doped TiO_2, (Nb, La) co-doped TiO_2 [13], (Ho, Ta) co-doped TiO_2 [14], (Bi, Nb) co-doped TiO_2, (Gd, Nb) co-doped TiO_2 $(A_{1/3}Nb_{2/3})_xTi_{1-x}O_2$ ($A = Ba^{2+}, Ca^{2+}, Zn^{2+}, Mg^{2+}$), and so on.

In another way, E_b can be improved via the sintering behavior modification of TiO_2-based ceramics, such as refining grain size [15, 16], adding glass phase [17], applying spark plasma sintering (SPS), and so on. Liu et al. [17], obtained a W_{rec} of 1.15 J/cm^3 for alkali-free glass-modified TiO_2 ceramics at 501.7 kV/cm. Besides that, using sol–gel process, Wei et al. [18] obtained denser and uniform TiO_2-based ceramics with DBS, and storage density increased from 374.5 kV/cm and 0.76 J/cm^3 to 482.9 kV/cm and 1.05 J/cm^3, respectively. Unfortunately, W_{rec} of those ceramic bulks after different modification ways is still only at a scale of 1 J/cm^3, which is less than that of other energy storage dielectric materials.

It should be noticed that high-quality TiO_2 film is a good direction owing to its high E_b. For instance, Chao and Dogan [19] fabricated 0.1-mm-thick TiO_2 films using

Table 14.1 Energy storage properties of TiO_2-based linear dielectric materials.

Composition	Type	E_b (kV/cm)	W_{rec} (J/cm^3)	References
TiO_2	Film	1800	15	[15]
TiO_2	Film	1400	14	[19]
TiO_2	Bulk	350	0.76	[18]
TiO_2–CaO–MgO–Al_2O_3–SiO_2	Bulk	482.9	1.05	[18]
Mn-doped TiO_2–BaO–B_2O_3–Al_2O_3–SiO_2	Bulk	501.7	1.15	[17]

tape-casting method, which achieved a W_{rec} of 14 J/cm^3 at 1400 kV/cm. The W_{rec} and E_b of all TiO_2-based dielectric materials are listed in Table 14.1.

14.2.1.2 $SrTiO_3$-Based Ceramics

$SrTiO_3$ (ST) is a cubic phase with ABO_3-type perovskite structure, accompanying to space group of Pm_3m and lattice constant of $a = 3.905$ Å [20, 21]. ST ceramics possess moderate ε_r of ~300, E_b of ~100 kV/cm, and low tan δ of ~10^{-3}, as well as good temperature- and frequency-independent dielectric properties, bias voltage stability, and thermoelectric energy conversion efficiency [22–24]. Therefore, it is considered to be a potential energy storage and conversion candidate. Compared with TiO_2 ceramic (~110) and polymer LD (<10), ST has an advantage of relatively high ε_r, and is thus more suitable for pulse capacitor.

Firstly, in order to increase ε_r and thus increase W_{rec}. In a similar approach to the co-doped TiO_2 ceramics, Chen et al. [25–27] reported (Bi, Sr)TiO_3 ceramics with giant dielectric constant (~5000), discussed the related physical mechanisms, and proposed that the first and second ionization of oxygen vacancies as well as corresponding thermal movement were the main reasons. To avoid the problem of Bi-containing oxides volatilizing at high temperature, Shen et al. [22, 28–30] used trivalent nonvolatile rare earth ions (Re^{3+} = La, Sm, Gd, Er, Nd, etc.) to replace Sr^{2+} and designed three composition formulas based on three possible charge compensation mechanisms such as equimolar substitution, introducing Sr or Ti vacancy in advance, and successfully fabricated ceramics with perovskite structure. It is experimentally verified the feasibility of introducing ion vacancies in advance for charge compensation, and then the concept of "forced charge compensation mechanism" is summarized and proposed.

To increase the E_b, microstructure regulation for enhancing E_b of ST ceramics has been widely studied. In 2014, Song et al. [31] prepared $Ba_{0.4}Sr_{0.6}TiO_3$ ceramics with various grain sizes (0.5–5.6 μm) and observed that the dielectric peak gradually depressed and broadened and E_b gradually increased with decreasing grain size, which should be closely related to the ratio of grain/grain boundary and polar nanoregions (PNRs). $Ba_{0.4}Sr_{0.6}TiO_3$ ceramic bulk with a grain size of 0.5 μm achieves a high W_{rec} = 1.28 J/cm^3 measured at the highest E_b of 243 kV/cm. In addition, the addition of suitable glass compositions is also an effective method to enhance E_b and reduce the sintering temperature of ST-based ceramics [32, 33]. In 2019, Shen et al.

[34] used a homemade glass frit to modify BST, enhancing the E_b (160 kV/cm) and reducing the high-temperature resistivity, which expanded the working temperature range for energy storage ceramic capacitor applications.

In terms of ST-based ceramic films, "Defect engineering" is an effective tool to enhance W_{rec} by strengthening relaxor characteristics for ST-based ceramic films. Yang et al. [35] fabricated $(Sr_{1-1.5x}Bi_x)Ti_{0.99}Mn_{0.01}O_3$ (SBTM, $x = 0.01, 0.05, 0.1$) thin films with a thickness of 217 nm using sol–gel method. As x value increases, relaxor behaviors are gradually strengthened due to a slight rotation of the (TiO_6) octahedra induced by the formation of $Bi^{3+}–V''_{sr}$ defect complex. The introduction of $Bi_{0.5}Na_{0.5}TiO_3$ (BNT) [36] and $BiFeO_3$ (BF) [37] has a similar effect. In the related mechanism studies, Hou et al. [38] investigated the influence of interface difference and thickness on the energy storage performances of ST thin films, and observed the existence of ionic diffusion layers and oxygen vacancies using high resolution transmission electron microscope (HR-TEM). Moreover, they observed that E_b and P_{max} (up to 102 μC/cm^2) along the positive direction were higher than the negative direction. Therefore, the maximum W_{rec} of ST thin films reaches 307 J/cm^3 for positive direction, which may be related to local electric field and redistribution of oxygen vacancy, as shown in Figure 14.3.

Various meaningful and interesting works have been done in the optimization of macro/micro structures to improve the energy storage properties of ST-based

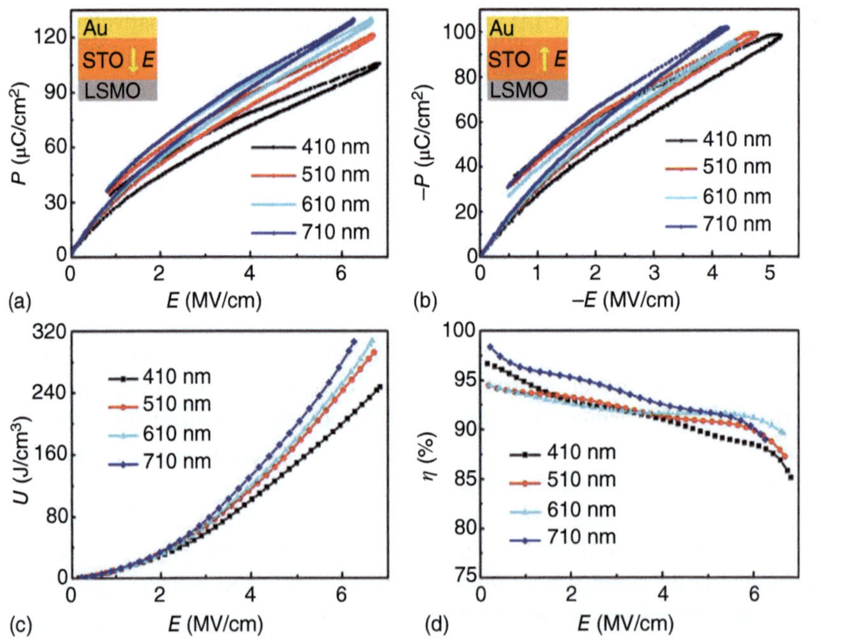

Figure 14.3 *P–E* loops of Au/STO/LSMO capacitors under (a) positive fields and (b) negative fields, the insets show the orientation of *E*. Variations of (c) recoverable energy density and (d) energy storage efficiency under positive electric fields of Au/STO/LSMO capacitors. Source: Hou et al. [38]. Reproduced with permission of American Chemical Society.

Table 14.2 Energy storage properties of ST-based ceramics.

Composition	Type	E_b (kV/cm)	W_{rec} (J/cm^3)	References
$Ba_{0.3}Sr_{0.7}TiO_3$	Bulk	90	0.23	[40]
$Ba_{0.4}Sr_{0.6}TiO_3$	Bulk	90	0.33	[40]
$Ba_{0.3}Sr_{0.7}TiO_3$	Bulk	140	0.57	[19]
$Ba_{0.3}Sr_{0.7}TiO_3$ (SPS)	Bulk	230	1.13	[19]
$Ba_{0.3}Sr_{0.7}TiO_3$ + 2% BBSZ	Bulk	160	0.63	[34]
$0.9(Sr_{0.7}Bi_{0.2})TiO_3$–$0.1Bi(Mg_{0.5}Hf_{0.5})O_3$	Bulk	360	3.1	[24]
$SrTiO_3$ (STO) films grown on $La_{0.67}Sr_{0.33}MnO_3$ (LSMO)	Film	6800	307	[38]
With Au and Al being deposited on $SrTiO_3$ thin films as top electrode	Film	—	15.7	[39]
Mn-doped $0.4BiFeO_3$–$0.6SrTiO_3$ (BFSTO) thin film	Film	3600	51	[37]
$SrTiO_3$ film with Al_2O_3 blocking layer	Film	75 423	39.49	[41]
$(Sr_{0.85}Bi_{0.1})Ti_{0.99}Mn_{0.01}O_3$ thin film	Film	—	24.4	[35]
$Sr_{0.995}(Na_{0.5}Bi_{0.5})_{0.005}(Ti_{0.99}Mn_{0.01})O_3$ amorphous thin film	Film	—	65.3	[36]

ceramic films. It is well known that the amorphous phase usually possesses higher E_b but lower ε_r than its crystalline counterparts. Gao et al. [39] studied energy storage behaviors of amorphous ST thin films with different top electrodes and proposed "self-healing" mechanism based on the anodic oxidation reaction in aluminum electrolytic capacitors. At a relative humidity of 60%, amorphous ST films with Al top electrode achieve a W_{rec} of 15.7 J/cm^3 at 3500 kV/cm, which approaches to eight times that of the samples with Au electrode. The energy storage properties of ST-based ceramics are listed in Table 14.2.

14.2.1.3 Glass Ceramics

Compared with ceramics or polymer-based dielectrics, glass–ceramics have high glass transition temperature and submicron grains. The microstructure of a glass ceramic is characterized by the presence of crystallites, which are held together by an amorphous glass matrix. So it has the superiorities of high thermal stability and low dielectric loss, as well as low electric field dependence of the relative dielectric constant. Thus, it is recognized that glass–ceramics are considered to be one of the most prospective materials for the use of energy-storage devices [42].

Over the past decade, niobate-based glass–ceramics with tetragonal tungsten bronze structure $(A_1)_4$-$(A_2)_2C_{4B10}O_{30}$ and cubic perovskite structure ABO_3 have attracted much attention owing to their superior E_b. Many efforts have been taken to improve the dielectric properties and energy storage performances of these glass–ceramics. For instance, Xiu et al. [43] reported that the theoretical energy storage was optimized when replacing a small amount of SiO_2 with B_2O_3 in niobate-based glass–ceramics. When the B_2O_3–SiO_2 proportion is 1.5–33.5, the optimal theoretical energy storage density reaches 7.4 J/cm^3. Wang et al. [44] reported that the content of SiO_2 and Nb_2O_5 has a great influence on the discharge properties of Na_2O–BaO–PbO–Nb_2O_5–SiO_2–Al_2O_3 glass–ceramics, and the maximum actual discharge energy density of 0.74 J/cm^3 was achieved when the ratio of SiO_2/Nb_2O_5 is 30 : 32.

A large number of results have also been achieved by adding ceramic substrates to energy-storage amorphous films. Trolier-Mckinstry and Michael [45] explored composite thin films of nanocrystalline PbO particles in an amorphous $Pb_{1.1}TiO_{3.1}$ matrix in order to increase the stored energy density of dielectrics. Excess lead was added in the precursor solution to promote the formation of nanocrystallites in the amorphous matrix. The presence of PbO nanocrystals enhanced the moderately high permittivity of the film, and the amorphous matrix facilitated the high breakdown field of 5 MV/cm. The lack of structural periodicity of the amorphous network and the isolated, randomly distributed lead oxide nanocrystals did not present a conduction pathway through the film, leading to an extremely low leakage current (on the order of 10^{-8} A/cm^2) and a good W_{rec} of 28 J/cm^3. Energy storage properties of glass ceramics are listed in Table 14.3.

In summary, for LD ceramic bulks, giant dielectric constants can be observed in TiO_2-based bulks with donor/acceptor co-doping at B-site, and ST-based bulks with donor/acceptor co-doping at B-site or aliovalent doping of at A-site bulks by chemical modification. Related physical mechanisms; some controversies, however, still exist. Energy storage properties of ceramic bulks are limited at the

Table 14.3 Energy storage properties of glass ceramics.

Composition	Type	E_b (kV/cm)	W_{rec} (J/cm^3)	References
$SrOBaONb_2O_5B_2O_3SiO_2$ with CeO_2 added	Bulk	1250	3.39	[46]
BaO–SrO–Nb_2O_5–B_2O_3 + 0.5 mol%Gd_2O_3	Bulk	1075	6.94	[47]
SrO–BaO–Nb_2O_5-based glass ceramics with B_2O_3/SiO_2	Bulk	—	7.4	[43]
$Sr_{0.5}Ba_{0.5}Nb_2O_6$ + La_2O_3	Bulk	1127	7.1	[48]
Na_2O–BaO–Nb_2O_5–SiO_2	Bulk	100	0.17	[49]
Na_2O–BaO–PbO–Nb_2O_5–SiO_2–Al_2O_3	Bulk	300	17.9	[44]
BaO–Na_2O–Nb_2O_5 + CaF_2	Bulk	—	14.3	[50]
Bi_2O_3–Nb_2O_5–SiO_2–Al_2O_3 + CeO_2	Bulk	300	20.9	[51]
10$PbTiO_3$–10Fe_2O_3–30V_2O_5–50B_2O_3	Bulk	—	0.16	[52]
26.5BaO–7.5K_2O–30Nb_2O_5–6Al_2O_3–30SiO_2–0.6PbO–$x$$ZrO_2$ (x = 1.5)	Bulk	1949	21.62	[53]
$Pb_{1.1}TiO_{3.1}$	Film	5000	28	[45]
$Bi_{1.5}Zn_{0.9}Nb_{1.35}Ta_{0.15}O_{6.9}$	Film	3800	40.2	[54]

expense of a rapid decrease in E_b. Adding a suitable glass phase, special sintering technology and refining grain size are both able to enhance E_b of ceramic bulks. For ceramic films, adjusting suitable ratio of amorphous and crystalline or introducing a high-insulating layer would be a good way to improve their breakdown behavior.

14.2.2 Relaxor Ferroelectric Ceramics

FE is a special dielectric material that possesses spontaneous polarization (Ps) at a certain temperature range, and the direction of Ps can be changed with an external electric field. Compared with LD, FE displays an obvious nonlinear characteristic since the domain cannot quickly respond to electric field stimulation. Generally, the polarization behavior of dielectric material can be characterized by *P*–*E* loop, and thus FE can be classified into normal FE and RFE. Normal FE possesses high P_{max} while its high P_r means that most of energy is dissipated during the discharge process. By contrast, RFE exhibits slim *P*–*E* loop with high P_{max} and low P_r (i.e. high $\Delta P = P_{max} - P_r$) meaning that electric energy can be effectively released, and thus obtain better energy storage performances. Note that strengthening the relaxor characteristics and enhancing E_b have become important factors for enhancing W_{rec}. Furthermore, FE here discussed mainly refers to RFE.

14.2.2.1 Pb-Based Relaxor Ferroelectric Ceramics

$PbZr_{1-x}Ti_xO_3$ (PZT, $0 \leq x \leq 1$) is a ceramic located at morphotropic phase boundary (MPB) where Zr:Ti is 52 : 48, possesses a high piezoelectric activity (d_{33} up to 300 pC/N) and good temperature stability, and becomes an extremely popular dielectric material [55, 56]. In addition, other ceramic compositions, such as PZT 65/35 and 70/30 also received more attention and had no limitation by piezoelectric properties [57, 58]. For example, the researches on actuator in PZT ceramics are increasing due to the large electrostrain under low electric field. And a small amount of La^{3+} (about 7–10 mol%) replacing Pb^{2+} would effectively strengthen the relaxor characteristics of PZT-based ceramics owing to a disrupted long-range FE order and diffuse phase transition [59, 60]. $(Pb,La)(Zr,Ti)O_3$ (PLZT)-based RFE is therefore considered to be a promising energy storage ceramic capacitor candidate.

In terms of the Pb-based ceramic films, Hu et al. [61] investigated the effect of different Zr/Ti ratios on the energy storage properties of PLZT thin films at a fixed La content of 8 mol%. As Ti/(Zr + Ti) ratio gradually increases, ε_r enhances while $\tan\delta$ shows an opposite trend, indicating phase structure gradually transforms from RFE into normal FE. At $E_b = 2180$ kV/cm, W_{rec} and η of the PLZT 8/52/48 RFE thin films are 30 J/cm^3 and 78%, respectively. In addition, the addition of excess Pb is a common chemical compensation method to solve the problems of Pb volatilization during the annealing process and suppressing the formation of pyrochlore phase [62].

Reasonable design and selection of heterostructure for Pb-based thin films is an important step in optimizing energy storage properties. In 2013, Zhang et al. [63] prepared compositionally gradient $(Pb_{1-x}La_x)(Zr_{0.65}Ti_{0.35})O_3$ (PLZT, $x = 0.08, 0.09, 0.1$) thick films using sol–gel method. Up-graded PLZT films possess a W_{rec} of 12.4 J/cm^3 at 800 kV/cm, a down-graded one of 8.9 J/cm^3, and the lowest single composition

Table 14.4 Energy storage performances of Pb-based relaxor ferroelectric materials.

Composition	Type	E_b (kV/cm)	W_{rec} (J/cm^3)	η (%)	References
$(1-x)Pb(Mg_{1/3}Nb_{2/3})O_3-xPbTiO_3$	Bulk	—	0.47	—	[58]
$(Pb_{0.89}La_{0.11})(Zr_{0.70}Ti_{0.30})_{0.9725}O_3$	Bulk	—	0.85	92.9	[64]
$(Pb_{1-x}La_x)(Zr_{0.65}Ti_{0.35})O_3$ (PLZT)	Film	800	12.4	—	[63]
$(Pb_{0.91}La_{0.09})(Zr_{0.65}Ti_{0.35})O_3$ + Mn	Film	—	30.8	—	[65]
4 mol% Nb-doped $PbZr_{0.4}Ti_{0.6}O_3$	Film	1351	20	70	[62]
$(Pb_{1-x}La_x)(Zr_{0.65}Ti_{0.35})O_3$ (La/Zr/Ti = 8/52/48)	Film	25.9	30	78	[61]
PLZT film deposited on the LNO/Ni	Film	2600	85	65	[66]
$(PbZrO_3/PbZr_{0.52}Ti_{0.48}O_3)$ bilayer thin films	Film	2410	28.2	—	[67]
$Pb_{0.9}La_{0.1}(Zr_{0.52}Ti_{0.48})O_3$ (PLZT)	Film	2500	40.2	—	[68]
$PbZr_{0.52}Ti_{0.48}O_3/Al_2O_3/PbZr_{0.52}Ti_{0.48}O_3$ opposite double-heterojunction	Film	5711	63.7	81.3	[69]

one of 7.1 J/cm^3. Energy storage performances of Pb-based RFE materials are summarized and listed in Table 14.4.

14.2.2.2 $Bi_{0.5}Na_{0.5}TiO_3$-Based Ceramics

$Bi_{0.5}Na_{0.5}TiO_3$ (BNT) is a FE material first discovered by Smolenskii et al. [70], which possesses complicated phase structure and good dielectric, piezoelectric, and FE properties, especially high P_{max} (~40 μC/cm^2) [71–74]. And so it becomes a popular research topic on fundamental theories and practical studies for FE materials [75–77]. However, the characteristics of high P_r (~38 μC/cm^2), high E_c (~73 kV/cm), and poor sintering behavior for pure BNT ceramics hinder its energy storage applications.

With the addition of $BaTiO_3$, $Bi_{0.5}Na_{0.5}TiO_3$–$xBaTiO_3$ (x = 6–7%) binary solid solution near MPB exhibits excellent electrical properties and is a most promising candidate for replacing Pb-based ceramics [78–80]. Since then, extensive energy storage studies have been done on this system. It is particularly important that strengthening dynamic of PNRs through disturbing long-range FE ordering or expanding nonergodic–ergodic phase transition range both could optimize polarization behavior to obtain high W_{rec}. In 2011, Gao et al. [81] first reported that $0.89Bi_{0.5}Na_{0.5}TiO_3$–$0.06BaTiO_3$–$0.05K_{0.5}Na_{0.5}NbO_3$ (0.89BNT–0.06BT–0.05KNN) ceramics possessed a W_{rec} of 0.46 J/cm^3 at 56 kV/cm. To simultaneously increase the W_{rec} and η, Fan et al. [82] prepared a lead-free RE/AFE $0.94(Na_{0.82}K_{0.18})_{1/2}Bi_{1/2}TiO_3$–$0.06FeNbO_4/0.96NaNbO_3$–$0.04CaZrO_3$ (NKBT–FN/NN–CZ) composite ceramic with multiphase structure of an AFE embedded in a RFE through heat treatment of powder and the traditional solid phase reaction method, as shown in Figure 14.4. A high W_{rec} of 2.85 J/cm^3 and a high η of 80%

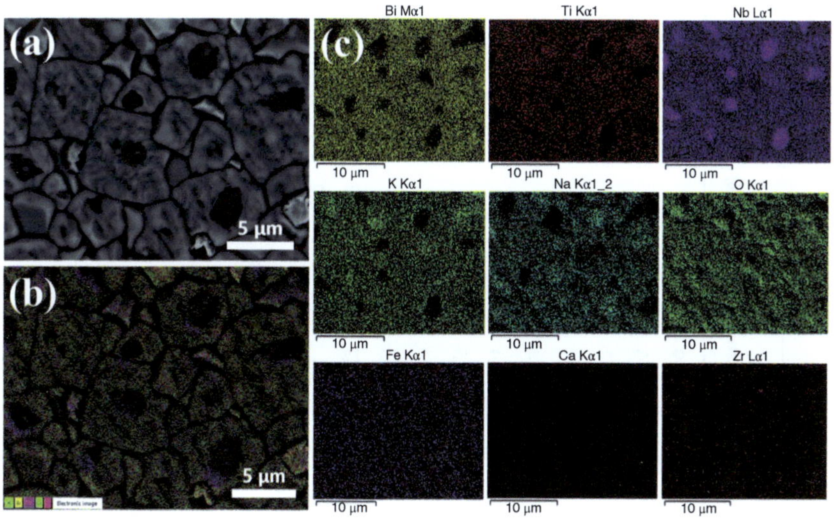

Figure 14.4 The backscattered electron (BEI) and element mappings of micrographs of the 80/20 composites: (a) and (b) polished and thermally etched fracture surface. (c) The corresponding element distribution and energy-dispersive spectrum patterns recorded from the local area. Source: Fan et al. [82]/Reproduced with permission of Royal Society of Chemistry.

were obtained in the NKBT–FN/NN–CZ composites at a composition ratio of 70/30 at room temperature. Moreover, at a high temperature of 120 °C, the energy storage performance of this composition ratio becomes more excellent ($W_{rec} \sim 3.5\,J/cm^3$ and $\eta \sim 91\%$). This study demonstrates that the design of a relaxor/antiferroelectric composite provides a highly effective method to improve the energy storage performance of lead-free ceramics.

Distinguished from single composition of MLCC, macrostructure modification of BNT-based films mainly focuses on gradient composition. It is widely accepted that BNT-based ceramic film is p-type conductivity due to many vacancies generating acceptor states in the band gap, and thus p–n junctions and block layers are applied to inhibit charge transportation. For instance, Guo et al. [83] reported the introduction of $Bi_{3.25}La_{0.75}Ti_3O_{12}$ (BLT) and $Pb(Zr_{0.4}Ti_{0.6})O_3$ (PZT) dielectric layer on pyroelectric and FE properties of 0.94BNT–0.06BT ceramic films and observed leakage current reduction of about 3 orders of magnitude. Energy storage performances of BNT-based RFE materials are summarized and listed in Table 14.5.

14.2.2.3 Relaxor $BaTiO_3$-Based Ceramics

$BaTiO_3$ (BT) with simple perovskite structure possesses a moderate Curie temperature (T_C) of 120 °C and a corresponding high ε_r of 104 [95, 96], and becomes a common dielectric material in passive capacitors. BT-based materials also received more attention in memory and memristors for information storage and transfer. In addition, the structure and physical property research of BT-based materials still attracts more attentions [97–99]. However, pure BT ceramic has some problems,

Table 14.5 Energy storage performances of BNT-based relaxor ferroelectric materials.

Composition	Type	E_b (kV/cm)	W_{rec} (J/cm^3)	η (%)	References
$0.89Bi_{0.5}Na_{0.5}TiO_3$–$0.06BaTiO_3$–$0.05K_{0.5}Na_{0.5}NbO_3$	Bulk	56	0.59	—	[81]
$Na_{0.5}Bi_{0.5}TiO_3$ + 11 mol%Mn	Bulk	95	1.06	—	[84]
($SrTiO_3$ + 0.5 wt%Li_2CO_3)/($0.93Bi_{0.5}Na_{0.5}TiO_3$–$0.07Ba_{0.94}La_{0.04}Zr_{0.02}Ti_{0.98}O_3$)	Bulk	294	2.72	—	[85]
$(Na_{0.25}Bi_{0.25}Sr_{0.5})(Ti_{0.8}Sn_{0.2})O_3$	Bulk		3.4	90	[86]
$(1-x)Bi_{0.5}Na_{0.5}TiO_3$–$xSr_{0.7}La_{0.2}TiO_3$	Bulk	315	4.14	92	[74]
$0.95Bi_{0.5}Na_{0.5}TiO_3$–$0.05SrZrO_3$	Bulk	230	3.14	79	[77]
$(Ba_{0.3}Sr_{0.7})_{0.35}(Bi_{0.5}Na_{0.5})_{0.65}TiO_3$	Bulk		1.25		[87]
$(Ba_{0.3}Sr_{0.7})_x(Bi_{0.5}Na_{0.5})_{1-x}TiO_3$ ($x = 0.5$)	Bulk	100	1.04	77	[88]
$Sr_{0.85}Bi_{0.1}TiO_3$-modified $Bi_{0.44}65Na_{0.44}65Ba_{0.05}7La_{0.05}TiO_3$	Bulk		4.55	90	[80]
$0.85[(1-x)Bi_{0.5}Na_{0.5}TiO_3$–$xBi_{0.1}Sr_{0.85}TiO_3]$–$0.15KNbO_3$ ($x = 0.3$)	Bulk	290	3.72	90.7	[76]
$(1-x)[0.8Bi_{0.5}Na_{0.5}TiO_3$–$0.2Ba(Zr_{0.3}Ti_{0.7})O_3]$–$xSr_{0.7}La_{0.2}TiO_3$ ($x = 0.3$)	Bulk	210	2.6	92.2	[75]
$0.94(Bi_{0.5}Na_{0.5})0.94TiO_3$–$0.06BaTiO_3/BiFeO_3$	film	2400	31.96	61	[89]
$(Na_{0.5}Bi_{0.5})TiO_3$–$x(Sr_{0.7}Bi_{0.2})TiO_3$	MLCC		9.5	92	[90]
$(1-x)$ $[0.94(0.75Bi_{0.5}Na_{0.5}TiO_3$–$0.25NaNbO_3)$–$0.06BaTiO_3]$–$xCaZrO_3$	MLCC	120	0.31–0.35	77	[91]
$(Na_{0.8}K_{0.2})0.5Bi_{0.5}TiO_3/0.6(Na_{0.8}K_{0.2})0.5Bi_{0.5}TiO_3$–$0.4SrTiO_3$	MLCC		73.7	68.1	[92]
$0.87BaTiO_3$–$0.13Bi(Zn_{2/3}(Nb_{0.85}Ta_{0.15})1/3)O_3$	MLCC	400	14.08		[93]
⟨111⟩-Textured $Na_{0.5}Bi_{0.5}TiO_3$–$Sr_{0.7}Bi_{0.2}TiO_3$	MLCC	1030	21.5		[94]

such as high Pr, the reduction from Ti^{4+} to Ti^{3+} at high temperature, which restricts it to achieve high W_{rec} [100].

In 2009, Ogihara et al. [101] synthesized a temperature-stable $0.7BaTiO_3$–$0.3BiScO_3$ (0.7BT–0.3BS) RFE ceramic with a thickness of 0.2 mm and obtained a W_{rec} of 2.3 J/cm^3 at 225 kV/cm. When thickness reduces to 15 μm, W_{rec} enhances to 6.1 J/cm^3 at 730 kV/cm. It should be mentioned that other systems such as $BaTiO_3$–$Bi(Mg,Ti)O_3$ (BT–BMT) also display good energy storage capabilities due to the low tolerance factor of Bi-based compound [102, 103]. In recent years, $Ba_{0.85}Ca_{0.15}Zr_{0.10}Ti_{0.90}O_3$ (BCZT)-based RFE has gained more attention in FE and pyroelectric properties because of its high ε_r at room temperature[104–107]. However, W_{rec} of BCZT-based ceramic bulks is commonly less than 3 J/cm^3 limited to a low E_b. If the trade-off relationship between ε_r and E_b gets a good balance, BCZT-based ceramics would be a promising energy storage candidate.

"Core–shell" structure is a common way of modifying BT-based ceramics, especially in enhancing temperature stability for ceramic capacitor applications. In 2014, Su et al. [108] coated BT nanocrystals with $65PbO$–$20B_2O_3$–$15SiO_2$ and $65Bi_2O_3$–$20B_2O_3$–$15SiO_2$ glass phases, reducing sintering temperature to 900 °C, as shown in Figure 14.5. Meanwhile, the authors estimated W_{rec} approaching to

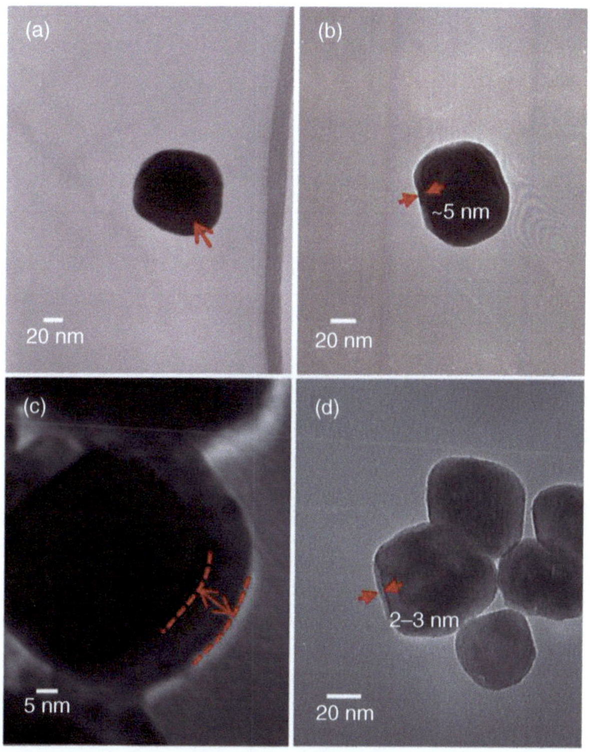

Figure 14.5 TEM micrographs of (a,b) as prepared $BaTiO_3/PbO$–B_2O_3–SiO_2 core–shell nanoparticles with various shell thicknesses (the arrows in (a) and (b) indicate the coating layer), (c,d) as prepared $BaTiO_3/Bi_2O_3$–B_2O_3–SiO_2 core–shell nanoparticles with the shell region defined by arrows and dashed lines. Source: Su et al. [108]/Reproduced with permission of Royal Society of Chemistry.

10 J/cm³ by equation for Bi-based glass phase. Similarly, high-E_b materials such as SiO_2 [109] and $SrTiO_3$ (ST) [110] are both used as shell materials to improve breakdown behavior. It should be pointed out that high-E_b materials usually possess low ε_r, and so W_{rec} of coated BT ceramics cannot be effectively enhanced at the sacrifice of ε_r.

$Ba(Zr,Ti)O_3$ (BZT)-based ceramic films are considered to be a potential energy storage material since Zr^{4+} has a stable valence and similar ion radius to Ti^{4+}, as well as good relaxor characteristics. As Zr^{4+} content increases, relaxor degree gradually enhances, and a weak diffuse phase transition accompanying slim P–E loop can be founded when Zr^{4+} is near 15 mol%, which is attributed to the inhomogeneous distribution of Zr and mechanical stress [111, 112]. Therefore, energy storage performances of BZT-based ceramic systems receive more attention.

"Interface engineering," including interface compatibility, periodic number, and space charge, plays a critical role in enhancing E_b and optimizing polarization behavior of BT-based ceramic films. Sun et al. [113] studied energy storage properties of laminated $Ba_{0.7}Ca_{0.3}TiO_3/BaZr_{0.2}Ti_{0.8}O_3$ (BCT/BZT) thin films with two layers as a period, and the number is 2, 4, 8. With increasing period, E_b enhances from 3 to 4.5 MV/cm, and the maximum W_{rec} is 52.4 J/cm³ for $N = 8$. Meanwhile, a series of multilayer structures consisting of two materials or a stack of different dielectric layers have become the most charming model systems since some unique properties can be enhanced. Energy storage performances of the BT-based RFE materials are summarized and listed in Table 14.6.

Table 14.6 Energy storage performances of the BT-based relaxor ferroelectric materials.

Composition	Type	E_b (kV/cm)	W_{rec} (J/cm³)	η (%)	References
$0.7BaTiO_3$–$0.3BiScO_3$	Bulk	730	6.1	—	[101]
SiO_2-coated $BaTiO_3$	Bulk		1.2	53.8	[109]
$Bi(Mg_{0.5}Zr_{0.5})O_3$–modified $BaTiO_3$	Bulk	301.4	2.9	86.8	[114]
$Ba_{0.85}Ca_{0.15}Zr_{0.1}Ti_{0.9}O_3 + Bi_2O_3$–$B_2O_3$–$SiO_2$	Bulk	330	2.12	90.5	[105]
$0.9(Sr_{0.7}Bi_{0.2})TiO_3$–$0.1Bi(Mg_{0.5}Zr_{0.5})O_3$–modified $BaTiO_3$	Bulk	370	4.03	96.2	[100]
BT@3 mol% BS	Bulk	120	0.68	—	[115]
BT@$65PbO$–$20B_2O_3$–$15SiO_2$ and $65Bi_2O_3$–$20B_2O_3$–$15SiO_2$	Bulk	1140	10	—	[108]
BT@ST	Bulk	47	0.22	90	[110]
$BaTiO_3/Ba_{0.3}Sr_{0.7}TiO_3$	Film	6000	46		[116]
(100) highly textured $BaZr_xTi_{1-x}O_3$	Film	3000	214	72.8	[117]
Laminated $Ba_{0.7}Ca_{0.3}TiO_3/BaZr_{0.2}Ti_{0.8}O_3$	Film	4500	52.4	—	[113]

14.2.2.4 High-Entropy Relaxor Ferroelectric Ceramics

The concept of high entropy was first introduced in 2003 by Ranganathan of the Indian Institute of Science and in 2004 by Ye Junwei and Cantor of the University of Oxford [118, 119]. The presence of different characteristic elements gives high-entropy materials four core effects: high-entropy effect, lattice distortion, hysteresis diffusion, and cocktail effect. The presence of these effects makes high-entropy ceramics a relatively promising ceramic material with excellent stability and energy storage properties [120–122].

In 2019, Pu et al. [123] synthesized a single-phase homogeneous A-site high-entropy $(Na_{0.2}Bi_{0.2}Ba_{0.2}Sr_{0.2}Ca_{0.2})TiO_3$ ceramic by using a solid-state method. It exhibited relaxation behavior because of the disorder of the microscopic composition. In addition, an applied electric field of 145 kV/cm produces a discharge energy density of 1.02 J/cm^3. Recently, an important breakthrough in energy storage ceramics based on the high entropy concept was achieved by Chen et al. [124]. By introducing (Li+, Ba^{2+}, Bi^{3+}, Sc^{3+}, Hf^{4+}, Zr^{4+}, Ta^{5+}, Sb^{5+}) ions into the KNN lattice, a local structure of R–O–T–C multiphase coexistence was obtained (Figure 14.6), further refining the PNRs and obtaining a huge improvement in energy storage performance and efficiency, respectively 10.06 J/cm^3 and 90.8%. Similarly, Yang et al. [125] designed a $(Bi_{3.25}La_{0.75})(Ti_{3-3x}Zr_xHf_xSn_x)O_{12}$ high-entropy system based on $Bi_4Ti_3O_{12}$ ceramic thin film to obtain a microstructure with the coexistence of nanograins and amorphous phases and obtained a huge energy storage density of 182 J/cm^3 under a 6360 kV/cm electric field and 78% efficiency. These two works are of epoch-making significance for the research and application of high-entropy energy storage ceramics.

The energy storage properties of high-entropy RFE ceramics are summarized in Table 14.7.

In summary, for RFE ceramics, the formation of PNRs due to disturbed long-range FE order or nonergodic–ergodic phase transition can strengthen relaxor characteristics. Reflecting on macroscopic FE properties, *P–E* loop goes slim. In addition, delay-saturated polarization is also used to optimize polarization behavior. Designing multilayer ceramic capacitor (MLCC), "core–shell" structure with nonferroelectric phase as shell material is commonly accepted method to improve electric breakdown behavior of ceramic bulks. By comparison, chemical modification of ceramic films functions in two roles: For one, suppress the generation and transportation of vacancy defect to reduce leakage current; for the other one, strengthen their relaxor characteristics and optimize polarization behavior. In the heterostructure of ceramic films, match degree of physical parameters such as lattice constant, thermal expansion coefficient, etc., gradient sequence, template, or new inert layers are all important factors to influence energy storage performances.

14.2.3 Antiferroelectric Ceramics

As a special group of FE material, AFE has many similarities with FE, whereas still exists obvious differences. Generally speaking, AFE materials possess Ps in a

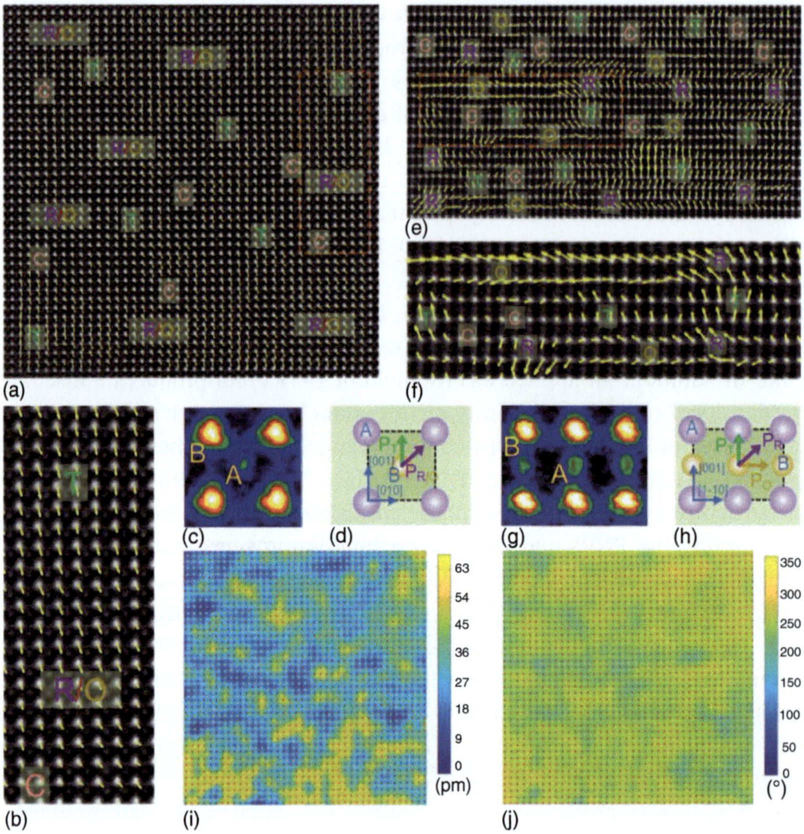

Figure 14.6 (a) Atomic-resolution HAADF STEM polarization vector image along $[100]_c$. (b) Enlarged image of the marked area (dark red rectangle) in a showing the transition of polarization vectors from T to R/O to C. (c) Magnified image and (d) schematic projection of the unit cell along $[100]_c$. (e) Atomic resolution HAADF STEM polarization vector image along $[110]_c$. (f) Enlarged image of the marked area (dark red rectangle) in (e) showing the transition of polarization vectors from T to R to O to C. (g) Magnified image and (h) schematic projection of the unit cell along $[110]_c$. (i) Polarization magnitude mapping, and (j) polarization angle mapping along $[100]_c$. Source: Chen et al. [124]. Reproduced with permission of Springer Nature.

Table 14.7 The energy storage properties of high-entropy RFE ceramics.

Composition	Type	E_b (kV/cm)	W_{rec} (J/cm^3)	η (%)	References
$(Na_{0.2}Bi_{0.2}Ba_{0.2}Sr_{0.2}Ca_{0.2})TiO_3$	Bulk	145	1.02		[123]
$(Bi_{0.2}Na_{0.2}K_{0.2}Ba_{0.2}Ca_{0.2})TiO_3$	Bulk	129	0.684	87.5	[126]
$(Bi_{0.2}Na_{0.2}K_{0.2}La_{0.2}Sr_{0.2})TiO_3$	Bulk	180	0.959	67	[127]
$0.75BaTiO_3$–$0.25Na_{0.5}Bi_{0.5}TiO_3$ + 0.1 mol%$Bi(Zn_{0.2}Mg_{0.2}Al_{0.2}Sn_{0.2}Zr_{0.2})O_3$	Bulk	273	3.74	82.2	[128]
KNN+ (Li^+, Ba^{2+}, Bi^{3+}, Sc^{3+}, Hf^{4+}, Zr^{4+}, Ta^{5+}, Sb^{5+})	Bulk	740	10.06	90.8	[124]
$(Bi_{3.25}La_{0.75})(Ti_{3-3x}Zr_xHf_xSn_x)O_{12}$	Film	6350	182	78	[125]

unit cell, but the direction of Ps is antiparallel to that of neighboring unit cell. AFE, therefore, does not exhibit polarization in macroscopic characteristics. It is particularly important that AFE exists a unique feature: double hysteresis loop under an external field. P is linearly proportional to E at a low electric field. When E exceeds the forward switching (AFE-to-FE) field EA–F, AFE displays an obvious FE behavior, and P fastly increases and gradually reaches P_{max}. Note that AFE can undergo a ferroelectric–antiferroelectric (FE–AFE) phase transition field (EF–A) after removing E. Consequently, P gradually reduces and returns to the initial state (that is, $E = 0$, $P_r = 0$). A similar variation in P can be observed as E continuously increases in an opposite direction.

It is hard to see a double hysteresis loop for most pure AFE materials at a low electric field, which usually requires special conditions of high temperature and high electric field to stimulate. Thereby, it is a crucial challenge for AFE to obtain double P–E loop, especially for ceramic bulks. In addition, a typical high-squareness P–E loop, and corresponding internal strain induced by AFE–FE phase transition both result in low W_{rec} and reduced device life [129]. In this regard, how to stabilize and relax the AFE phase (corresponding to enhancing EA–F, reducing the difference of EA–F–EF–A) and enhance E_b has become a particularly important issue to acquire good energy storage performances.

14.2.3.1 $PbZrO_3$-Based Antiferroelectric Ceramics

$PbZrO_3$ (PZ) is the prototype AFE and serves as a model system to be thoroughly studied so far. The origin of the phase transition mechanism, however, is not well understood [130]. Despite this, a characteristic double P–E loop with high P_{max} and low P_r makes it suitable for energy storage capacitors. PZ-based energy storage ceramics have gained increasing attention since Chen et al. [131] reported W_{rec} of 7.1 J/cm^3 for PZ thin films during phase transition, but basically around ceramic films.

It is widely accepted that tolerance factor t of perovskite structure is closely related to phase stability, and the equation is represented as:

$$t = \frac{r_A + r_O}{\sqrt{2}(r_A + r_O)}$$

where r_A, r_B, r_O denote ion radii of A, B, and O, respectively. In general, AFE phase can be stabilized for $t \approx 1$, and reducing t can enhance to some degree the stability of AFE phase. For example, Hao et al. [132] substituted Pb^{2+} (1.20 Å) with smaller Sr^{2+} (1.12 Å) to reduce t, and P–E loop goes slim together with increased EA–F and decreased ΔE. As a consequence, $(Pb_{0.95}Sr_{0.05})ZrO_3$ (PSZ_5) thin films obtain a W_{rec} of 14.5 J/cm^3. Choosing smaller ion radius while aliovalent to La^{3+} as donor dopant has a similar effect [133]. In addition, controlling the orientation of PZ thin film also achieves the purpose of stabilizing the AFE phase. PZ thin film with (100) orientation needs a higher electric field than with (111) one to finish AFE–FE phase transition [134].

Tailoring local electric fields and stress to enhance W_{rec} is hard but interesting work. As is well known, introducing a new nanoparticle in materials often brings novel properties and improved properties. For example, Sa et al. [135] used α-Fe_2O_3

Table 14.8 The energy storage performance of PZ-based AFE materials.

Composition	Type	E_b (kV/cm)	W_{rec} (J/cm^3)	η (%)	References
$Pb_{0.98}La_{0.02}(Hf_{0.45}Sn_{0.55})_{0.995}O_3$	Bulk	380	7.63	94	[136]
$PbZrO_3$	Film		7.1		[131]
$PbZrO_3$ + 5%La	Film	6000	14.9		[133]
$PbZrO_3$ + Sr	Film		14.5		[132]
$PbZrO_3$ + Eu	Film		16.3		[137]
(110) or (111) oriented $PbZrO_3$	Film	700	5.3		[138]
$PbZrO_3$ + Fe_2O_3NPs	Film	600	17.4		[135]
NiO nano-columns embedded $PbZrO_3$	Film		24.6		[139]

nanoparticles to modify PZ thin films and obtained a W_{rec} of 17.4 J/cm^3 and a P_{max} as high as 78 μC/cm^2 attributing to the local field effect. To our best knowledge, it is difficult to directly measure and characterize local electric field and stress, and thus corresponding methods need to be developed. Furthermore, the gradient sequence of thin film is significant for enhancing W_{rec}. The energy storage performance of PZ-based AFE materials is listed in Table 14.8.

14.2.3.2 (Pb, La)(Zr, Sn, Ti)O_3-Based Antiferroelectric Ceramics

As discussed in Section 14.2.2.1, $Pb(Zr_{1-x}Ti_x)O_3$ (PZT, $x \approx 0.05$) solid solutions located at rich Zr regions of FE–AFE phase boundary, exhibit rich phase structures and well-defined electrical properties [140]. It is worth noting that PZT systems at FE–AFE phase boundary differ from MPB in electric properties. For instance, FE and AFE cannot be transformed into each other under force or electric field, while temperature, stress, or other factors induced by FE–AFE phase transition can occur. Meanwhile, the region of the AFE phase for PZT 95/5 is relatively narrow, and a slight composition fluctuation would easily cause deviation from FE–AFE phase boundary.

La^{3+} and Sn^{4+} are very popular A/B site dopants in PZT 95/5 ceramics, where their functions are similar [141–145]. La^{3+}, as a donor dopant, substitutes Pb^{2+} to disturb long-range ordering of domain by vacancy defect, which would strengthen the relaxor characteristic and expand the stabilized region of the AFE phase. Sn^{4+} as an equal-valence dopant of Ti^{4+} functions not only expands the stabilized region of AFE but also adjusts Zr/Ti ratio to enable Ti up to 10 mol%. Currently, the related works of stabilizing the Pb-based AFE phase mainly focus on adjusting suitable Zr/Sn/Ti ratio [146–148]. Liu et al. [149] found that E_{A-F} linearly increased and the squareness of P–E loop slightly improved when Ti content reduces from 0.11 to 0.07 mol at a fixed Zr of 0.58 mol. W_{rec} enhances, thereby, from 0.28 to 2.35 J/cm^3 for $Pb_{0.97}La_{0.02}(Zr_{0.58}Sn_{0.35}Ti_{0.07})O_3$ AFE ceramic bulks. In addition, it should be particularly noted that a high W_{rec} is difficult to obtain when E_{A-F} exceeds E_b.

In improving the breakdown behavior of Pb-based AFE ceramic bulks, Zhang et al. [150–152] did a series of works using some special sintering technology. For instance, Zhang et al. [151] reported that $(Pb_{0.87}Ba_{0.1}La_{0.02})(Zr_{0.68}Sn_{0.24}Ti_{0.08})O_3$ (PBLZST) ceramics using hot-press (HP) possessed a high W_{rec} of 3.2 J/cm^3 at $E_b = 180$ kV/cm due to smaller grain size and well insulation. Considering that special sintering technology requires expensive equipment, this method is not suitable for large-scale production. Some low-cost and convenient solutions are proposed, such as Bian et al. [153] used amorphous SiO_2 to coat $Pb_{0.97}La_{0.02}(Zr_{0.33}Sn_{0.55}Ti_{0.12})O_3$ (PLZST 2/33/55/12), and measured a high W_{rec} of 2.68 J/cm^3 due to E_b enhancing from 12.2 to 23.8 kV/mm. It should be noticed that the introduction of non-AFE phase will reduce the content of the original AFE phase, even that E_{A-F} would disappear.

With the development of micro-electric devices, flexible substrate, such as Ti, Si, and Ni foils, is required to meet future application scenes. Ma et al. [154, 155] did a series of meaningful works in the low-cost integration development of PLZT-based thin films. For instance, $Pb_{0.92}La_{0.08}Zr_{0.95}Ti_{0.05}O_3$ (PLZT 8/95/5) AFE ceramic films possess a high W_{rec} of 53 J/cm^3 at mental foil by CSD, and effective work time would maintain 5000 hours at room temperature [154]. Energy storage properties of PLZST-based AFE ceramics and films are summarized and listed in Table 14.9.

14.2.3.3 $NaNbO_3$-Based Antiferroelectric Ceramics

$NaNbO_3$ (NN) is a well-documented nonpolar AFE phase and possesses a complicated crystal structure due to the rotation of oxygen octahedron and off-centered displacement of Nb^{5+} [160, 161]. In addition to temperature-induced transition,

Table 14.9 Energy storage properties of PLZST-based antiferroelectric materials.

Composition	Type	E_b (kV/cm)	W_{rec} (J/cm^3)	η (%)	References
$(Pb_{0.858}Ba_{0.1}La_{0.02}Y_{0.008})(Zr_{0.65}Sn_{0.3}Ti_{0.05})O_3$–$(Pb_{0.97}La_{0.02})(Zr_{0.9}Sn_{0.05}Ti_{0.05})O_3$(SPS)	Bulk	162	6.5		[150]
$(Pb_{0.87}Ba_{0.1}La_{0.02})(Zr_{0.68}Sn_{0.24}Ti_{0.08})O_3$(HP)	Bulk	180	3.2		[151]
$Pb_{0.97}La_{0.02}(Zr_{0.50}Sn_{0.46}Ti_{0.04})O_3$	Bulk		3.2	86.5	[144]
$Pb_{0.97}La_{0.02}(Zr_{0.33}Sn_{0.55}Ti_{0.12})O_3$@$SiO_2$	Bulk	238	2.68		[153]
$(Pb_{0.98}La_{0.02})(Zr_{0.55}Sn_{0.45})_{0.995}O_3$	Bulk	400	10.4	87	[156]
$(Pb_{0.98-x}La_{0.02}Sr_x)(Zr_{0.9}Sn_{0.1})$	Bulk		11.18	82.2	[157]
$(Pb_{0.97}La_{0.02})(Zr_{0.93}Sn_{0.03}Ti_{0.04})O_3$	Bulk		2.0–3.4		[145]
$Pb_{0.92}La_{0.08}Zr_{0.95}Ti_{0.05}O_3$	Film	300	37		[154]
$(Pb_{0.97}La_{0.02})(Zr_{1-x-y}Sn_xTi_y)O_3$	Film		13.7		[142]
$(Pb_{0.97}La_{0.02})(Zr_{0.66}Sn_{0.23}Ti_{0.11})O_3$	Film	4000	46.3	84	[158]
$(Pb_{0.98}La_{0.02})(Zr_{0.95}Ti_{0.05})O_3$	Film		30		[159]
$Pb_{0.97}La_{0.02}(Zr_{0.95}Ti_{0.05})O_3 + SrTiO_3$	Film		101	62	[143]

AFE–FE phase transition can also be induced by mechanical stress [162], grain size [163], and electric field [164]. Note that an electric field-induced AFE–FE phase transition is usually irreversible at room temperature because of the small free energy difference between AFE and FE.

Generally, reducing tolerance factor t is still a widely accepted method to stabilize AFE phase of NN ceramics. For instance, Shimizu et al. [165] used $CaZrO_3$ (CZ), with similar electronegativity but smaller tolerance factor t, to modify NN ceramics. As the content of CZ increases, an obvious double P–E loop accompanying E_{A-F} gradually increases while P_{max} decreases. Since then, other compounds, such as $SrZrO_3$ (SZ) [166], $CaHfO_3$ (CH) [167], and $BiScO_3$ (BS) [168], have both been utilized to stabilize AFE phase, but few report their energy storage performances. It should be pointed out that double P–E loop of NN-based ceramics still has more hysteresis loops at room temperature, resulting in low W_{rec}. Recently, Jiang et al. [169] discovered NN-BF system, whose W_{rec} and η were improved to 18.5 J/cm^3 and 83.3%, which set the highest record for ceramic bulks of AFE materials.

A double P–E hysteresis loop can be observed in high-quality NN single crystal and ceramic film compared with ceramic bulk. However, a few studies reported energy storage properties of NN-based ceramic films and concentrated on piezoelectric and dielectric tunability. Luo et al. [170] fabricated Mn-doped $0.96NaNbO_3$–$0.04CaZrO_3$ (0.96NN–0.04CZ) thin films using sol–gel method, and a leakage current of 1 mol% Mn reduces the magnitude of 103–104 compared to pure compositions. W_{rec} and η of 1 mol% Mn thin films are 19.64 J/cm^3 and 64.5%, respectively. Nevertheless, its simple composition and absence of toxic metal elements make it have broad prospects for energy storage research in the future. The energy storage properties of NN-based dielectric materials are summarized in Table 14.10.

14.2.3.4 $AgNbO_3$-Based Antiferroelectric Ceramics

$AgNbO_3$ (AN) has a complicated phase structure and relatively low bandgap (~2.8 eV) and still has difficulty in observing double hysteresis loop at room

Table 14.10 The energy storage properties of NN-based dielectric materials.

Composition	Type	E_b (kV/cm)	W_{rec} (J/cm^3)	η (%)	References
$Na_{0.7}Bi_{0.1}NbO_3$	Bulk	250	4.03	85.4	[171]
$NaNbO_3$–$0.1Bi(Mg_{2/3}Nb_{1/3})NbO_3$	Bulk	300	2.8	82	[172]
$NaNbO_3$–$0.15Bi(Mg_{2/3}Nb_{1/3})NbO_3$	Bulk	300	2.4	90	[172]
$(1-x)NaNbO_3$–$xBi(Ni_{1/2}Sn_{1/2})O_3$	Bulk		5	68	[173]
$0.76NaNbO_3$–$0.24(Bi_{0.5}Na_{0.5})TiO_3$	Bulk	680	12.2	69	[6]
$NaNbO_3$–$Sr_{0.7}Bi_{0.2}TiO_3$	Bulk	288	4.5	90.3	[174]
$0.76NaNbO_3$-$0.24Sr_{0.7}Bi_{0.2}TiO_3$(SPS)	Bulk	660	12.2	88	[5]
$0.90NaNbO_3$–$0.10BiFeO_3$	Bulk	995	18.5	83.3	[169]
$NaNbO_3$–$0.04CaZrO_3$ + Mn	Film		19.64		[170]

temperature. Early studies of AN ceramics therefore concentrated on microwave communication and photocatalysis [175, 176]. In addition, Ag_2O decomposition at a high temperature requires the fabrication of AN-based ceramics in an oxygen-rich environment. Similar to NN, AN-based ceramic bulk is a main research direction for energy storage capacitors.

In 2007, Fu et al. [177] successfully fabricated AN ceramics with a double hysteresis loop, and P_{max} can up to 52 μC/cm^2 at 220 kV/cm exceeding other dielectric materials at the same electric field. The phenomenon strongly stimulates the studies of AN ceramics on energy storage applications. Tian et al. [178] synthesized pure AN AFE ceramic bulks, found two polarization structures by TEM and variable temperature *P–I–E* loops, and attained a high W_{rec} of 2.1 J/cm^3. In general, AN ceramics experience a series-phase transition [179]:

$$M_1 \xrightarrow{67\,^\circ C} M_2 \xrightarrow{267\,^\circ C} M_3 \xrightarrow{353\,^\circ C}$$

$$O_1 \xrightarrow{361\,^\circ C} O_2 \xrightarrow{387\,^\circ C} T \xrightarrow{579\,^\circ C} C$$

where M_1, M_2, and M_3 denote orthorhombic phases in rhombic orientation, O_1 and O_2 are the orthorhombic phases in a parallel orientation, while T and C denote the tetragonal and cubic phases, respectively. It is known that the stability of AFE phase of AN ceramic is closely related to phase transition among M_1, M_2, and M_3. Therefore, reducing phase transition temperature among M_1, M_2, and M_3 to low temperature would enhance W_{rec} of AN-based ceramics [179–182].

Note that fine-grain ceramics are a good choice to enhance E_b due to an inverse relationship between E_b and grain size. Recently, Wang et al. [183] synthesized AN ceramics using hydrothermal process and obtained a maximum E_b = 250 kV/cm among pure AN ceramics so far. However, P_{max} has an obvious decrease compared

Table 14.11 Energy storage behaviors of AN antiferroelectric materials.

Composition	Type	E_b (kV/cm)	W_{rec} (J/cm^3)	η (%)	References
$AgNbO_3$	Bulk		2.1		[178]
Ta-modified $AgNbO_3$	Bulk	240	4.2		[179]
$Ag_{1-3x}Bi_xNbO_3$	Bulk		2.6	86	[184]
W-modified $AgNbO_3$	Bulk		3.3		[185]
$Sm_{0.03}Ag_{0.91}NbO_3$	Bulk	290	5.2	68.5	[182]
Bi^{3+} on the A-site and Zn^{2+} on the B-site in $AgNbO_3$	Bulk		4.6		[181]
$Ag_{0.9}Sr_{0.05}NbO_3$	Bulk	190	2.9	56	[186]
2 mol% La-doped $AgNbO_3$	Bulk		4.4	70	[187]
$Ag_{0.92}Ca_{0.04}NbO_3$	Bulk	220	3.55		[188]
$(1-x)AgNbO_3$–$xAgTaO_3$	Bulk		6.3	90	[180]
$Ag_{0.97}Nd_{0.01}Ta_{0.20}Nb_{0.80}O_3$	Bulk	350	6.5		[189]

with other AN ceramics using solid-state method, and thus W_{rec} and η are only 1.8 J/cm^3 and 40%, respectively. It is noteworthy that little attention has been shared on enhancing E_b of AN ceramics by other different methods apart from ion-doping. Energy storage behaviors of AN AFE ceramic bulks and films are summarized and listed in Table 14.11.

In summary, for AFE ceramics, tolerance factor t, electronegative difference, and polarizability are all influencing the stability of AFE phase. Note that the characteristic double P–E loop of pure AFE phase is difficult to be observed, which needs to further explore the related reasons. Physical (special sintering technology, rolling process, etc.) and chemical (coating, hydrothermal process, etc.) methods can enhance E_b by improving sintering behavior or mechanical strength.

14.3 Organic Dielectric Materials

14.3.1 Polymer-Based Dielectrics for Energy Storage

14.3.1.1 Ferroelectric Polymers

FE polymers are a group of crystalline polar polymers that are also FE, meaning that they maintain a permanent electric polarization that can be reversed or switched in an external electric field. FE polymers are crystalline polar polymers that are also FE, meaning they have a permanent electric polarization that may be altered in an external electric field. A hysteresis loop is created by monitoring the electric polarization of a FE material as the applied electric field is altered. The material must be polar and crystalline, and polarization may be reversed in response to an electric field to produce ferroelectricity. Till present, approximately 1000 solid materials have been identified as FE, with fewer than 10 of the 32 crystalline types contributing. FEs have FE domain structures made up of dipoles that spontaneously align.

Ferroelectricity in polymers was first proven in poly(vinylidene fluoride) (PVDF) in the early 1970s, almost 50 years after the discovery of the FE phenomenon, and it quickly drew the interest of solid-state physicists and polymer research groups. These materials have several benefits over their inorganic FE equivalents, including simplicity of production, low weight, and flexibility. As a result, they open up new possibilities in a wide variety of sectors, including memory, capacitors, piezoelectric, and pyroelectric applications, among many others, sparking widespread interest throughout the world [190].

PVDF, having the chemical formula -$(CH_2$–$CF_2)_n$-, is a flexible polymer made from vinylidene chloride copolymerization [191]. VDF has fluorine and hydrogen contents of 59.4 and 3 wt%, respectively. The generated C—F bond has a more significant dipole moment (μ = 1.92 D) due to the electronegativity difference between fluorine and hydrogen, resulting in PVDF having a higher ε_r and U_d than typical polymers. A typical PVDF film has an ε_r, breakdown strength, dissipation factor at 1 kHz, and energy density of 12, 590 MV/m, 0.018, and >2 J/cm^3, respectively [192]. PVDF films with varied crystallinities (50–70%) may be made by modifying the preparation method [193]. PVDF tends to generate stable gauche (G) and trans (T) connections with bond angles of 60° or

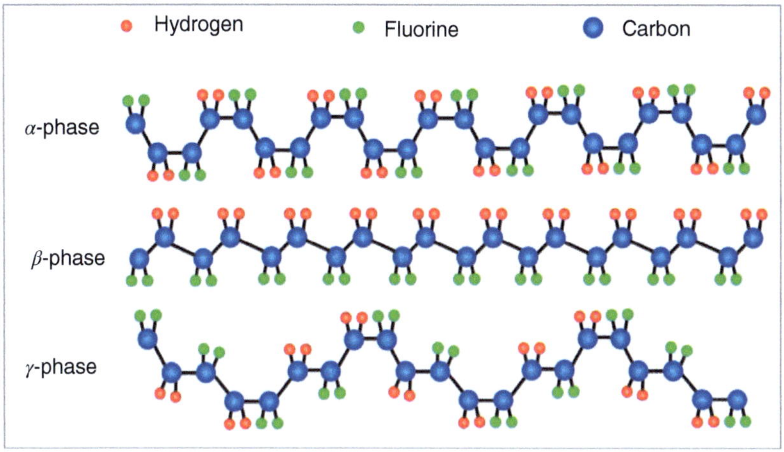

Figure 14.7 Schematic representation of the chain conformation for the α, β, and γ-phases of polyvinylidene fluoride (PVDF). Source: Wang et al. [194]. Reproduced with permission of MDPI.

180° between neighboring CH_2 and CF_2 units. PVDF is available in four distinct crystal shapes. The δ-phase and α-phase have identical 3D dimensions and a trans-gauche-trans-gauche′ (TGTG) chain conformation. The all-trans chain conformation of the β-phase is trans-trans-trans-trans (TTTT). γ-phase possesses the trans-trans-trans-gauche-trans-trans-gauche′ (TTTGTTTG) chain conformation, as indicated in Figure 14.7 [195].

The trans-gauche conformation is seen in three crystal forms of PVDF, i.e. α, γ, and δ phases. The dipole moment is 1.02 D in the direction parallel to the chain axis and 1.20 D in the direction perpendicular to the chain axis. The polarization is minimal because the polar CF_2 groups in this conformation are orientated differently. Since the dipole moments of trans-gauche conformation cancel one another, the α-phase has a zero macroscopic net dipole moment. The β-phase, on the other hand, has the highest polarization [196]. The polar groups are oriented in the same direction in the all-trans conformation, which has a strong dipole moment of 2.10 D in the direction perpendicular to the chain axis [193].

PVDF may be produced in a variety of crystal phases by using various preparation methods [197]. Direct cooling at normal speed following PVDF melting yields the lowest energy α-phase. To produce the δ-phase, a 100–200 MV/m electric field can be used to polarize and convert the α-phase into the δ-phase. Substantial temperatures or high electric fields may offer β- and γ-phases. The rearrangement of chain segment movement and the transition of distinct crystal phases may be facilitated by high-temperature annealing [198]. Crystal phase formation may be influenced by highly polar solvents as well. For example, cyclohexanone solvents can significantly develop α- and β-phases, whereas *N*,*N*-Dimethylformamide (DMF) solvent is easier to generate α- or γ-phase [199].

As shown in Figure 14.8, Li et al. investigated the energy storage capacity of three crystalline types of α-phase, β-phase, and γ-phase [200]. The energy loss (U_l)

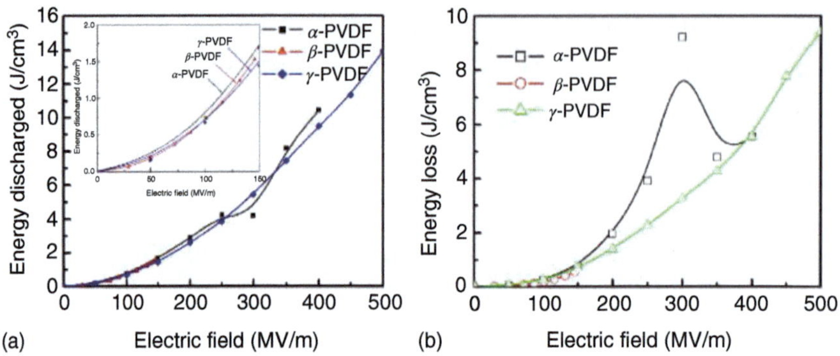

Figure 14.8 Various PVDF phases' (a) energy density and (b) loss as a function of an electric field. Source: Li et al. [200]. Reproduced with permission of American Institute of Physics.

and U_d of distinct PVDF crystalline phases were shown. At low electric fields, the three crystalline phases were shown to have identical energy storage capability. With polarization at 150 MV/m, the β-phase is saturated, and the maximal U_d was reported at 1.5 J/cm^3. The electric field is less than 300 MV/m, and the α-phase's U_d was somewhat higher than the γ-phase's. When the electric field is greater than 300 MV/m, the rate of α-phase U_d development slows dramatically. When the electric field is higher than 300 MV/m, the *D–E* loop of the α-phase shows a distinct tendency to switch to the γ-phase. The α-phase showed considerable energy loss between 150 and 350 MV/m. This means that a portion of the energy is spent transitioning from the α-phase TGTG′ to the γ-phase TTTG conformation. The loss curves of the α- and γ-phases match nicely when the transition is completed. Only the γ-phase can exist under a high electric field (500 MV/m) and has the greatest U_d (14 J/cm^3) of the three crystal phases of PVDF.

Ren et al. [201] reported a new Press and Folding (P&F) technique to increase the β-phase, as shown in Figure 14.9a. The increase in the number of P&F cycles results in a significant increase in PVDF polarization, which is related to the content of β-phase in PVDF (Figure 14.9b). After six P&F cycles, the β-phase content in PVDF increases from 8% to 98%. The reduction of remnant polarization and the RFE behavior at a high electric field are related to the internal stress generated during pressing (Figure 14.9c). The internal stress increases the β-phase molecular chain spacing, which makes residual stresses exist inside the material. These residual stresses promote domain flipping when the electric field changes, forming reversible highly polarized nanostructures. Finally, after six P&F cycles at a temperature of 140 °C, the prepared PVDF film achieved an ultrahigh U_d of 39.8 J/cm^3 at 880 MV/m (Figure 14.9d).

By combining PVDF with additional monomers to make PVDF-based binary copolymers, it is possible to prevent the early saturation of electric displacement caused by low electric fields. Modulation of the copolymer group content in the copolymer may modify the composition of each phase in PVDF, which can enhance the maximum polarization while simultaneously reducing the remnant

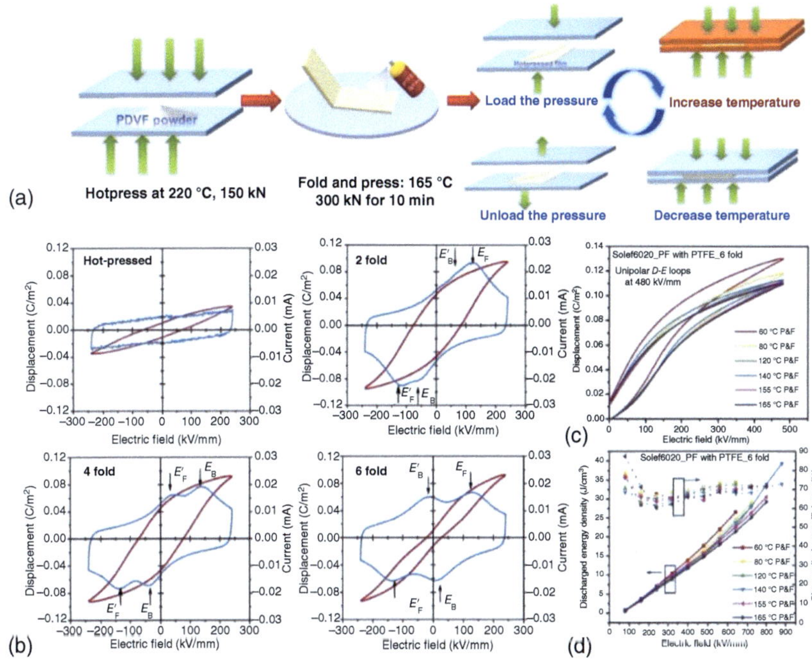

Figure 14.9 (a) Schematic demonstration of the P&F process of spraying PTFE release agent on the surface. (b) *D*–*E* and current–electric field (*I*–*E*) loops of PVDF films with different P&F cycles. (c) *D*–*E* loops of sixfold films for P&F operation at different temperatures. (d) U_d and η of P&F films. Source: Ren et al. [201]/Elsevier/CC BY 4.0.

polarization of the dielectric. Simultaneously, it is possible for the RFE and AFE dielectric characteristics to be obtained in the copolymers. These features are distinct from those of typical FEs. Some of the renowned PVDF-based binary copolymers are PVDF-TrFE, PVDF-HFP, and PVDF-CTFE, which are discussed below.

The copolymerization of PVDF with the TrFE unit disrupts TGTG' conformation and improves the β-phase's crystal stability, allowing for the transformation to the all-trans conformation. Xie et al. prepared P(VDF-TrFE) binary copolymers and concluded that due to the high content of the β-phase in P(VDF-TrFE), the resulting copolymers achieved high remnant polarization of $3\,\mu C/cm^2$ and low energy storage properties, i.e. U_d of about $10\,J/cm^3$ at an electric field of 370 MV/m (Figure 14.10) [202].

A small quantity of large-sized CTFE enhances chain spacing, stabilizing the TGTG′ conformation in the alpha-phase and preventing phase transitions from the α- to the γ-phase at high electric fields. Zhou et al. studied the dielectric and energy storage characteristics of P(VDF-CTFE). They reported $\tan\delta$ and ε_r of 0.03 and 13 at 1 kHz for P(VDF-CTFE), respectively. Also, a high E_b of >700 MV/m for the P(VDF-CTFE) film was reached at a stretching ratio of 500% by further refining the film production method and polymer film quality [203]. With an U_d of $25\,J/cm^3$ and a 600 MV/m electric field, P(VDF-CTFE) films showed RFE activity. Moreover,

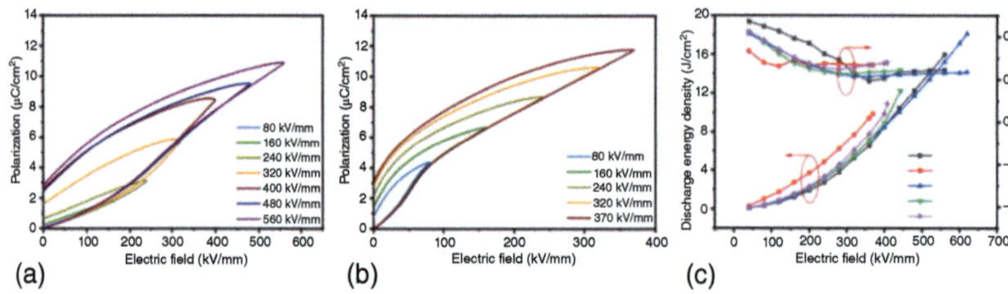

Figure 14.10 *D–E* loops of (a) P(VDF) and (b) P(VDF-TrFE). Discharge energy density and charge–discharge efficiency of P(VDF) and P(VDF-TrFE). Source: Xie et al. [202]. Reproduced with permission of American Chemical Society.

in P(VDF-HFP), the energy storage characteristics are greatly influenced by the HFP concentration in overall composition. At room temperature in P(VDF-HFP) (95.5/4.5 mol%), Zhou et al. improved the film processing conditions by studying the breakdown mechanism. They achieved an U_d of over 25 J/cm^3 at this temperature under a 700 MV/m electric field [204].

Besides the binary copolymers of P(VDF), the inclusion of the bulky comonomer in the ternary copolymerization process makes it possible to accomplish physical pinning and to improve interchain spacing. This pinning effect makes it possible to reduce the size of FE domains to the nanoscale range, which speeds up the process of dipole flipping and lowers coupling. At the same time, the dipole and domain wall between the pinning may be orientated at lower energy to provide low hysteresis RFE behavior, which is favorable to effective energy storage in dielectrics. This is possible because of the interaction between the pinning and the dipole. P(VDF-TrFE) is then added with a copolymer monomer that has a volume that is greater than that of TrFE in order to accomplish pinning in it; for example, monomers of chlorotrifluoroethylene (CFE), chlorotrifluoroethylene (CTFE), and hydrofluoropropylene (HFP) in order to fulfill the requirements. To mention some literature, P(VDF-TrFE-CFE) with 58.3/34.2/7.5 mol% compositions, respectively, achieved an U_d of about 9.5 J/cm^3 with an electric field of 425 MV/m, according to Chu et al. [205]. In addition, different P(VDF-TrFE-CTFE) compositions were examined by Zhang et al. [206] for their energy storage capabilities. The P(VDF-TrFE-CTFE) with 65.6/26.7/7.7 mol% compositions, respectively, showed a breakdown electric field of more than 500 MV/m. Also, it demonstrated an ε_r of about 60 at a Curie temperature of 35–40 °C and an U_d of >13 J/cm^3, as shown in Figure 14.11.

14.3.1.2 Nonferroelectric Polymers

Nonferroelectric polymers are further classified into two types, i.e. polar and nonpolar. For example, poly(methyl methacrylate) (PMMA) [207], polysulfones (PSF) [208], polythiourea (PTU) [209], and polyetherimide (PEI) [210], are categorized as polar polymers; while polystyrene (PS) and polypropylene (PP) are classified as nonpolar polymers [211]. Further, many polymers, for example, poly(phenylene sulfide) (PPS) [212], polycarbonate (PC)[213], poly(ethylene terephthalate) (PET) [214], and

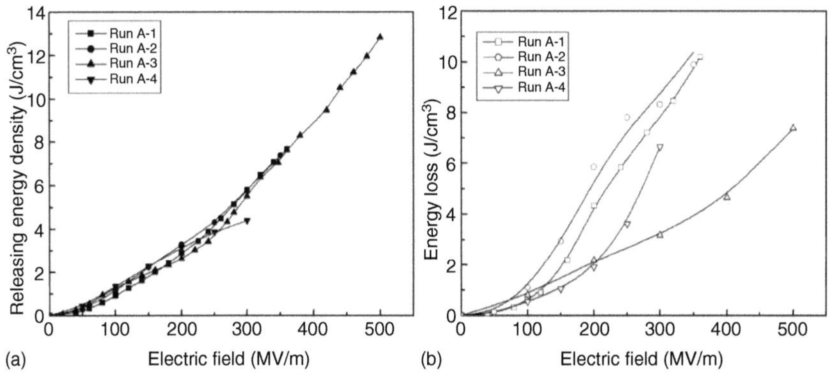

Figure 14.11 (a) Releasing energy density and (b) energy loss of VDF/CTFE copolymer (run A-1) and three VDF/TrFE/CTFE terpolymers (runs A-2, A-3, and A-4). Source: Zhang et al. [206]. Reproduced with permission of American Chemical Society.

Table 14.12 Nonferroelectric polymers' energy storage properties.

Polymers	Dielectric constant (at 1 kHz)	Dielectric loss (at 1 kHz)	Breakdown strength (MV/m)	Energy density (J/cm^3)	Efficiency (%)	References
PI	3–3.5	0.13	200	1.4	76	[197]
PEI	2.7	—	450	1.6	92%	[210]
PMMA	3.49	0.04	250	1.25	80	[216]
PC	2.8	0.001	520	0.5–1	95.8	[217]
PET	3.3	0.003	500	1–1.5	—	[212]
PP	2.25	0.0001	620	2.43	99	[218]
PS	2.7	—	450	5	—	[219]

poly(ethylene 2,6-naphthalate) (PEN) [212] are commercially available as dielectric capacitors. Treufeld et al. successfully prepared PI films and reported a dielectric constant and its respective losses of about 3.5 and 0.13 at 1 kHz [215]. They reported the highest E_b, U_d, and η of 200, 1.4, and 76, respectively. Further, Marwat et al. reported an energy density and efficiency of 1.6 J/cm^3 and >90% for PEI, respectively, at 450 MV/m [210]. Similarly, PMMA, PC, PET, PP, and PS are reported with U_d of 1.25, 1, 1.5, 5, and 2.43 J/cm^3. Table 14.12 summarizes the nonferroelectric polymers' energy storage properties.

14.3.1.3 Blended Polymers

To increase breakdown strength and maximum energy density, Rahimabady et al. found that adequate intermolecular interactions between polar groups in polymer blends were beneficial. A chemical solution method, followed by quenching, annealing, and hot pressing, was used to create α-phase P(VDF-HFP)/PVDF mix

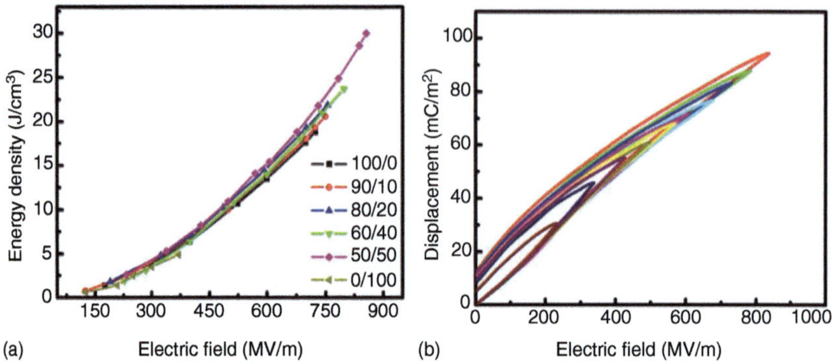

Figure 14.12 (a) Discharged energy density for P(VDF-HFP)/PVDF blended films as a function of the applied field. (b) Hysteresis loops for P(VDF-HFP)/PVDF (50/50 wt%), measured under various unipolar electric fields. Source: Rahimabady et al. [220]. Reproduced with permission of American Institute of Physics.

films. Compared to either of the two constituent polymers, PVDF or P(VDF-HFP), intermolecular electrostatic interactions between PVDF and P(VDF-HFP) in the highly uniform polymer blends led to a higher melting temperature, structural reconfiguration, higher crystallinity, larger elastic modulus, and an improved breakdown strength (>850 MV/m). P(VDF-HFP)/PVDF blended film also showed a higher dielectric constant, probably due to interfacial polarization, structural reconfiguration, and better crystallinity in the film. It was found that P(VDF-HFP), PVDF (50 : 50 wt%) blend films had an unusually high energy density of 30.1 J/cm^3, see Figure 14.12 [220].

Similarly, the ε_r and E_b of the PVDF-TrFE-CFE composite membranes were increased by adjusting the PVDF ratio, as shown in Figure 14.13. For this, Gong et al. prepared poly(vinylidene fluoride)/poly(vinylidene fluoride-trifluoroethylene-chlorofluoroethylene) (PVDF-PTC) blended films. The recombination and interfacial polarization effects were enhanced by adding terpolymer into P(VDF). As a result, a discharge energy density of 13.63 J/cm^3 was achieved at 480 MV/m [221]. Some of the polymer blends' energy storage properties are summarized

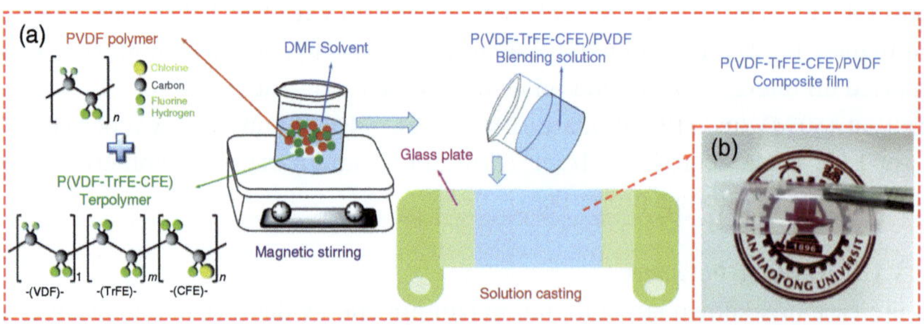

Figure 14.13 Schematic illustration of the fabrication process for the PTC/PVDF polymer composite films (a); photograph of the PVDF polymer composite films blended with 20 wt% PTC terpolymer (b). Source: Mao et al. [221]/Reproduced with permission of Royal Society of Chemistry.

Table 14.13 Energy storage properties of polymer blends.

Blends	Dielectric constant (at 1 kHz)	Dielectric loss (at 1 kHz)	Breakdown strength (MV/m)	Energy density (J/cm^3)	References
PVDF/P(VDF-TrFE-CFE) (20/80 wt%)	13	0.04	480	13.63	[221]
PVDF/P(VDF-TrFE) (70/30 vol%)	23	0.04	200	—	[222]
PVDF/P(VDF-TrFE-CTFE) (60/40 vol%)	30	0.08	640	19.6	[223]
P(VDF-TrFE)/P(VDF-TrFE-CTFE) (60/40 vol%)	38	0.1	400	—	[224]
PI/PEEU (15/75 wt%)	4.78	0.00299	495.65	5.14	[225]
PEI/PEMEU (50/50 wt%)	5.8	0.0087	—	—	[226]

in Table 14.13, showing one of the highest discharge energy densities of 19.6 for PVDF/P(VDF-TrFE-CTFE) (60/40 vol%) polymer blends [223].

14.3.2 Size Consideration in Polymer Dielectric Films

According to the literature, the dielectric breakdown process may be described using thermal [227], avalanche [228], partial discharge [229], electromechanical [230], and coexistence and coupling of these breakdown models [231]. Complicated and interdependent in nature, these processes are greatly influenced by the composition of materials as well as electrical and thermal conductivity, as well as glass transition temperature and operating temperature. There are other factors that contribute to polymer breakage strength, such as its thickness. For instance, PP films of thinner thicknesses from 8 to 76 μm were tested for dc breakdown by Laghari and coworker [232]. The study also found that the breakdown strength of dielectric samples increases with decreasing sample thickness. Because the thickness has a significantly greater impact on electric strength than the electrode area, both must be taken into account. Thick films' inhomogeneity is one possible explanation. Thick samples have an uneven electric field at the electrode's edge, increasing the likelihood of defects and limiting heat dissipation. Thicker samples also have a higher risk of faults. The breakdown field strength will be reduced by each of these factors (Figure 14.14) [233].

The dielectric constant is also affected by the thickness of films, in addition to the duration of measurement, test temperature, and applied electric field. For instance, vapor deposition polymerization was used by Kahouli et al. to create parylene thin films with the Si(100) index on a p-type Si substrate. With layer thickness, the dielectric constant grows following a power law with $\varepsilon' = d^{\beta}$ with 0.042 ± 0.04 for d less than 1000 nm and 0.0138 ± 0.02 for more than 1000 nm. As the thickness of the film increases, the number of voids and discontinuities in the material decreases,

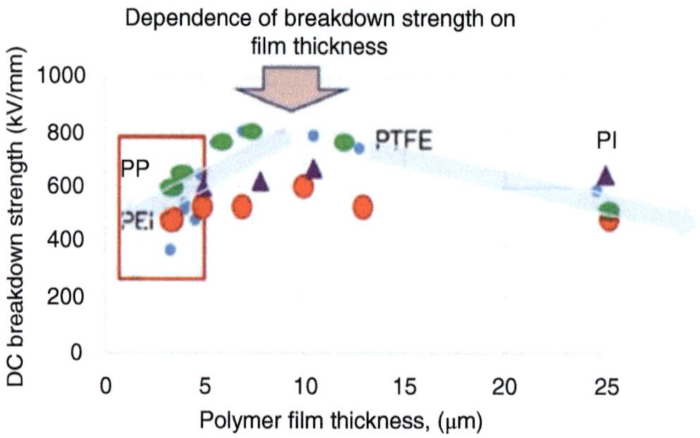

Figure 14.14 Dependence of electric breakdown strength on PEI, PP, PTFE, and PI polymer film thickness. Source: Tan [233]. Reproduced with permission of Wiley Periodicals LLC.

which could be a reason for such behavior. Similarly, using polymers, Zhang et al. discovered that FE P(VDF-TrFE) copolymer thin films had a critical crystalline thickness value. The crystallization process is severely impeded when the film thickness is less than 100 nm, resulting in poor crystallinity. Due to the film's poor crystallinity, dielectric constant and FE polarization vary significantly and discontinuously below critical thickness. The dielectric constant drops precipitously with decreasing film thickness, peaking at around 100 nm.

14.3.3 High-Temperature Stability Considerations in Polymer Dielectrics

Improvements in thermal stability (like a high T_g) and dielectric characteristics dominate polymer research for high-temperature dielectric applications (e.g. high dielectric constant). A selection of exemplary high-temperature polymer dielectrics is summarized in Table 14.14. The most promising high-temperature polymer dielectrics are PI, PEI, and FPE, based on their thermal and electrical characteristics.

There are two ways to make polyimide: polycondensation and imidization of diamines [234]. Pyromellitic dianhydride (PMDA) and 4,4′-oxydianiline (ODN) are two examples of PI (ODA). With a T_g of 360–410 °C and strong chemical resistance, PI's primary chain is made up of imides and aromatics, giving it remarkable thermal, chemical, and mechanical properties [235]. PEI has an ether (–O–) connection in its backbone, making it a modified form of PI. An ether linkage has been added to the PEI, making it easier to melt-extrude or solution cast. Even though PEI has less thermal stability than PI, the high-temperature capacitive performance of PEI is superior to that of PI. At 150 °C and a 300 MV/m electric field, the discharged energy density and charge/discharge efficiency of PEI, for example, are 1.14 J/cm^3 and 82%, respectively [227]. With a T_g of 330 °C and a steady dielectric constant until 300 °C, fluorene polyester (FPE) produced from fluorine bisphenol and phthaloyl chloride exhibits excellent thermal and dielectric stability [236].

Table 14.14 Energy storage properties of high-temperature dielectrics.

Polymer type	Thickness (μm)	ε_r	tan δ	T_g (°C)	λ (W/(m K))	E_b (MV/m) (thickness)
PET	25	3.3	0.005	75	0.15	300
PEN	75	3.2	0.005	120	0.15	160
PPS	9	3.2	0.002	120	0.30	490
PEEK	200	3.1	0.004	150	0.32	200
PC	10	3.2	0.002	150–210	0.21	500
PESU	25	3.5	0.001	225	0.18	180
PEI	25	3.2	0.001	217–260	0.22	200
FPE	8	2.8–3.2	0.006	330	0.20	430
PI	25	3.4–3.5	0.002	360–410	0.12	300
c-BCB	10	2.75	0.001	>350	0.30	300

For example, at 150 °C and 300 MV/m, FPE's charge/discharge efficiency is 1.04 J/cm^3 and 58%, which outperforms PI. The T_g of fluorenyl polyester (FDAPE) may be enhanced to 450 °C by chemical modification with a diamond-like hydrocarbon group, i.e. 4,9-diamantyl [237]. It is possible to obtain FDAPE with a low dissipation factor of 0.003–0.004 in the temperature range of 25–350 °C by adding heterocyclic rings. FDAPE is a good polymer dielectric for use in ultrawide-temperature power electronics because of these characteristics.

14.4 Composite Dielectric Materials

Electrostatic dielectric capacitors' primary shortcoming is an insufficient energy density for more complex uses [200]. This kind of capacitor often has a ceramic or polymer dielectric. Among the properties shared by ceramic dielectrics are high stiffness, exceptional thermal stability, and a high dielectric permittivity [201–203]. However, their low breakdown strength, poor flexibility, large specific mass, and challenging manufacturing conditions restrict their applicability in energy storage capacitors. In contrast, polymer dielectrics include not only low dielectric loss and high breakdown strength but also low specific mass, simple processing, applicability to large-scale production, and a unique self-healing capacity. A local dielectric breakdown clears away the material at the place of breakdown, thanks to the self-healing ability, preventing catastrophic failure of the whole capacitor. Polymer dielectrics' fundamental drawback is their low dielectric permittivity [238–243].

Polymer-based dielectric composite films, which include ceramic fillers into a polymer matrix, are one logical option to overcome these drawbacks and achieve high energy storage density in dielectric capacitors. Ceramic fillers are used to increase the dielectric constant and dielectric displacement of the polymer matrix, whereas the polymer matrix is included for its high breakdown strength. Micron-sized ceramic particle fillers have been the primary focus of research on polymer-based dielectric composites for energy storage applications [244, 245]. However, it was quickly discovered that these composites show deterioration in their dielectric characteristics owing to stress cracking and surface flaws. The size of the microparticles also sets a cap on the thickness of these composite films. However, dielectric coatings in the nanoscale range are needed to accommodate the ongoing structural downsizing trend in the electronics sector. Thinner films are better at storing the same amount of energy in less space. To reduce the size of these polymer-based composite dielectrics, nanofillers must be used instead of microfillers. Because of this, polymer nanocomposite film capacitors have been suggested and produced; they use composite dielectrics with negligible nanofiller loadings. Excellent flexibility, low density, and easy processability are maintained while still achieving the requisite dielectric properties. Polymer nanocomposite film capacitors have been shown to have superior power density, operating voltage, and cycle lifespan [246–250].

14.4.1 0D-Filled Nanocomposites

The insertion of 0D, i.e. spherical nanoparticles inside the polymer matrix, is one of the solutions in polymer nanocomposites for energy storage applications that has been the subject of many studies and is very simple to produce. Adding a relatively small quantity of 0D nanoparticles to a polymer matrix has been observed to improve the material's dielectric breakdown strength. On the other hand, adding a greater quantity of nanoparticles can increase the dielectric constant almost linearly, but it will drastically lower the material's breakdown strength. It is important to point out that the research that has been published on the incorporation of simply 0D spherical nanoparticles into the polymer matrix without any interface engineering or its coupling with other dimensional nanoparticles, i.e. 1D and 2D, has become obsolete. Therefore, the research focuses on determining the optimal number of spherical nanoparticles in polymer matrices to produce high breakdown strength and good energy storage capabilities. Some of the research about 0D polymer nanocomposites is summarized below.

Synthesis of the novel polymer/ceramic nanocomposite by Huang et al. [251] included core–shell $BaTiO_3@SiO_2$ (BT@SO) structures with a diameter of less than 10 nm, see Figure 14.15. Since the interfaces in this ultrafine nanostructure are almost ten times larger than those in conventional 100 nm fillers, the composite's breakdown strength and electrical displacement may be increased without sacrificing either. This is due to the fact that a composite's breakdown strength and electrical displacement may both benefit from high polarization. When tested at a 420 kV/mm electric field, the BT@SO/PVDF nanocomposite showed remarkable energy storage ability, with an U_d value of 11.5 J/cm^3.

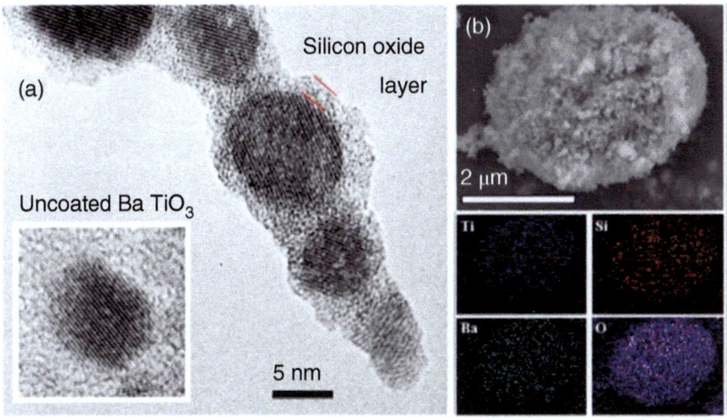

Figure 14.15 (a) TEM image of the 30 wt% SiO_2-coated core–shell BT@SO structure, the inserted TEM image in (a) presents the original uncoated BT particles, (b) SEM and EDS mapping images. Source: Bi et al. [251]. Reproduced with permission of Elsevier Ltd.

14.4.2 1D-Filled Nanocomposites

Recently, it has been suggested that 1D ceramic nanofillers with a bigger aspect ratio should be used instead of spherical nanoparticles (NPs) to solve the problems caused by a high NPs volume fraction in the polymer matrix [252]. (i) One of the benefits of using 1D nanofillers is that, due to their strong dipole moments, they may boost the nanocomposites' permittivity at a lower nanofiller concentration [253]. (ii) Their reduced surface energy feature as polymer nanocomposites is another benefit of their one-dimensional structure. (iii) 1D ceramic fillers cause less agglomeration inside the polymer matrix due to their low specific surface areas. Energy density may be increased by including several kinds of 1D nanofillers into polymers, such as linear (1D TiO_2 [TO]-based), FE (1D $BaTiO_3$ [BT]-based), PE (1D $SrTiO_3$ [ST]- and $BaSrTiO_3$ [BST]-based), and RFE (1D $BaZrTiO_3$ [BZT]-based).

In order to increase the breakdown strength of the final nanocomposites, numerous studies have hypothesized that a smaller dielectric difference between the linear fillers, such as TiO_2, and polymers like PVDF is necessary. The leakage current densities and dielectric loss properties of the anatase-phased TiO_2 nanowires (NWs) incorporated with P(VDF-HFP) FE polymer were greatly reduced in comparison to those of the raw TO NWs/P(VDF-HFP) nanocomposites. The surface of the NWs was modified with strong adhesion performance dopamine derivative (h-DA) containing elongated-chain tails with brush-like. At an electric field of 5200 kV/cm, 2.5 vol% h-DA-modified TO NWs/P(VDF-HFP) discharged the ultrahigh energy density of around 11 J/cm^3 [254].

One of the most popular nanofillers employed in polymer nanocomposites is 1D FE nanofillers because of their strong polarization responses and, therefore, their high maximum dielectric displacements. BT is an example of a material

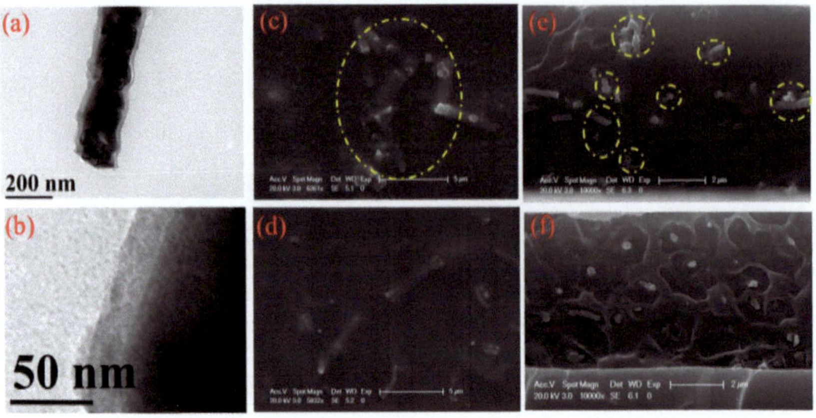

Figure 14.16 Typical TEM images (a) and enlarged view (b) of BCZT@PATP NFs. SEM images of 4.5 vol% BCZT NFs/PVDF (c: surface; e: cross-section) and 4.5 vol% BCZT@PATP NFs/PVDF nanocomposites (d: surface; f: cross-section). Source: Pan et al. [256]. Reproduced with permission of American Chemical Society.

with a high dielectric constant and a high D_{max}. The high aspect ratio BT NWs were synthesized using a novel synthesis method proposed by Tang et al. At an electric field of 3000 kV/cm, the energy storage density of the P(VDF-TrFE-CFE) nanocomposite functionalized with 17.5 vol% BT NWs were about 45.3% higher than the pristine P(VDF-TrFE-CFE) polymer (7.2 J/cm^3). Roughly 1.52 μS after the discharge began, the same nanocomposite showed a maximum power density of about 1200 kW/cm^3 [255].

The wide *D–E* loops of FE materials limit their energy density, so it is necessary to prefer one-dimensional relaxors and AFE nanofillers with large maximum polarization and small residual polarization. Researchers mixed 1D RFE nanofillers, which inherently have narrow *D–E* loops compared to 1D FE nanofillers, with diverse polymer matrices to achieve the desired features. Pan et al. synthesized PATP-modified BCT-BZT NFs by electrospinning, grafted them to the surface of the NFs, and then integrated the NFs into a PVDF polymer matrix. Chemical bonding and confinement of charge carrier motion at the filler-matrix interface area successfully improved NFs' distribution and compatibility in this way, see Figure 14.16. Thus, small nanofiller loading, i.e. 2.1 vol %, increased the energy storage density to 8.2 J/cm^3 at a relatively lower electric field of 3800 kV/cm [256].

14.4.3 2D-Filled Nanocomposites

In polymer nanocomposites, 2D fillers with a high aspect ratio and lateral dimensions may significantly improve ε_r and E_b at a low filling content by forming percolation systems or constructing effective conduction barriers. Adjustments to these characteristics are possible by using various filler stages and developing multilayer designs [239, 257, 258]. Polymer nanocomposites reinforced with $BaTiO_3$ (BT) plates were described by Wen et al. In comparison to unmodified PVDF, the composite with

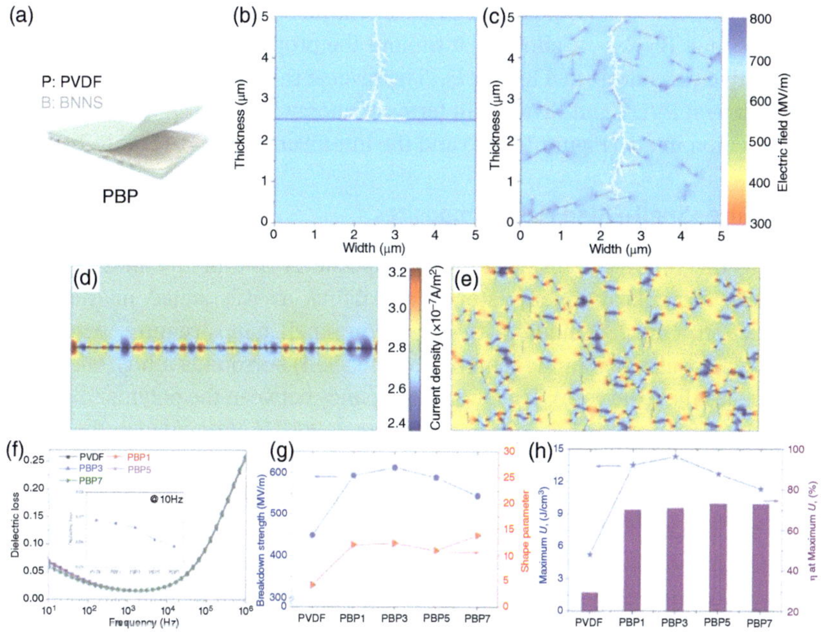

Figure 14.17 Schematic representation of (a) polymer nanocomposites with a sandwich structure; (b,c) the electric field distribution and the propagation route of an electrical tree in polymer nanocomposites with 0.2 vol% BNNS; its 2D current density maps (d,e); polymer nanocomposites' dielectric, breakdown, and energy storage characteristics (f–h). Source: Zhu et al. [260]. Reproduced with permission of John Wiley & Sons.

1 wt% loading has a maximum U_d of 9.7 J/cm^3, which represents an improvement of 100% [259].

Films with a BNNS layer in the center, as shown in Figure 14.17a, were the focus of an investigation by Zhu et al. [260]. The current density in the PVDF matrix near the in-plane BNNSs is considerably lower than that near the through-plane BNNSs for the sample with 0.2vol% randomly distributed BNNSs ($BNNS_{0.2}$) (Figure 14.17e). Thus, it is possible that high-current paths arise along the field direction, diminishing the effectiveness of the method used to reduce the leakage current throughout the composite material. By contrast, in PBP_3 (a sandwich structure with 0.16 vol% BNNSs in the intermediate layer), all BNNSs lie in the in-plane direction and create a tight topological barrier to restrict the total leakage current (Figure 14.17d). The dielectric characteristics of PBP_3 are enhanced by the fact that its compact BNNS interlayer may serve as a line of effective scattering centers under high electric fields, hence reducing the mobility of free charge carriers. To further understand the electrical breakdown process, a modified dielectric breakdown model was utilized to simulate the electrical tree propagation state of the two samples, taking into account the impact of space charge on the electric field distribution of composites. The outcomes are shown in Figure 14.17b,c. In PBP_3, the BNNS barrier layer causes the electrical tree to be redirected or perhaps cut off entirely. Because BNNSs do not create a continuous topological barrier, the electrical tree may spread unhindered over the $BNNS_{0.2}$ sample. Consequently, PBP_3's interlayer BNNS is more compact and continuous than $BNNS_{0.2}$'s, allowing it to limit the propagation of electric damage more effectively, resulting in a higher E_b. The reverse-sandwich-structured samples outperform neat PVDF and $BNNS_{0.2}$ in terms of energy storage (Figure 14.17h) due to the reduction in tan (Figure 14.17f) and the improvement in E_b (Figure 14.17g).

14.4.4 Core–Shell-Filled Nanocomposites

Enhancing the interfaces inside a nanocomposite is one of the most effective strategies to increase its energy storage capability, as was stated in the polarization section. One technique is to use a core–shell nanostructure, whereby a low-dielectric-constant ceramic nano-layer (the shell) surrounds a high-dielectric-constant nanofiller (the core). The huge discrepancy between the polymer dielectric constant ($\varepsilon_r = 10$) and the ceramic nanofillers ($\varepsilon_r > 200$) is attenuated in the former scenario. This results in increased dielectric breakdown strength and less internal distortion in polymer nanocomposites. This is also one of the methods used to increase the volume fraction of conductive fillers that can conduct electricity, such as metallic Al with a native oxide Al_2O_3 in the form of nanoparticles or NWs, or semi-conductive Si with a coherent surface oxide SiO_2, and so on. Despite rapid increases in dielectric constant, dielectric loss is kept to a minimum by the insulating sheath. So far, SiO_2 (SO), Al_2O_3 (AO), TiO_2 (TO), etc., have been the most effective nano-shells used in core–shell designs. Though the ε_r-mismatch is small, the large difference in electrical conductivity between ceramic nanofillers and polymers causes significant interfacial polarization due to the buildup of space charges at the interface. However, with an increase in interfacial polarization comes an increase in dielectric loss property and a decrease in dielectric breakdown strength in the polymer nanocomposites, which in turn leads to a decrease in the polymers' capacity to store energy. Due to polymer-nanofiller mismatch and distortions followed by poor breakdown strength, researchers have seldom studied the encapsulation of a relatively high dielectric constant shell on the surface of a core-nanofiller displaying a moderate ε_r [261–266]. To create high-energy storage nanocomposites, Wang et al. used the high-ε_r core @ low-ε_r shell method, preparing BT NWs encased by TO shells of varying thicknesses. At an electric field of 4400 kV/cm, it was found that 5 wt% TO@BT NWs-modified P(VDF-HFP), i.e. core–shell strategized nanocomposites, displayed ultrahigh-energy discharge density and high efficiency of around 9.5 J/cm^3 and 63%, respectively. The energy storage capability of the polymer nanocomposites is not linearly impacted by altering shell thickness, as shown by a comparison of core–shell strategized NWs with varied thicknesses. Using the finite element approach, we found that the higher energy storage capacity was mostly attributable to a reduction in the leakage current in the P(VDF-HFP) polymer matrix and an increase in the electric field concentration [263].

14.4.5 High-Temperature Nanocomposites

Biaxially orientated polypropylene (BOPP) is the best dielectric polymer currently on the market; however, it can only function at temperatures below 105 °C. As a result, dielectric polymers can never be used in high-temperature applications without first implementing some kind of heat control system. For instance, cooling systems must be used to reduce the ambient temperature from about 140 to about 70 °C in order to accommodate BOPP film capacitors in the power inverters of hybrid and electric vehicles, which are used to control and convert direct current from batteries into the alternating current required to power the motor. The integrated power system loses efficiency and reliability as a result of the additional mass and size. Due to the exponential growth in the amount of heat produced by electronic devices and circuitry as they get smaller and more useful, the demand for high-temperature dielectric polymers has skyrocketed with the popularity of lightweight and flexible electronic devices.

High-temperature uses of polymer-based capacitors are constrained by the polymer's low glass transition temperature and higher losses with increasing temperature. Li et al. [267] evaluated the dielectric and energy storage characteristics of the composite crosslinked divinyl tetramethyl disiloxane-bis(benzocyclobutene)/boron nitride nanosheets (c-BCB/BNNS) to those of several high-temperature polymer films, they discovered that the c-BCB/BNNS demonstrated superior energy density and efficiency with excellent temperature stability. Due to the increased thermal conductivity provided by the BNNSs, superior temperature stability can be ascribed to their usage in nanocomposites. Also, the insertion of BNNSs results in a boost to Young's modulus. Due to the increased Young's modulus, electromechanical failure occurs less often. For this reason, expanding the use of ceramic/polymer composites in energy storage may be as simple as combining the two types of materials.

Marwat et al. [196] developed heterogeneous sandwich-structured nanocomposites consisting of externally pure PEI and $x = 1–11$ wt% of xDA@Ag@BT/PVDF. Based on the results, PEI helped insulate charge injection from electrodes, decreased polymer-free volume, increased electrical insulation, and enhanced thermal and mechanical stability. Using Ag-grafts on top of BT NPs helped DA@Ag@BT/PVDF achieve more polarizability, and the well-known Coulomb blockade effect synergistically increased E_b. Finite element modeling supported a decrease in the local electric field concentration inside the polymer matrices of the outer and inner layers, indicating improved E_b and U_d performance. The ideal heterogeneous sandwich-structured nanocomposites showed an ultrahigh U_e of 21.03 J/cm^3 at 592.1 MV/m, see Figure 14.18, which is the greatest U_d recorded to date when the comparable NPs content and an analogous electric field are taken into account. Moreover, essential heterogeneous sandwich-structured nanocomposites demonstrated remarkable thermal stability up to 170 °C in high-temperature energy storage experiments.

In conclusion, a lot of research is still being done to improve the thermal stability of polymer nanocomposites for energy storage applications. For this, various strategies are incorporated, including using high-T_g polymers, introducing thermally conductive nanoparticles, and engineering the interface between polymer and nanoparticles. All these strategies facilitate the high energy storage capability of polymer nanocomposites at high temperatures. However, to create an innovative, industrially feasible solution that can replace BOPP, it is still necessary to develop

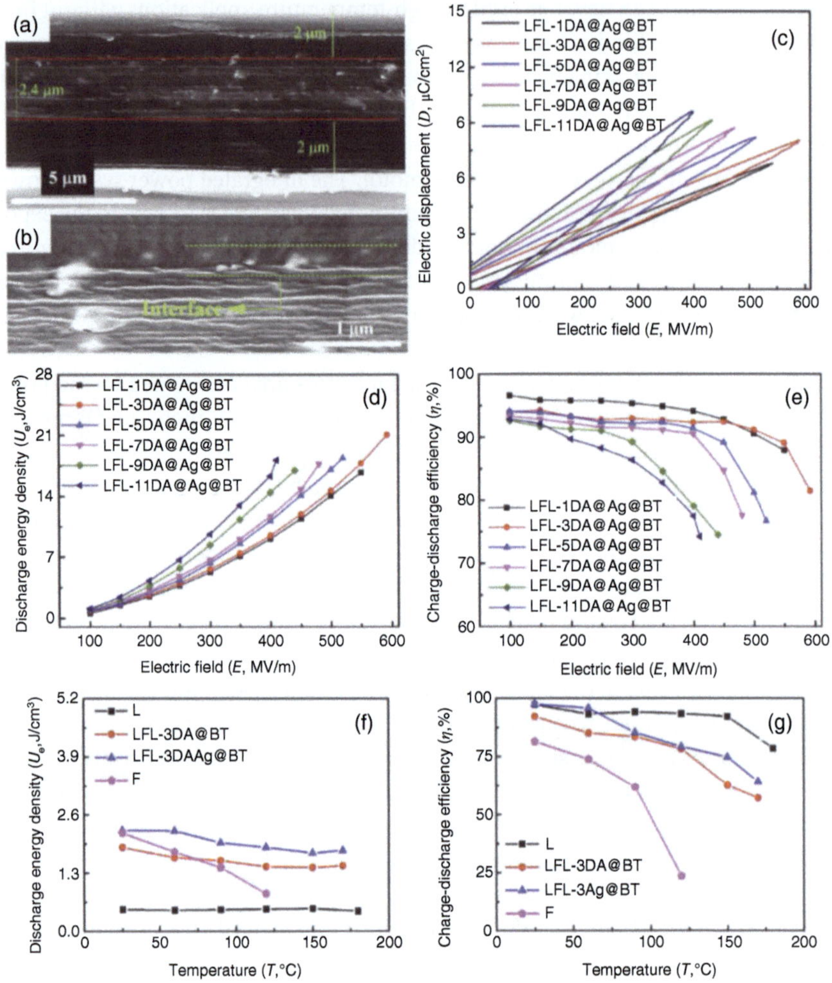

Figure 14.18 SEM images of (a) sandwich-structured polymer nanocomposites and (b) the magnified image of an interface between the layers of sandwich structure. (c) *D*–*E* loops, (d) energy density, and (e) charge–discharge efficiency of various nanocomposites. (f) High-temperature energy density and (g) charge–discharge efficiency of polymers and the sandwich-structured material of interest. Source: Marwat et al. [196]/Reproduced with permission of Elsevier.

novel engineered polymers, nanoparticles, interface engineering, and innovative methodologies to obtain thermally stable high-energy storage dielectrics.

14.5 Conclusions

Dielectric capacitors have lower cost, shorter charge–discharge times (<0.01 seconds), higher power density (>10^5 W/kg), lower loss, and higher operating voltages. Thus, they are the preferred choice for energy storage applications with high voltage, low cost, and large-scale fabrication [268]. For example, dielectric capacitors are the core components of electric vehicle inverters, medical pacemakers, HVDC transmission systems, and electromagnetic catapults on aircraft carriers [269, 270]. In this chapter, we detail the energy storage theory of dielectric capacitors and the research progress on different kinds of dielectric materials. First, the current state of energy storage in the world and the advantages of dielectric capacitors in energy storage fields are presented. Then, we introduce the definitions and calculation formulas for energy storage parameters, such as ε_r, η, W_{rec}, and P. The influences of several polarization mechanisms on the ε_r and polarization properties are investigated. Moreover, the causes of breakdown and loss in dielectrics are explained by understanding several different types of breakdown mechanisms and conduction mechanisms. In Section 14.3, we classify dielectrics for energy storage into inorganic, organic, and composite dielectrics and investigate the energy storage properties of various dielectric materials as well as performance improvement strategies. In Section 14.4, we summarize the industrial preparation of MLCCs and polymer thin films. As can be seen in the paper, a great improvement in the energy storage dielectric was achieved, which realized a simultaneous increase of ε_r and E_b. Even though energy storage dielectrics have achieved some results, there are still some difficult problems to be solved. The problems and future research priorities of dielectric materials are summarized as follows:

(1) Low ε_r or low E_b are the main problems that limit the energy storage applications of most dielectrics. LDs have a high E_b and high η, but their ε_r is low. Normal FE dielectrics are the opposite. This limits the improvement of both U_d. Modification of ceramics by chemical modification, defect engineering, microstructure modulation, or modification of polymers by modification strategies such as blend, composite, and multilayer have proven to be effective methods to achieve energy storage density in dielectric materials. Meanwhile, RFE and AFE dielectrics with narrow hysteresis loops, which exhibit low remnant polarization, are also expected to achieve high ε_r and high η. However, the preparation and theory of RFEs and AFEs are complicated. How to prepare special FE polymers by simple methods may be a key direction for future research.
(2) When capacitors are used in high-temperature environments such as electric vehicles and oil and gas exploration, there are very strict requirements for the thermal stability of capacitors. Based on studies on breakdown and conduction mechanisms, researchers have come up with feasible solutions. The introduction of linear and paraelectric dielectrics to reduce losses, or the preparation of multilayer structures and the use of differences in dielectric properties between layers to impede charge and breakdown paths, are both feasible approaches. In addition, there are two key directions for researchers to solve the high-temperature problem of polymers in the future. Adding a high thermal conductivity second phase or coating the dielectric surface with an insulating nanolayer to increase the Schottky barrier to inhibit charge injection to improve capacitor performance.

(3) Although some dielectrics with ultrahigh levels of energy storage properties have been successfully fabricated in the laboratory, they are not feasible for large-scale commercial production. Dielectrics applied to capacitors need to consider not only the energy storage properties of the material but also the mechanical, thermal stability, service life, and other comprehensive performance improvements to achieve the preparation of devices in industry.

In the future, the development of dielectric capacitors will be more rapid, and the energy storage requirements of dielectric materials will be more stringent. Based on computer simulations and analysis, it will become feasible for researchers to select suitable dielectric materials and design suitable structures to solve these problems through continuous experiments. The preparation of dielectric materials with better performance through various modification methods will contribute to the innovative development of the energy storage industry and effectively solve global energy problems. This requires the joint efforts and cooperation of staff and researchers in related fields.

References

1 Palneedi, H., Peddigari, M., Hwang, G.-T.T. et al. (2018). High-performance dielectric ceramic films for energy storage capacitors: progress and outlook. *Adv. Funct. Mater.* 28: 1803665. https://doi.org/10.1002/adfm.201803665.

2 Berberich, L.J. and Bell, M.E. (1940). The dielectric properties of the rutile form of TiO_2. *J. Appl. Phys.* 11: 681–692. https://doi.org/10.1063/1.1712721.

3 Müller, K.A. and Burkard, H. (1979). $SrTiO_3$: an intrinsic quantum paraelectric below 4 K. *Phys. Rev. B* 19: 3593–3602. https://doi.org/10.1103/PhysRevB.19.3593.

4 Silva, J.P.B., Silva, J.M.B., Oliveira, M.J.S. et al. (2019). High-performance ferroelectric–dielectric multilayered thin films for energy storage capacitors. *Adv. Funct. Mater.* 29: 1807196. https://doi.org/10.1002/adfm.201807196.

5 Tan, H., Yan, Z., Chen, S.G. et al. (2022). SPS prepared NN-24SBT lead-free relaxor-antiferroelectric ceramics with ultrahigh energy-storage density and efficiency. *Scr. Mater.* 210: 114428. https://doi.org/10.1016/j.scriptamat.2021.114428.

6 Qi, H., Zuo, R., Xie, A. et al. (2019). Ultrahigh energy-storage density in $NaNbO_3$-based lead-free relaxor antiferroelectric ceramics with nanoscale domains. *Adv. Funct. Mater.* 29: 1903877. https://doi.org/10.1002/adfm.201903877.

7 Pan, H., Lan, S., Xu, S. et al. (2021). Ultrahigh energy storage in superparaelectric relaxor ferroelectrics. *Science* 374: 100–104. https://doi.org/10.1126/science.abi7687.

8 Huebner, W., Zhang, S.C., Gilmore, B., et al. (1999). High breakdown strength, multilayer ceramics for compact pulsed power applications. *Digest of Technical Papers. 12th IEEE International Pulsed Power Conference*, Monterey, CA,

USA (27–30 June 1999). IEEE. pp. 1242–1245. https://doi.org/10.1109/ppc.1999.823749.

9 Gupta, S.M. and Tripathi, M. (2011). A review of TiO_2 nanoparticles. *Chin. Sci. Bull.* 56: 1639–1657. https://doi.org/10.1007/s11434-011-4476-1.

10 Parker, R.A. and Wasilik, J.H. (1960). Dielectric constant and dielectric loss of TiO_2 (rutile) at low frequencies. *Phys. Rev.* 120: 1631–1637. https://doi.org/10.1103/PhysRev.120.1631.

11 Guo, D., Ito, A., Goto, T. et al. (2013). Preparation of rutile TiO_2 thin films by laser chemical vapor deposition method. *J. Adv. Ceram.* 2: 162–166. https://doi.org/10.1007/s40145-013-0056-y.

12 Hu, W., Liu, Y., Withers, R.L. et al. (2013). Electron-pinned defect-dipoles for high-performance colossal permittivity materials. *Nat. Mater.* 12: 821–826. https://doi.org/10.1038/nmat3691.

13 Yu, Y., Zhao, Y., Qiao, Y.L. et al. (2021). Defect engineering of rutile TiO_2 ceramics: route to high voltage stability of colossal permittivity. *J. Mater. Sci. Technol.* 84: 10–15. https://doi.org/10.1016/j.jmst.2020.12.046.

14 Fan, J., Chen, Y., Long, Z. et al. (2022). Grain-boundary engineering inducing thermal stability, low dielectric loss and high energy storage in Ta + Ho co-doped TiO_2 ceramics. *Ceram. Int.* 48: 21584–21592. https://doi.org/10.1016/j.ceramint.2022.04.124.

15 Chao, S., Petrovsky, V., and Dogan, F. (2010). Effects of sintering temperature on the microstructure and dielectric properties of titanium dioxide ceramics. *J. Mater. Sci.* 45: 6685–6693. https://doi.org/10.1007/s10853-010-4761-4.

16 Ye, Y., Zhang, S.C., Dogan, F., et al. (2003). Influence of nanocrystalline grain size on the breakdown strength of ceramic dielectrics. *Digest of Technical Papers. 12th IEEE International Pulsed Power Conference*, Dallas, TX, USA (15–18 June 2003). IEEE. pp. 719–722. https://doi.org/10.1109/ppc.2003.1277809.

17 Liu, J., Zhang, J., Wei, M. et al. (2016). Dielectric properties of manganese-doped TiO_2 with different alkali-free glass contents for energy storage application. *J. Mater. Sci.: Mater. Electron.* 27: 7680–7684. https://doi.org/10.1007/s10854-016-4753-5.

18 Wei, M., Zhang, J., Huang, J. et al. (2016). Microstructure and electrical properties of TiO_2–CaO–MgO–Al_2O_3–SiO_2 glass-ceramic with sol–gel method. *J. Mater. Sci.: Mater. Electron.* 27: 11623–11627. https://doi.org/10.1007/s10854-016-5294-7.

19 Chao, S. and Dogan, F. (2011). Processing and dielectric properties of TiO_2 thick films for high-energy density capacitor applications. *Int. J. Appl. Ceram. Technol.* 8: 1363–1373. https://doi.org/10.1111/j.1744-7402.2010.02592.x.

20 Yadav, A.K. and Gautam, C.R. (2014). A review on crystallisation behaviour of perovskite glass ceramics. *Adv. Appl. Ceram.* 113: 193–207. https://doi.org/10.1179/1743676113Y.0000000134.

21 Pai, Y.-Y., Tylan-Tyler, A., Irvin, P., and Levy, J. (2018). Physics of $SrTiO_3$-based heterostructures and nanostructures: a review. *Rep. Prog. Phys.* 81: 36503. https://doi.org/10.1088/1361-6633/aa892d.

22 Hu, Q.G., Shen, Z.Y., Li, Y.M. et al. (2014). Enhanced energy storage properties of dysprosium doped strontium titanate ceramics. *Ceram. Int.* 40: 2529–2534. https://doi.org/10.1016/j.ceramint.2013.07.126.

23 Fergus, J.W. (2012). Oxide materials for high temperature thermoelectric energy conversion. *J. Eur. Ceram. Soc.* 32: 525–540. https://doi.org/10.1016/j.jeurceramsoc.2011.10.007.

24 Kong, X., Yang, L., Cheng, Z., and Zhang, S. (2020). Bi-modified $SrTiO_3$-based ceramics for high-temperature energy storage applications. *J. Am. Ceram. Soc.* 103: 1722–1731. https://doi.org/10.1111/jace.16844.

25 Chen, A., Zhi, Y., and Cross, L. (2000). Oxygen-vacancy-related low-frequency dielectric relaxation and electrical conduction. *Phys. Rev. B: Condens. Matter Mater. Phys.* 62: 228–236. https://doi.org/10.1103/PhysRevB.62.228.

26 Chen, A. and Zhi, Y. (2002). Dielectric relaxor and ferroelectric relaxor: Bi-doped paraelectric $SrTiO_3$. *J. Appl. Phys.* 91: 1487–1494. https://doi.org/10.1063/1.1428799.

27 Chen, A. and Zhi, Y. (2007). High capacitance-temperature sensitivity and "giant" dielectric constant in $SrTiO_3$. *Appl. Phys. Lett.* 90: 202903. https://doi.org/10.1063/1.2736298.

28 Shen, Z.Y., Hu, Q.G., Li, Y.M. et al. (2013). Structure and dielectric properties of $Re_{0.02}Sr_{0.97}TiO_3$ (Re = La, Sm, Gd, Er) ceramics for high-voltage capacitor applications. *J. Am. Ceram. Soc.* 96: 2551–2555. https://doi.org/10.1111/jace.12364.

29 Shen, Z.Y., Li, Y.M., Luo, W.Q. et al. (2013). Structure and dielectric properties of $Nd_xSr_{1-x}TiO_3$ ceramics for energy storage application. *J. Mater. Sci.: Mater. Electron.* 24: 704–710. https://doi.org/10.1007/s10854-012-0798-2.

30 Shen, Z.Y., Luo, W.Q., Li, Y.M. et al. (2013). Electrical hetero-structure of $Nd_{0.1}Sr_{0.9}TiO_3$ ceramic for energy storage applications. *J. Mater. Sci.: Mater. Electron.* 24: 607–612. https://doi.org/10.1007/s10854-012-0871-x.

31 Song, Z., Liu, H., Zhang, S. et al. (2014). Effect of grain size on the energy storage properties of $(Ba_{0.4}Sr_{0.6})TiO_3$ paraelectric ceramics. *J. Eur. Ceram. Soc.* 34: 1209–1217. https://doi.org/10.1016/j.jeurceramsoc.2013.11.039.

32 Zhang, Q., Wang, L., Luo, J. et al. (2009). Improved energy storage density in barium strontium titanate by addition of $BaO-SiO_2-B_2O_3$ glass. *J. Am. Ceram. Soc.* 92: 1871–1873. https://doi.org/10.1111/j.1551-2916.2009.03109.x.

33 Kim, S.H. and Koh, J.H. (2008). ZnBO-doped (Ba, Sr)TiO_3 ceramics for the low-temperature sintering process. *J. Eur. Ceram. Soc.* 28: 2969–2973. https://doi.org/10.1016/j.jeurceramsoc.2008.04.034.

34 Shen, Z.Y., Wang, Y., Tang, Y. et al. (2019). Glass modified barium strontium titanate ceramics for energy storage capacitor at elevated temperatures. *J. Mater.* 5: 641–648. https://doi.org/10.1016/j.jmat.2019.06.003.

35 Yang, X., Li, W., Qiao, Y. et al. (2019). High energy-storage density of lead-free $(Sr_{1-1.5x}Bi_x)Ti_{0.99}Mn_{0.01}O_3$ thin films induced by Bi^{3+}-VSr dipolar defects. *Phys. Chem. Chem. Phys.* 21: 16359–16366. https://doi.org/10.1039/c9cp01368g.

36 Zhang, Y., Li, W., Wang, Z. et al. (2019). Ultrahigh energy storage and electrocaloric performance achieved in $SrTiO_3$ amorphous thin films: via polar

cluster engineering. *J. Mater. Chem. A* 7: 17797–17805. https://doi.org/10.1039/c9ta05446d.

37 Pan, H., Zeng, Y., Shen, Y. et al. (2017). $BiFeO_3$-$SrTiO_3$ thin film as a new lead-free relaxor-ferroelectric capacitor with ultrahigh energy storage performance. *J. Mater. Chem. A* 5: 5920–5926. https://doi.org/10.1039/c7ta00665a.

38 Hou, C., Huang, W., Zhao, W. et al. (2017). Ultrahigh energy density in $SrTiO_3$ film capacitors. *ACS Appl. Mater. Interfaces* 9: 20484–20490. https://doi.org/10.1021/acsami.7b02225.

39 Gao, W., Yao, M., and Yao, X. (2017). Improvement of energy density in $SrTiO_3$ film capacitor via self-repairing behavior. *Ceram. Int.* 43: 13069–13074. https://doi.org/10.1016/j.ceramint.2017.06.162.

40 Wang, Y., Shen, Z.Y., Li, Y.M. et al. (2015). Optimization of energy storage density and efficiency in $Ba_xSr_{1-x}TiO_3$ ($x \leq 0.4$) paraelectric ceramics. *Ceram. Int.* 41: 8252–8256. https://doi.org/10.1016/j.ceramint.2015.02.156.

41 Gao, W., Yao, M., and Yao, X. (2018). Achieving ultrahigh breakdown strength and energy storage performance through periodic interface modification in $SrTiO_3$ thin film. *ACS Appl. Mater. Interfaces* 10: 28745–28753. https://doi.org/10.1021/acsami.8b07151.

42 Hao, X., Zhai, J., Kong, L.B., and Xu, Z. (2014). A comprehensive review on the progress of lead zirconate-based antiferroelectric materials. *Prog. Mater Sci.* 63: 1–57. https://doi.org/10.1016/j.pmatsci.2014.01.002.

43 Xiu, S., Xiao, S., Xue, S. et al. (2016). Crystallization kinetics behaviour and dielectric properties of strontium barium niobate-based glass–ceramics. *J. Mater. Sci.: Mater. Electron.* 27: 5324–5330. https://doi.org/10.1007/s10854-016-4431-7.

44 Wang, S., Tian, J., Jiang, T. et al. (2018). Effect of phase structures on dielectric properties and energy storage performances in Na_2O-BaO-PbO-Nb_2O_5-SiO_2-Al_2O_3 glass-ceramics. *Ceram. Int.* 44: 23109–23115. https://doi.org/10.1016/j.ceramint.2018.09.118.

45 Michael, E.K. and Trolier-Mckinstry, S. (2014). Amorphous-nanocrystalline lead titanate thin films for dielectric energy storage. *J. Ceram. Soc. Jpn.* 122: 250–255. https://doi.org/10.2109/jcersj2.122.250.

46 Liu, T.Y., Chen, G.H., Song, J., and Yuan, C.L. (2013). Crystallization kinetics and dielectric characterization of CeO_2-added BaO-SrO-Nb_2O_5-B_2O_3-SiO_2 glass-ceramics. *Ceram. Int.* 39: 5553–5559. https://doi.org/10.1016/j.ceramint.2012.12.069.

47 Song, J. and Chen, G. (2014). Dielectric behavior and energy storage properties in BaO-SrO-Nb_2O_5-B_2O_3 system glass-ceramics with Gd_2O_3 addition. *J. Mater. Sci.: Mater. Electron.* 25: 349–354. https://doi.org/10.1007/s10854-013-1593-4.

48 Zheng, J., Chen, G.H., Yuan, C.L. et al. (2016). Dielectric characterization and energy-storage performance of lead-free niobate glass-ceramics added with La_2O_3. *Ceram. Int.* 42: 1827–1832. https://doi.org/10.1016/j.ceramint.2015.09.146.

49 Zhou, Y., Li, Y., Qiao, Y., and Feng, R. (2018). Investigation of structure, dielectric and energy-storage properties of lead-free niobate glass and glass-ceramics. *J. Alloys Compd.* 747: 55–59. https://doi.org/10.1016/j.jallcom.2018.03.016.

50 Jiang, T., Chen, K., Shen, B., and Zhai, J. (2019). Enhanced energy-storage density in sodium-barium-niobate based glass-ceramics realized by doping CaF_2 nucleating agent. *J. Mater. Sci.: Mater. Electron.* 30: 15277–15284. https://doi.org/10.1007/s10854-019-01900-1.

51 Tian, J., Wang, S., Jiang, T. et al. (2019). Ultrahigh energy storage density and excellent charge–discharge properties of Bi_2O_3-Nb_2O_5-SiO_2-Al_2O_3 glass ceramic with CeO_2 doping. *J. Electron. Mater.* 48: 6183–6188. https://doi.org/10.1007/s11664-019-07432-y.

52 EI-Desoky, M.M., Wally, N.K., Ali, A.M. et al. (2021). Relaxor ferroelectric-like behavior in 10$PbTiO_3$–10Fe_2O_3–30V_2O_5–50B_2O_3 glass for energy storage applications. *J. Mater. Sci.: Mater. Electron.* 32: 22408–22416. https://doi.org/10.1007/s10854-021-06727-3.

53 Liu, C., Xie, S., Bai, H. et al. (2022). Excellent energy storage performance of niobate-based glass-ceramics via introduction of nucleating agent. *J. Mater.* 1–9. https://doi.org/10.1016/j.jmat.2022.03.001.

54 Michael-Sapia, E.K., Li, H.U., Jackson, T.N., and Trolier-Mckinstry, S. (2015). Nanocomposite bismuth zinc niobate tantalate for flexible energy storage applications. *J. Appl. Phys.* 118: 234102. https://doi.org/10.1063/1.4937560.

55 Chen, Y., Wang, S., Zhou, H. et al. (2020). A systematic analysis of the radial resonance frequency spectra of the PZT-based (Zr/Ti = 52/48) piezoceramic thin disks. *J. Adv. Ceram.* 9: 380–392. https://doi.org/10.1007/s40145-020-0378-5.

56 Shrout, T.R. and Zhang, S.J. (2007). Lead-free piezoelectric ceramics: alternatives for PZT? *J. Electroceram.* 19: 111–124. https://doi.org/10.1007/s10832-007-9047-0.

57 Gao, J., Liu, Y., Wang, Y. et al. (2017). High temperature-stability of $(Pb_{0.9}La_{0.1})(Zr_{0.65}Ti_{0.35})O_3$ ceramic for energy-storage applications at finite electric field strength. *Scr. Mater.* 137: 114–118. https://doi.org/10.1016/j.scriptamat.2017.05.011.

58 Zhang, T.F., Tang, X.G., Liu, Q.X. et al. (2016). Energy-storage properties and high-temperature dielectric relaxation behaviors of relaxor ferroelectric $Pb(Mg_{1/3}Nb_{2/3})O_3$-$PbTiO_3$ ceramics. *J. Phys. D: Appl. Phys.* 49: 95302. https://doi.org/10.1088/0022-3727/49/9/095302.

59 Dai, X. and Viehland, D. (1994). Effects of lanthanum modification on the antiferroelectric-ferroelectric stability of high zirconium-content lead zirconate titanate. *J. Appl. Phys.* 76: 3701–3709. https://doi.org/10.1063/1.357439.

60 Gupta, S.M., Li, J.F., and Viehland, D. (1998). Coexistence of relaxor and normal ferroelectric phases in morphotropic phase boundary compositions of lanthanum-modified lead zirconate titanate. *J. Am. Ceram. Soc.* 81: 557–564. https://doi.org/10.1111/j.1151-2916.1998.tb02374.x.

61 Hu, Z., Ma, B., Liu, S. et al. (2014). Relaxor behavior and energy storage performance of ferroelectric PLZT thin films with different Zr/Ti ratios. *Ceram. Int.* 40: 557–562. https://doi.org/10.1016/j.ceramint.2013.05.139.

62 Peng, B., Xie, Z., Yue, Z., and Li, L. (2014). Improvement of the recoverable energy storage density and efficiency by utilizing the linear dielectric response in ferroelectric capacitors. *Appl. Phys. Lett.* 105: 52904. https://doi.org/10.1063/1.4892454.

63 Zhang, L., Hao, X., Yang, J. et al. (2013). Large enhancement of energy-storage properties of compositional graded $(Pb_{1-x}La_x)(Zr_{0.65}Ti_{0.35})O_3$ relaxor ferroelectric thick films. *Appl. Phys. Lett.* 103: 113902. https://doi.org/10.1063/1.4821209.

64 Kumar, A., Kim, S.H., Peddigari, M. et al. (2019). High energy storage properties and electrical field stability of energy efficiency of $(Pb_{0.89}La_{0.11})(Zr_{0.70}Ti_{0.30})0.9725O_3$ relaxor ferroelectric ceramics. *Electron. Mater. Lett.* 15: 323–330. https://doi.org/10.1007/s13391-019-00124-z.

65 Liu, Y., Hao, X., and An, S. (2013). Significant enhancement of energy-storage performance of $(Pb_{0.91}La_{0.09})(Zr_{0.65}Ti_{0.35})O_3$ relaxor ferroelectric thin films by Mn doping. *J. Appl. Phys.* 114: 174102. https://doi.org/10.1063/1.4829029.

66 Ma, B., Hu, Z., Koritala, R.E. et al. (2015). PLZT film capacitors for power electronics and energy storage applications. *J. Mater. Sci.: Mater. Electron.* 26: 9279–9287. https://doi.org/10.1007/s10854-015-3025-0.

67 Zhang, T., Li, W., Hou, Y. et al. (2017). High-energy storage density and excellent temperature stability in antiferroelectric/ferroelectric bilayer thin films. *J. Am. Ceram. Soc.* 100: 3080–3087. https://doi.org/10.1111/jace.14876.

68 Nguyen, M.D., Nguyen, C.T.Q., Vu, H.N., and Rijnders, G. (2018). Controlling microstructure and film growth of relaxor-ferroelectric thin films for high break-down strength and energy-storage performance. *J. Eur. Ceram. Soc.* 38: 95–103. https://doi.org/10.1016/j.jeurceramsoc.2017.08.027.

69 Zhang, T., Li, W., Zhao, Y. et al. (2018). High energy storage performance of opposite double-heterojunction ferroelectricity–insulators. *Adv. Funct. Mater.* 28: 1706211. https://doi.org/10.1002/adfm.201706211.

70 Smolenskii, G.A., Isupov, V.A., Agranovskaya, A.I., and Krainik, N.N. (1961). New ferroelectrics of complex composition. *Sov. Phys. Solid State* 2 (11): 2651–2654.

71 Rao, B.N., Datta, R., Chandrashekaran, S.S. et al. (2013). Local structural disorder and its influence on the average global structure and polar properties in $Na_{0.5}Bi_{0.5}TiO_3$. *Phys. Rev. B: Condens. Matter Mater. Phys.* 88: 224103. https://doi.org/10.1103/PhysRevB.88.224103.

72 Reichmann, K., Feteira, A., and Li, M. (2015). Bismuth sodium titanate based materials for piezoelectric actuators. *Materials* 8: 8467–8495. https://doi.org/10.3390/ma8125469.

73 Suchanicz, J., Kluczewska-Chmielarz, K., Sitko, D., and Jagło, G. (2021). Electrical transport in lead-free $Na_{0.5}Bi_{0.5}TiO_3$ ceramics. *J. Adv. Ceram.* 10: 152–165. https://doi.org/10.1007/s40145-020-0430-5.

74 Qiao, X., Zhang, F., Wu, D. et al. (2020). Superior comprehensive energy storage properties in $Bi_{0.5}Na_{0.5}TiO_3$-based relaxor ferroelectric ceramics. *Chem. Eng. J.* 388: 124158. https://doi.org/10.1016/j.cej.2020.124158.

75 Yang, F., Pan, Z., Ling, Z. et al. (2021). Realizing high comprehensive energy storage performances of BNT-based ceramics for application in pulse power capacitors. *J. Eur. Ceram. Soc.* 41: 2548–2558. https://doi.org/10.1016/j.jeurceramsoc.2020.11.049.

76 Zhang, X., Hu, D., Pan, Z. et al. (2021). Enhancement of recoverable energy density and efficiency of lead-free relaxor-ferroelectric BNT-based ceramics. *Chem. Eng. J.* 406: 126818. https://doi.org/10.1016/j.cej.2020.126818.

77 Zhu, C., Cai, Z., Luo, B. et al. (2020). High temperature lead-free BNT-based ceramics with stable energy storage and dielectric properties. *J. Mater. Chem. A* 8: 683–692. https://doi.org/10.1039/c9ta10347c.

78 Ma, C. and Tan, X. (2011). In situ transmission electron microscopy study on the phase transitions in lead-free $(1-x)(Bi_{1/2}Na_{1/2})TiO_3$-$xBaTiO_3$ ceramics. *J. Am. Ceram. Soc.* 94: 4040–4044. https://doi.org/10.1111/j.1551-2916.2011.04670.x.

79 Jo, W., Schaab, S., Sapper, E. et al. (2011). On the phase identity and its thermal evolution of lead free $(Bi_{1/2}Na_{1/2})TiO_3$-6 mol $BaTiO_3$. *J. Appl. Phys.* 110: 74106. https://doi.org/10.1063/1.3645054.

80 Ye, H., Yang, F., Pan, Z. et al. (2021). Significantly improvement of comprehensive energy storage performances with lead-free relaxor ferroelectric ceramics for high-temperature capacitors applications. *Acta Mater.* 203: 116484. https://doi.org/10.1016/j.actamat.2020.116484.

81 Gao, F., Dong, X., Mao, C. et al. (2011). Energy-storage properties of $0.89Bi_{0.5}Na_{0.5}TiO_3$-$0.06BaTiO_3$-$0.05K_{0.5}Na_{0.5}NbO_3$ lead-free anti-ferroelectric ceramics. *J. Am. Ceram. Soc.* 94: 4382–4386. https://doi.org/10.1111/j.1551-2916.2011.04731.x.

82 Fan, P., Zhang, S.T., Xu, J. et al. (2020). Relaxor/antiferroelectric composites: a solution to achieve high energy storage performance in lead-free dielectric ceramics. *J. Mater. Chem. C* 8: 5681–5691. https://doi.org/10.1039/d0tc00589d.

83 Guo, Y., Li, M., Zhao, W. et al. (2009). Ferroelectric and pyroelectric properties of $(Na_{0.5}Bi_{0.5})TiO_3$-$BaTiO_3$ based trilayered thin films. *Thin Solid Films* 517: 2974–2978. https://doi.org/10.1016/j.tsf.2008.11.100.

84 Cao, W., Li, W., Feng, Y. et al. (2016). Defect dipole induced large recoverable strain and high energy-storage density in lead-free $Na_{0.5}Bi_{0.5}TiO_3$-based systems. *Appl. Phys. Lett.* 108: 202902. https://doi.org/10.1063/1.4950974.

85 Yang, H., Liu, P., Yan, F. et al. (2019). A novel lead-free ceramic with layered structure for high energy storage applications. *J. Alloys Compd.* 773: 244–249. https://doi.org/10.1016/j.jallcom.2018.09.252.

86 Yang, L., Kong, X., Cheng, Z., and Zhang, S. (2019). Ultra-high energy storage performance with mitigated polarization saturation in lead-free relaxors. *J. Mater. Chem. A* 7: 8573–8580. https://doi.org/10.1039/c9ta01165j.

87 Li, D., Shen, Z.Y., Li, Z. et al. (2020). Optimization of polarization behavior in $(1-x)$BSBNT-xNN ceramics for pulsed power capacitors. *J. Mater. Chem. C* 8: 7650–7657. https://doi.org/10.1039/d0tc01699c.

88 Li, D., Shen, Z.Y., Li, Z. et al. (2020). *P–E* hysteresis loop going slim in $Ba_{0.3}Sr_{0.7}TiO_3$-modified $Bi_{0.5}Na_{0.5}TiO_3$ ceramics for energy storage applications. *J. Adv. Ceram.* 9: 183–192. https://doi.org/10.1007/s40145-020-0358-9.

89 Chen, P., Wu, S., Li, P. et al. (2018). Great enhancement of energy storage density and power density in BNBT/xBFO multilayer thin film hetero-structures. *Inorg. Chem. Front.* 5: 2300–2305. https://doi.org/10.1039/c8qi00487k.

90 Li, J., Li, F., Xu, Z., and Zhang, S. (2018). Multilayer lead-free ceramic capacitors with ultrahigh energy density and efficiency. *Adv. Mater.* 30: 1802155. https://doi.org/10.1002/adma.201802155.

91 Jia, W., Hou, Y., Zheng, M. et al. (2018). Superior temperature-stable dielectrics for MLCCs based on $Bi_{0.5}Na_{0.5}TiO_3$-$NaNbO_3$ system modified by $CaZrO_3$. *J. Am. Ceram. Soc.* 101: 3468–3479. https://doi.org/10.1111/jace.15519.

92 Qian, J., Han, Y., Yang, C. et al. (2020). Energy storage performance of flexible NKBT/NKBT-ST multilayer film capacitor by interface engineering. *Nano Energy* 74: 104862. https://doi.org/10.1016/j.nanoen.2020.104862.

93 Wang, H., Zhao, P., Chen, L. et al. (2020). Energy storage properties of $0.87BaTiO_3$-$0.13Bi(Zn_{2/3}(Nb_{0.85}Ta_{0.15})_{1/3})O_3$ multilayer ceramic capacitors with thin dielectric layers. *J. Adv. Ceram.* 9: 292–302. https://doi.org/10.1007/s40145-020-0367-8.

94 Li, J., Shen, Z., Chen, X. et al. (2020). Grain-orientation-engineered multilayer ceramic capacitors for energy storage applications. *Nat. Mater.* 19: 999–1005. https://doi.org/10.1038/s41563-020-0704-x.

95 Pan, M.J. and Randall, C. (2010). A brief introduction to ceramic capacitors. *IEEE Electr. Insul. Mag.* 26: 44–50. https://doi.org/10.1109/MEI.2010.5482787.

96 Zheng, L., Yuan, L., Liang, G. et al. (2019). *Mater. Chem. Front.* 3: 726–736. https://doi.org/10.1039/c9qm00006b.

97 Matias, A., Nikola, N., Virginia, R. et al. (2017). $BaTiO_3$-based piezoelectrics: fundamentals, current status, and perspectives. *Appl. Phys. Rev.* 4: 1–53.

98 Gao, W., Zhu, Y., Wang, Y. et al. (2020). A review of flexible perovskite oxide ferroelectric films and their application. *J. Mater.* 6: 1–16. https://doi.org/10.1016/j.jmat.2019.11.001.

99 Bai, F., Viehland, D., Jia, Y. et al. (2004). Multiferroic $BaTiO_3$-$CoFe_2O_4$ nanostructures. *Science* 303: 661–663.

100 Hu, D., Pan, Z., Tan, X. et al. (2021). Optimization the energy density and efficiency of $BaTiO_3$-based ceramics for capacitor applications. *Chem. Eng. J.* 409: 127375. https://doi.org/10.1016/j.cej.2020.127375.

101 Ogihara, H., Randall, C.A., and Trolier-Mckinstry, S. (2009). High-energy density capacitors utilizing $0.7BaTiO_3$-$0.3BiScO_3$ ceramics. *J. Am. Ceram. Soc.* 92: 1719–1724. https://doi.org/10.1111/j.1551-2916.2009.03104.x.

102 Choi, D.H., Baker, A., Lanagan, M. et al. (2013). Structural and dielectric properties in $(1-x)BaTiO_3$-$x Bi(Mg_{1/2}Ti_{1/2})O_3$ ceramics ($0.1 \leq x \leq 0.5$) and potential for high-voltage multilayer capacitors. *J. Am. Ceram. Soc.* 96: 2197–2202. https://doi.org/10.1111/jace.12312.

103 Li, M.D., Tang, X.G., Zeng, S.M. et al. (2018). Oxygen-vacancy-related dielectric relaxation behaviours and impedance spectroscopy of $Bi(Mg_{1/2}Ti_{1/2})O_3$ modified $BaTiO_3$ ferroelectric ceramics. *J. Mater.* 4: 194–201. https://doi.org/10.1016/j.jmat.2018.03.001.

104 Hanani, Z., Mezzane, D., Amjoud, M. et al. (2019). Phase transitions, energy storage performances and electrocaloric effect of the lead-free $Ba_{0.85}Ca_{0.15}Zr_{0.10}Ti_{0.90}O_3$ ceramic relaxor. *J. Mater. Sci.: Mater. Electron.* 30: 6430–6438. https://doi.org/10.1007/s10854-019-00946-5.

105 Wang, X.W., Zhang, B.H., Shi, Y.C. et al. (2020). Enhanced energy storage properties in $Ba_{0.85}Ca_{0.15}Zr_{0.1}Ti_{0.9}O_3$ ceramics with glass additives. *J. Appl. Phys.* 127: 74103. https://doi.org/10.1063/1.5138948.

106 Hanani, Z., Merselmiz, S., Danine, A. et al. (2020). Enhanced dielectric and electrocaloric properties in lead-free rod-like BCZT ceramics. *J. Adv. Ceram.* 9: 210–219. https://doi.org/10.1007/s40145-020-0361-1.

107 Patel, S., Sharma, D., Singh, A., and Vaish, R. (2016). Enhanced thermal energy conversion and dynamic hysteresis behavior of Sr-added $Ba_{0.85}Ca_{0.15}Ti_{0.9}Zr_{0.1}O_3$ ferroelectric ceramics. *J. Mater.* 2: 75–86. https://doi.org/10.1016/j.jmat.2016.01.002.

108 Su, X., Riggs, B.C., Tomozawa, M. et al. (2014). Preparation of $BaTiO_3$/low melting glass core-shell nanoparticles for energy storage capacitor applications. *J. Mater. Chem. A* 2: 18087–18096. https://doi.org/10.1039/c4ta04282d.

109 Zhang, Y., Cao, M., Yao, Z. et al. (2015). Effects of silica coating on the microstructures and energy storage properties of $BaTiO_3$ ceramics. *Mater. Res. Bull.* 67: 70–76. https://doi.org/10.1016/j.materresbull.2015.01.056.

110 Wu, L., Wang, X., Gong, H. et al. (2015). Core-satellite $BaTiO_3$@$SrTiO_3$ assemblies for a local compositionally graded relaxor ferroelectric capacitor with enhanced energy storage density and high energy efficiency. *J. Mater. Chem. C* 3: 750–758. https://doi.org/10.1039/c4tc02291b.

111 Yu, Z., Ang, C., Guo, R., and Bhalla, A.S. (2002). Ferroelectric-relaxor behavior of $Ba(Ti_{0.7}Zr_{0.3})O_3$ ceramics. *J. Appl. Phys.* 92: 2655–2657. https://doi.org/10.1063/1.1495069.

112 Hennings, D., Schnell, A., and Simon, G. (1982). Diffuse ferroelectric phase transitions in $Ba(Ti_{1-y}Zr_y)O_3$ ceramics. *J. Am. Ceram. Soc.* 65: 539–544. https://doi.org/10.1111/j.1151-2916.1982.tb10778.x.

113 Sun, Z., Ma, C., Liu, M. et al. (2017). Ultrahigh energy storage performance of lead-free oxide multilayer film capacitors via interface engineering. *Adv. Mater.* 29: 1604427. https://doi.org/10.1002/adma.201604427.

114 Yuan, Q., Li, G., Yao, F.Z. et al. (2018). Simultaneously achieved temperature-insensitive high energy density and efficiency in domain engineered $BaTiO_3$-$Bi(Mg_{0.5}Zr_{0.5})O_3$ lead-free relaxor ferroelectrics. *Nano Energy* 52: 203–210. https://doi.org/10.1016/j.nanoen.2018.07.055.

115 Wu, L., Wang, X., and Li, L. (2016). Core-shell $BaTiO_3$@$BiScO_3$ particles for local graded dielectric ceramics with enhanced temperature stability and energy storage capability. *J. Alloys Compd.* 688: 113–121. https://doi.org/10.1016/j.jallcom.2016.07.057.

116 Ortega, N., Kumar, A., Scott, J.F. et al. (2012). Relaxor-ferroelectric superlattices: high energy density capacitors. *J. Phys.: Condens. Matter* 24: 445901. https://doi.org/10.1088/0953-8984/24/44/445901.

117 Instan, A.A., Pavunny, S.P., Bhattarai, M.K., and Katiyar, R.S. (2017). Ultrahigh capacitive energy storage in highly oriented $Ba(Zr_xTi_{1-x})O_3$ thin films prepared by pulsed laser deposition. *Appl. Phys. Lett.* 111: 142903. https://doi.org/10.1063/1.4986238.

118 Yeh, J.W., Chen, S.K., Lin, S.J. et al. (2004). Nanostructured high-entropy alloys with multiple principal elements: novel alloy design concepts and outcomes. *Adv. Eng. Mater.* 6: 299–303. https://doi.org/10.1002/adem.200300567.

119 Cantor, B., Chang, I.T.H., Knight, P., and Vincent, A.J.B. (2004). Microstructural development in equiatomic multicomponent alloys. *Mater. Sci. Eng., A* 375–377: 213–218. https://doi.org/10.1016/j.msea.2003.10.257.

120 Yeh, J.W. (2006). Recent progress in high-entropy alloys. *Ann. Chim. Sci. Des Mater.* 31: 633–648. https://doi.org/10.3166/acsm.31.633-648.

121 Yeh, J.W., Chen, S.K., Gan, J.Y. et al. (2004). Formation of simple crystal structures in Cu-Co-Ni-Cr-Al-Fe-Ti-V alloys with multiprincipal metallic elements. *Metall. Mater. Trans. A Phys. Metall. Mater. Sci.* 35 A: 2533–2536. https://doi.org/10.1007/s11661-006-0234-4.

122 Miracle, D.B. and Senkov, O.N. (2017). A critical review of high entropy alloys and related concepts. *Acta Mater.* 122: 448–511. https://doi.org/10.1016/j.actamat.2016.08.081.

123 Pu, Y., Zhang, Q., Li, R. et al. (2019). Dielectric properties and electrocaloric effect of high-entropy $(Na_{0.2}Bi_{0.2}Ba_{0.2}Sr_{0.2}Ca_{0.2})TiO_3$ ceramic. *Appl. Phys. Lett.* 115: 223901. https://doi.org/10.1063/1.5126652.

124 Chen, L., Deng, S., Liu, H. et al. (2022). Giant energy-storage density with ultrahigh efficiency in lead-free relaxors via high-entropy design. *Nat. Commun.* 13: 3089. https://doi.org/10.1038/s41467-022-30821-7.

125 Yang, B., Zhang, Y., Pan, H. et al. (2022). High-entropy enhanced capacitive energy storage. *Nat. Mater.* https://doi.org/10.1038/s41563-022-01274-6.

126 Liu, J., Ren, K., Ma, C. et al. (2020). Dielectric and energy storage properties of flash-sintered high-entropy $(Bi_{0.2}Na_{0.2}K_{0.2}Ba_{0.2}Ca_{0.2})TiO_3$ ceramic. *Ceram. Int.* 46: 20576–20581. https://doi.org/10.1016/j.ceramint.2020.05.090.

127 Yang, W. and Zheng, G. (2022). High energy storage density and efficiency in nanostructured $(Bi_{0.2}Na_{0.2}K_{0.2}La_{0.2}Sr_{0.2})TiO_3$ high-entropy ceramics. *J. Am. Ceram. Soc.* 105: 1083–1094. https://doi.org/10.1111/jace.18129.

128 Zhou, S., Pu, Y., Zhang, X. et al. (2022). High energy density, temperature stable lead-free ceramics by introducing high entropy perovskite oxide. *Chem. Eng. J.* 427: 131684. https://doi.org/10.1016/j.cej.2021.131684.

129 Chen, X., Zhang, H., Cao, F. et al. (2009). Charge-discharge properties of lead zirconate stannate titanate ceramics. *J. Appl. Phys.* 106: 34105. https://doi.org/10.1063/1.3187778.

130 Tagantsev, A.K., Vaideeswaran, K., Vakhrushev, S.B. et al. (2013). The origin of antiferroelectricity in $PbZrO_3$. *Nat. Commun.* 4: 2229. https://doi.org/10.1038/ncomms3229.

131 Chen, N., Bai, G.R., Auciello, O. et al. (1999). Properties and orientation of antiferroelectric lead zirconate thin films grown by MOCVD. *Mater. Res. Soc. Symp. Proc.* 541: 345–350. https://doi.org/10.1557/proc-541-345.

132 Hao, X., Zhai, J., and Yao, X. (2009). Improved energy storage performance and fatigue endurance of Sr-doped $PbZrO_3$ antiferroelectric thin films. *J. Am. Ceram. Soc.* 92: 1133–1135. https://doi.org/10.1111/j.1551-2916.2009.03015.x.

133 Parui, J. and Krupanidhi, S.B. (2008). Enhancement of charge and energy storage in sol-gel derived pure and La-modified $PbZrO_3$ thin films. *Appl. Phys. Lett.* 92: 192901. https://doi.org/10.1063/1.2928230.

134 Tani, T., Li, J.F., Viehland, D., and Payne, D.A. (1994). Antiferroelectric-ferroelectric switching and induced strains for sol-gel derived lead zirconate thin layers. *J. Appl. Phys.* 75: 3017–3023. https://doi.org/10.1063/1.356146.

135 Sa, T., Cao, Z., Wang, Y., and Zhu, H. (2014). Enhancement of charge and energy storage in $PbZrO_3$ thin films by local field engineering. *Appl. Phys. Lett.* 105: 43902. https://doi.org/10.1063/1.4891768.

136 Chao, W., Yang, T., and Li, Y. (2020). Achieving high energy efficiency and energy density in $PbHfO_3$-based antiferroelectric ceramics. *J. Mater. Chem. C* 8: 17016–17024. https://doi.org/10.1039/d0tc04617e.

137 Ye, M., Sun, Q., Chen, X. et al. (2012). Electrical and energy storage performance of Eu-doped $PbZrO_3$ thin films with different gradient sequences. *J. Am. Ceram. Soc.* 95: 1486–1488. https://doi.org/10.1111/j.1551-2916.2012.05130.x.

138 Ge, J., Remiens, D., Dong, X. et al. (2014). Enhancement of energy storage in epitaxial $PbZrO_3$ antiferroelectric films using strain engineering. *Appl. Phys. Lett.* 105: 112908. https://doi.org/10.1063/1.4896156.

139 Chen, M.J., Ning, X.K., Wang, S.F., and Fu, G.S. (2019). Significant enhancement of energy storage density and polarization in self-assembled $PbZrO_3$:NiO nano-columnar composite films. *Nanoscale* 11: 1914–1920. https://doi.org/10.1039/c8nr08887j.

140 Shrout, T.R. and Zhang, S. (2007). Erratum: Lead-free piezoelectric ceramics: alternatives for PZT? (Journal of Electroceramics (2007) 19 (183)). *J. Electroceram.* 19: 183. https://doi.org/10.1007/s10832-007-9095-5.

141 Xu, B., Moses, P., Pai, N.G., and Cross, L.E. (1998). Charge release of lanthanum-doped lead zirconate titanate stannate antiferroelectric thin films. *Appl. Phys. Lett.* 72: 593–595. https://doi.org/10.1063/1.120817.

142 Sharifzadeh Mirshekarloo, M., Yao, K., and Sritharan, T. (2010). Large strain and high energy storage density in orthorhombic perovskite $(Pb_{0.97}La_{0.02})(Zr_{1-x-y}Sn_xTi_y)O_3$ antiferroelectric thin films. *Appl. Phys. Lett.* 97: 142902. https://doi.org/10.1063/1.3497193.

143 Zhang, A.H., Wang, W., Li, Q.J. et al. (2020). Internal-strain release and remarkably enhanced energy storage performance in PLZT–$SrTiO_3$ multilayered films. *Appl. Phys. Lett.* 117: 252901. https://doi.org/10.1063/5.0030279.

144 Dan, Y., Xu, H., Zou, K. et al. (2018). Energy storage characteristics of $(Pb,La)(Zr,Sn,Ti)O_3$ antiferroelectric ceramics with high Sn content. *Appl. Phys. Lett.* 113: 63902. https://doi.org/10.1063/1.5044712.

145 Liu, P., Fan, B., Yang, G. et al. (2019). High energy density at high temperature in PLZST antiferroelectric ceramics. *J. Mater. Chem. C* 7: 4587–4594. https://doi.org/10.1039/c9tc00944b.

146 Xu, B., Ye, Y., and Cross, L.E. (2000). Dielectric properties and field-induced phase switching of lead zirconate titanate stannate antiferroelectric thick films on silicon substrates. *J. Appl. Phys.* 87: 2507–2515. https://doi.org/10.1063/1.372211.

147 Markowski, K., Park, S.E., Yoshikawa, S., and Cross, L.E. (1996). Effect of compositional variations in the lead lanthanum zirconate stannate titanate system on electrical properties. *J. Am. Ceram. Soc.* 79: 3297–3304. https://doi.org/10.1111/j.1151-2916.1996.tb08108.x.

148 Zheng, Q., Yang, T., Wei, K. et al. (2012). Effect of Sn:Ti variations on electric filed induced AFE–FE phase transition in PLZST antiferroelectric ceramics. *Ceram. Int.* 38: S9–S12. https://doi.org/10.1016/j.ceramint.2011.04.037.

149 Liu, Z., Bai, Y., Chen, X. et al. (2017). Linear composition-dependent phase transition behavior and energy storage performance of tetragonal PLZST antiferroelectric ceramics. *J. Alloys Compd.* 691: 721–725. https://doi.org/10.1016/j.jallcom.2016.08.328.

150 Zhang, L., Jiang, S., Fan, B., and Zhang, G. (2015). Enhanced energy storage performance in $(Pb_{0.858}Ba_{0.1}La_{0.02}Y_{0.008})(Zr_{0.65}Sn_{0.3}Ti_{0.05})O_3$-$(Pb_{0.97}La_{0.02})(Zr_{0.9}Sn_{0.05}Ti_{0.05})O_3$ anti-ferroelectric composite ceramics by spark plasma sintering. *J. Alloys Compd.* 622: 162–165. https://doi.org/10.1016/j.jallcom.2014.09.171.

151 Zhang, G., Zhu, D., Zhang, X. et al. (2015). High-energy storage performance of $(Pb_{0.87}Ba_{0.1}La_{0.02})(Zr_{0.68}Sn_{0.24}Ti_{0.08})O_3$ antiferroelectric ceramics fabricated by the hot-press sintering method. *J. Am. Ceram. Soc.* 98: 1175–1181. https://doi.org/10.1111/jace.13412.

152 Zhang, G., Liu, S., Yu, Y. et al. (2013). Microstructure and electrical properties of $(Pb_{0.87}Ba_{0.1}La_{0.02})(Zr_{0.68}Sn_{0.24}Ti_{0.08})O_3$ anti-ferroelectric ceramics fabricated by the hot-press sintering method. *J. Eur. Ceram. Soc.* 33: 113–121. https://doi.org/10.1016/j.jeurceramsoc.2012.08.011.

153 Bian, F., Yan, S., Xu, C. et al. (2018). Enhanced breakdown strength and energy density of antiferroelectric Pb,La(Zr,Sn,Ti)O_3 ceramic by forming core-shell structure. *J. Eur. Ceram. Soc.* 38: 3170–3176. https://doi.org/10.1016/j.jeurceramsoc.2018.03.028.

154 Ma, B., Kwon, D.K., Narayanan, M., and Balachandran, U. (2009). Dielectric properties and energy storage capability of antiferroelectric $Pb_{0.92}La_{0.08}Zr_{0.95}Ti_{0.05}O_3$ film-on-foil capacitors. *J. Mater. Res.* 24: 2993–2996. https://doi.org/10.1557/jmr.2009.0349.

155 Ma, B., Kwon, D.K., Narayanan, M., and (Balu) Balachandran, U. (2009). Fabrication of antiferroelectric PLZT films on metal foils. *Mater. Res. Bull.* 44: 11–14. https://doi.org/10.1016/j.materresbull.2008.09.006.

156 Wang, H., Liu, Y., Yang, T., and Zhang, S. (2019). Ultrahigh energy-storage density in antiferroelectric ceramics with field-induced multiphase transitions. *Adv. Funct. Mater.* 29: 1807321. https://doi.org/10.1002/adfm.201807321.

157 Liu, X., Li, Y., and Hao, X. (2019). Ultra-high energy-storage density and fast discharge speed of $(Pb_{0.98}$-$x La_{0.02}Sr_x)(Zr_{0.9}Sn_{0.1})_{0.995}O_3$ antiferroelectric ceramics prepared via the tape-casting method. *J. Mater. Chem. A* 7: 11858–11866. https://doi.org/10.1039/c9ta02149c.

158 Lin, Z., Chen, Y., Liu, Z. et al. (2018). Large energy storage density, low energy loss and highly stable $(Pb_{0.97}La_{0.02})(Zr_{0.66}Sn_{0.23}Ti_{0.11})O_3$ antiferroelectric thin-film capacitors. *J. Eur. Ceram. Soc.* 38: 3177–3181. https://doi.org/10.1016/j.jeurceramsoc.2018.03.004.

159 Gao, M., Tang, X., Leung, C.M. et al. (2019). Phase transition and energy storage behavior of antiferroelectric PLZT thin films epitaxially deposited on SRO buffered STO single crystal substrates. *J. Am. Ceram. Soc.* 102: 5180–5191. https://doi.org/10.1111/jace.16380.

160 Zhang, M.H., Fulanović, L., Egert, S. et al. (2020). Electric-field-induced antiferroelectric to ferroelectric phase transition in polycrystalline $NaNbO_3$. *Acta Mater.* 200: 127–135. https://doi.org/10.1016/j.actamat.2020.09.002.

161 Chen, J. and Feng, D. (1988). TEM study of phases and domains in $NaNbO_3$ at room temperature. *Phys. Status Solidi* 109: 171–185. https://doi.org/10.1002/pssa.2211090117.

162 Saito, T., Adachi, H., Wada, T. et al. (2005). *Jpn. J. Appl. Phys., Part 1* 44: 6969–6972. https://doi.org/10.1143/JJAP.44.6969.

163 Koruza, J., Groszewicz, P., Breitzke, H. et al. (2017). Grain-size-induced ferroelectricity in $NaNbO_3$. *Acta Mater.* 126: 77–85. https://doi.org/10.1016/j.actamat.2016.12.049.

164 Antipin, M.Y., Lindeman, S.V., and Struchkov, Y.T. (1993). Crystal structure of the electric-field-induced ferroelectric phase of $NaNbO_3$. *Ferroelectrics* 141: 307–311. https://doi.org/10.1080/00150199308223458.

165 Shimizu, H., Guo, H., Reyes-Lillo, S.E. et al. (2015). Lead-free antiferroelectric: xCaZrO$_3$-(1 − x)NaNbO$_3$ system ($0 \leq x \leq 0.10$). *Dalton Trans.* 44: 10763–10772. https://doi.org/10.1039/c4dt03919j.

166 Guo, H., Shimizu, H., Mizuno, Y., and Randall, C.A. (2015). Strategy for stabilization of the antiferroelectric phase (*Pbma*) over the metastable ferroelectric phase ($P2_1ma$) to establish double loop hysteresis in lead-free $(1-x)$NaNbO$_3$–xSrZrO$_3$ solid solution. *J. Appl. Phys.* 117: 214103. https://doi.org/10.1063/1.4921876.

167 Gao, L., Guo, H., Zhang, S., and Randall, C.A. (2016). A perovskite lead-free antiferroelectric xCaHfO$_3$-(1 − x)NaNbO$_3$ with induced double hysteresis loops at room temperature. *J. Appl. Phys.* 120: 204102. https://doi.org/10.1063/1.4968790.

168 Gao, L., Guo, H., Zhang, S., and Randall, C.A. (2018). Stabilized antiferroelectricity in xBiScO$_3$-(1 − x)NaNbO$_3$ lead-free ceramics with established double hysteresis loops. *Appl. Phys. Lett.* 112: 92905. https://doi.org/10.1063/1.5017697.

169 Jiang, J., Meng, X., Li, L. et al. (2021). Ultrahigh energy storage density in lead-free relaxor antiferroelectric ceramics via domain engineering. *Energy Storage Mater.* 43: 383–390. https://doi.org/10.1016/j.ensm.2021.09.018.

170 Luo, B., Dong, H., Wang, D., and Jin, K. (2018). Large recoverable energy density with excellent thermal stability in Mn-modified $NaNbO_3$-$CaZrO_3$ lead-free thin films. *J. Am. Ceram. Soc.* 101: 3460–3467. https://doi.org/10.1111/jace.15528.

171 Zhou, M., Liang, R., Zhou, Z., and Dong, X. (2018). Superior energy storage properties and excellent stability of novel $NaNbO_3$-based lead-free ceramics with A-site vacancy obtained: via a Bi_2O_3 substitution strategy. *J. Mater. Chem. A* 6: 17896–17904. https://doi.org/10.1039/c8ta07303a.

172 Ye, J., Wang, G., Zhou, M. et al. (2019). Excellent comprehensive energy storage properties of novel lead-free $NaNbO_3$-based ceramics for dielectric capacitor applications. *J. Mater. Chem. C* 7: 5639–5645. https://doi.org/10.1039/c9tc01414d.

173 Dong, X., Li, X., Chen, X. et al. (2021). High energy storage density and power density achieved simultaneously in $NaNbO_3$-based lead-free ceramics via antiferroelectricity enhancement. *J. Mater.* 7: 629–639. https://doi.org/10.1016/j.jmat.2020.11.016.

174 Wei, T., Liu, K., Fan, P. et al. (2021). Novel $NaNbO_3$–$Sr_{0.7}Bi_{0.2}TiO_3$ lead-free dielectric ceramics with excellent energy storage properties. *Ceram. Int.* 47: 3713–3719. https://doi.org/10.1016/j.ceramint.2020.09.228.

175 Kania, A. and Kwapuliński, J. (1999). $Ag_{1-x}Na_xNbO_3$ (ANN) solid solutions: from disordered antiferroelectric $AgNbO_3$ to normal antiferroelectric $NaNbO_3$. *J. Phys.: Condens. Matter* 11: 8933–8946. https://doi.org/10.1088/0953-8984/11/45/316.

176 Wang, D., Kako, T., and Ye, J. (2009). New series of solid-solution semiconductors $(AgNbO_3)_{1-x}(SrTiO_3)_x$ with modulated band structure and enhanced visible-light photocatalytic activity. *J. Phys. Chem. C* 113: 3785–3792. https://doi.org/10.1021/jp807393a.

177 Fu, D., Endo, M., Taniguchi, H. et al. (2007). $AgNbO_3$: a lead-free material with large polarization and electromechanical response. *Appl. Phys. Lett.* 90: 252907. https://doi.org/10.1063/1.2751136.

178 Tian, Y., Jin, L., Zhang, H. et al. (2016). High energy density in silver niobate ceramics. *J. Mater. Chem. A* 4: 17279–17287. https://doi.org/10.1039/c6ta06353e.

179 Zhao, L., Liu, Q., Gao, J. et al. (2017). Lead-free antiferroelectric silver niobate tantalate with high energy storage performance. *Adv. Mater.* 29: 1701824. https://doi.org/10.1002/adma.201701824.

180 Luo, N., Han, K., Cabral, M.J. et al. (2020). Constructing phase boundary in $AgNbO_3$ antiferroelectrics: pathway simultaneously achieving high energy density and efficiency. *Nat. Commun.* 11: 4824. https://doi.org/10.1038/s41467-020-18665-5.

181 Yan, Z., Zhang, D., Zhou, X. et al. (2019). Silver niobate based lead-free ceramics with high energy storage density. *J. Mater. Chem. A* 7: 10702–10711. https://doi.org/10.1039/c9ta00995g.

182 Luo, N., Han, K., Zhuo, F. et al. (2019). Aliovalent A-site engineered $AgNbO_3$ lead-free antiferroelectric ceramics toward superior energy storage density. *J. Mater. Chem. A* 7: 14118–14128. https://doi.org/10.1039/c9ta02053e.

183 Wang, J., Wan, X., Rao, Y. et al. (2020). Hydrothermal synthesized $AgNbO_3$ powders: leading to greatly improved electric breakdown strength in ceramics. *J. Eur. Ceram. Soc.* 40: 5589–5596. https://doi.org/10.1016/j.jeurceramsoc.2020.06.031.

184 Tian, Y., Jin, L., Zhang, H. et al. (2017). Phase transitions in bismuth-modified silver niobate ceramics for high power energy storage. *J. Mater. Chem. A* 5: 17525–17531. https://doi.org/10.1039/c7ta03821f.

185 Zhao, L., Gao, J., Liu, Q. et al. (2018). Silver niobate lead-free antiferroelectric ceramics: enhancing energy storage density by B-site doping. *ACS Appl. Mater. Interfaces* 10: 819–826. https://doi.org/10.1021/acsami.7b17382.

186 Han, K., Luo, N., Mao, S. et al. (2019). Realizing high low-electric-field energy storage performance in $AgNbO_3$ ceramics by introducing relaxor behaviour. *J. Mater.* 5: 597–605. https://doi.org/10.1016/j.jmat.2019.07.006.

187 Gao, J., Zhang, Y., Zhao, L. et al. (2019). Enhanced antiferroelectric phase stability in La-doped $AgNbO_3$: perspectives from the microstructure to energy storage properties. *J. Mater. Chem. A* 7: 2225–2232. https://doi.org/10.1039/c8ta09353a.

188 Luo, N., Han, K., Zhuo, F. et al. (2019). Design for high energy storage density and temperature-insensitive lead-free antiferroelectric ceramics. *J. Mater. Chem. C* 7: 4999–5008. https://doi.org/10.1039/c8tc06549g.

189 Lu, Z., Bao, W., Wang, G. et al. (2021). Mechanism of enhanced energy storage density in $AgNbO_3$-based lead-free antiferroelectrics. *Nano Energy* 79: 105423. https://doi.org/10.1016/j.nanoen.2020.105423.

190 Li, Q. and Wang, Q. (2016). Ferroelectric polymers and their energy-related applications. *Macromol. Chem. Phys.* 217: 1228–1244. https://doi.org/10.1002/macp.201500503.

191 Prateek, V.K. and Thakur, R.K. (2016). Gupta, recent progress on ferroelectric polymer-based nanocomposites for high energy density capacitors: synthesis, dielectric properties, and future aspects. *Chem. Rev.* 116: 4260–4317. https://doi.org/10.1021/acs.chemrev.5b00495.

192 Rabuffi, M. and Picci, G. (2002). Status quo and future prospects for metallized polypropylene energy storage capacitors. *IEEE Trans. Plasma Sci.* 30: 1939–1942. https://doi.org/10.1109/TPS.2002.805318.

193 Zhu, L. and Wang, Q. (2012). Novel ferroelectric polymers for high energy density and low loss dielectrics. *Macromolecules* 45: 2937–2954. https://doi.org/10.1021/ma2024057.

194 Wang, X., Sun, F., Yin, G. et al. (2018). Tactile-sensing based on flexible PVDF nanofibers via electrospinning: a review. *Sensors* 18: https://doi.org/10.3390/s18020330.

195 Gong, H.H., Zhang, Y., Cheng, Y.P. et al. (2021). The application of controlled/living radical polymerization in modification of PVDF-based fluoropolymer. *Chin. J. Polym. Sci.* 39: 1110–1126. https://doi.org/10.1007/s10118-021-2616-x.

196 Marwat, M.A., Ma, W., Fan, P. et al. (2020). Ultrahigh energy density and thermal stability in sandwich-structured nanocomposites with dopamine@Ag@$BaTiO_3$. *Energy Storage Mater.* 31: 492–504. https://doi.org/10.1016/j.ensm.2020.06.030.

197 Li, J., Meng, Q., Li, W., and Zhang, Z. (2011). Influence of crystalline properties on the dielectric and energy storage properties of poly(vinylidene fluoride). *J. Appl. Polym. Sci.* 122: 1659–1668. https://doi.org/10.1002/app.34020.

198 Guan, F., Pan, J., Wang, J. et al. (2010). Crystal orientation effect on electric energy storage in poly(vinylidene fluoride-*co*-hexafluoropropylene) copolymers. *Macromolecules* 43: 384–392. https://doi.org/10.1021/ma901921h.

199 Feng, Q.K., Zhong, S.L., Pei, J.Y. et al. (2022). Recent progress and future prospects on all-organic polymer dielectrics for energy storage capacitors. *Chem. Rev.* 122: 3820–3878. https://doi.org/10.1021/acs.chemrev.1c00793.

200 Li, W., Meng, Q., Zheng, Y. et al. (2010). Electric energy storage properties of poly(vinylidene fluoride). *Appl. Phys. Lett.* 96: 192905. https://doi.org/10.1063/1.3428656.

201 Ren, X., Meng, N., Zhang, H. et al. (2020). Giant energy storage density in PVDF with internal stress engineered polar nanostructures. *Nano Energy* 72: 104662. https://doi.org/10.1016/j.nanoen.2020.104662.

202 Xie, H., Wang, L., Gao, X. et al. (2020). High breakdown strength and energy density in multilayer-structured ferroelectric composite. *ACS Omega* 5: 32660–32666. https://doi.org/10.1021/acsomega.0c05031.

203 Lovinger, A.J., Furukawa, T., Davis, G.T., and Broadhurst, M.G. (1983). Crystallographic changes characterizing the Curie transition in three ferroelectric copolymers of vinylidene fluoride and trifluoroethylene: 2. Oriented or poled samples. *Polymer* 24: 1233–1239. https://doi.org/10.1016/0032-3861(83)90051-4.

204 Zhou, X., Zhao, X., Suo, Z. et al. (2009). Electrical breakdown and ultrahigh electrical energy density in poly(vinylidene fluoride-hexafluoropropylene) copolymer. *Appl. Phys. Lett.* 94: 162901. https://doi.org/10.1063/1.3123001.

205 Chu, B., Zhou, X., Ren, K. et al. (2006). A dielectric polymer with high electric energy density and fast discharge speed. *Science* 313: 334–336. https://doi.org/10.1126/science.1127798.

206 Zhang, Z. and Chung, T.C.M. (2007). Study of VDF/TrFE/CTFE terpolymers for high pulsed capacitor with high energy density and low energy loss. *Macromolecules* 40: 783–785. https://doi.org/10.1021/ma0627119.

207 Marwat, M.A., Xie, B., Zhu, Y. et al. (2019). Largely enhanced discharge energy density in linear polymer nanocomposites by designing a sandwich structure. *Compos. Part A Appl. Sci. Manuf.* 121: 115–122. https://doi.org/10.1016/j.compositesa.2019.03.016.

208 Zhang, Z., Wang, D.H., Litt, M.H. et al. (2018). High-temperature and high-energy-density dipolar glass polymers based on sulfonylated poly(2,6-dimethyl-1,4-phenylene oxide). *Angew. Chem. Int. Ed.* 57: 1528–1531. https://doi.org/10.1002/anie.201710474.

209 Wu, S., Li, W., Lin, M. et al. (2013). Aromatic polythiourea dielectrics with ultrahigh breakdown field strength, low dielectric loss, and high electric energy density. *Adv. Mater.* 25: 1734–1738. https://doi.org/10.1002/adma.201204072.

210 Marwat, M.A., Xie, B., Zhu, Y. et al. (2019). Sandwich structure-assisted significantly improved discharge energy density in linear polymer nanocomposites with high thermal stability. *Colloids Surf. A Physicochem. Eng. Asp.* 581: 123802. https://doi.org/10.1016/j.colsurfa.2019.123802.

211 Zhang, H., Marwat, M.A., Xie, B. et al. (2020). Polymer matrix nanocomposites with 1D ceramic nanofillers for energy storage capacitor applications. *ACS Appl. Mater. Interfaces* 12: 1–37. https://doi.org/10.1021/acsami.9b15005.

212 Li, H., Zhou, Y., Liu, Y.Y. et al. (2021). Dielectric polymers for high-temperature capacitive energy storage. *Chem. Soc. Rev.* 50: 6369–6400. https://doi.org/10.1039/d0cs00765j.

213 Ju, T., Chen, X., Langhe, D. et al. (2022). Enhancing breakdown strength and lifetime of multilayer dielectric films by using high temperature polycarbonate skin layers. *Energy Storage Mater.* 45: 494–503. https://doi.org/10.1016/j.ensm.2021.12.009.

214 Wang, Q., Hou, Y., Ma, S. et al. (2022). Study on insulating properties of polyethylene terephthalate/montmorillonite nanocomposites. *AIP Adv.* 12: 15214. https://doi.org/10.1063/5.0077665.

215 Treufeld, I., Wang, D.H., Kurish, B.A. et al. (2014). Enhancing electrical energy storage using polar polyimides with nitrile groups directly attached to the main chain. *J. Mater. Chem. A* 2: 20683–20696. https://doi.org/10.1039/c4ta03260h.

216 Ali, U., Karim, K.J.B.A., and Buang, N.A. (2015). A review of the properties and applications of poly (methyl methacrylate) (PMMA). *Polym. Rev.* 55: 678–705. https://doi.org/10.1080/15583724.2015.1031377.

217 Zhang, X.M., Liu, J.G., and Yang, S.Y. (2016). A review on recent progress of R&D for high-temperature resistant polymer dielectrics and their applications in electrical and electronic insulation. *Rev. Adv. Mater. Sci.* 46: 22–38.

218 Ritamaki, M., Rytoluoto, I., and Lahti, K. (2019). Performance metrics for a modern BOPP capacitor film. *IEEE Trans. Dielectr. Electr. Insul.* 26: 1229–1237. https://doi.org/10.1109/TDEI.2019.007970.

219 Amaral, F., Rubinger, C.P.L., Henry, F. et al. (2008). Dielectric properties of polystyrene-CCTO composite. *J. Non-Cryst. Solids* 354: 5321–5322. https://doi.org/10.1016/j.jnoncrysol.2008.05.056.

220 Rahimabady, M., Yao, K., Arabnejad, S. et al. (2012). Intermolecular interactions and high dielectric energy storage density in poly(vinylidene fluoride-hexafluoropropylene)/poly(vinylidene fluoride) blend thin films. *Appl. Phys. Lett.* 100: 252907. https://doi.org/10.1063/1.4730603.

221 Mao, P., Wang, J., Zhang, L. et al. (2020). Tunable dielectric polarization and breakdown behavior for high energy storage capability in P(VDF-TrFE-CFE)/PVDF polymer blended composite films. *Phys. Chem. Chem. Phys.* 22: 13143–13153. https://doi.org/10.1039/d0cp01071e.

222 Meng, N., Zhu, X., Mao, R. et al. (2017). Nanoscale interfacial electroactivity in PVDF/PVDF-TrFE blended films with enhanced dielectric and ferroelectric properties. *J. Mater. Chem. C* 5: 3296–3305. https://doi.org/10.1039/c7tc00162b.

223 Zhang, X., Shen, Y., Shen, Z. et al. (2016). Achieving high energy density in PVDF-based polymer blends: suppression of early polarization saturation and enhancement of breakdown strength. *ACS Appl. Mater. Interfaces* 8: 27236–27242. https://doi.org/10.1021/acsami.6b10016.

224 Chen, X.Z., Li, X., Qian, X.S. et al. (2013). A polymer blend approach to tailor the ferroelectric responses in P(VDF-TrFE) based copolymers. *Polymer* 54: 2373–2381. https://doi.org/10.1016/j.polymer.2013.02.041.

225 Ahmad, A., Tong, H., Fan, T., and Xu, J. (2021). All-organic polymer blend dielectrics of poly (arylene ether urea) and polyimide: toward high energy density and high temperature applications. *J. Polym. Sci.* 59: 1414–1423. https://doi.org/10.1002/pol.20200725.

226 Zhang, Q.M., Chen, X., Zhang, T., and Zhang, Q.M. (2019). Giant permittivity materials with low dielectric loss over a broad temperature range enabled by weakening intermolecular hydrogen bonds. *Nano Energy* 64: 103916. https://doi.org/10.1016/j.nanoen.2019.103916.

227 Zhou, Y., Li, Q., Dang, B. et al. (2018). High-throughput, and environmentally benign approach to polymer dielectrics exhibiting significantly improved capacitive performance at high temperatures. *Adv. Mater.* 30: 1805672. https://doi.org/10.1002/adma.201805672.

228 Sparks, M., Mills, D.L., Warren, R. et al. (1981). Theory of electron-avalanche breakdown in solids. *Phys. Rev. B* 24: 3519–3536. https://doi.org/10.1103/PhysRevB.24.3519.

229 Tanaka, T. (1986). Internal partial discharge and material degradation. *IEEE Trans. Electr. Insul.* EI-21: 899–905. https://doi.org/10.1109/TEI.1986.348999.

230 Fothergill, J.C. (1991). Filamentary electromechanical breakdown. *IEEE Trans. Electr. Insul.* 26: 1124–1129. https://doi.org/10.1109/14.108149.

231 Shen, Z.H., Wang, J.J., Jiang, J.Y. et al. (2019). Phase-field modeling and machine learning of electric-thermal-mechanical breakdown of polymer-based dielectrics. *Nat. Commun.* 10: 1843. https://doi.org/10.1038/s41467-019-09874-8.

232 Cygan, S. and Laghari, J.R. (1987). Dependence of the electric strength on thickness area and volume of polypropylene. *IEEE Trans. Electr. Insul.* 2–4. https://doi.org/10.1109/TEI.1987.298949.

233 Tan, D.Q. (2020). The search for enhanced dielectric strength of polymer-based dielectrics: a focused review on polymer nanocomposites. *J. Appl. Polym. Sci.* 137: 49379. https://doi.org/10.1002/app.49379.

234 Kirby, A.J. (1992). Polyimide (PI) plastic: uses, structure, properties.

235 Vanherck, K., Koeckelberghs, G., and Vankelecom, I.F.J. (2013). Crosslinking polyimides for membrane applications: a review. *Prog. Polym. Sci.* 38: 874–896. https://doi.org/10.1016/j.progpolymsci.2012.11.001.

236 Ho, J.S. and Greenbaum, S.G. (2018). Polymer capacitor dielectrics for high temperature applications. *ACS Appl. Mater. Interfaces* 10: 29189–29218. https://doi.org/10.1021/acsami.8b07705.

237 Venkat, N., Dang, T.D., Bai, Z. et al. (2010). High temperature polymer film dielectrics for aerospace power conditioning capacitor applications. *Mater. Sci. Eng. B Solid-State Mater. Adv. Technol.* 168: 16–21. https://doi.org/10.1016/j.mseb.2009.12.038.

238 Guo, R., Luo, H., Yan, M. et al. (2021). Significantly enhanced breakdown strength and energy density in sandwich-structured nanocomposites with low-level $BaTiO_3$ nanowires. *Nano Energy* 79: 105412. https://doi.org/10.1016/j.nanoen.2020.105412.

239 Hu, J., Zhang, S., and Tang, B. (2021). 2D filler-reinforced polymer nanocomposite dielectrics for high-k dielectric and energy storage applications. *Energy Storage Mater.* 34: 260–281. https://doi.org/10.1016/j.ensm.2020.10.001.

240 Li, L., Cheng, J., Cheng, Y. et al. (2021). Significantly enhancing the dielectric constant and breakdown strength of linear dielectric polymers by utilizing ultralow loadings of nanofillers. *J. Mater. Chem. A* 9: 23028–23036. https://doi.org/10.1039/d1ta05408b.

241 Wang, C., He, G., Chen, S. et al. (2021). Enhanced performance of all-organic sandwich structured dielectrics with linear dielectric and ferroelectric polymers. *J. Mater. Chem. A* 9: 8674–8684. https://doi.org/10.1039/d1ta00974e.

242 Zhang, Y., Zhang, C., Feng, Y. et al. (2019). Energy storage enhancement of P(VDF-TrFE-CFE)-based composites with double-shell structured BZCT nanofibers of parallel and orthogonal configurations. *Nano Energy* 66: 104195. https://doi.org/10.1016/j.nanoen.2019.104195.

243 Qiao, R., Wang, C., Chen, S. et al. (2022). High-temperature dielectric polymers with high breakdown strength and energy density via constructing the electron traps in blends. *Compos. Part A Appl. Sci. Manuf.* 152: 106679. https://doi.org/10.1016/j.compositesa.2021.106679.

244 Sun, Q., Wang, J., Sun, H. et al. (2021). Simultaneously enhanced energy density and discharge efficiency of layer-structured nanocomposites by reasonably designing dielectric differences between $BaTiO_3$@SiO_2/PVDF layers and BNNSs/PVDF-PMMA layers. *Compos. Part A Appl. Sci. Manuf.* 149: 106546. https://doi.org/10.1016/j.compositesa.2021.106546.

245 Chi, Q., Zhou, Y., Feng, Y. et al. (2020). Excellent energy storage performance of polyetherimide filled by oriented nanofibers with optimized diameters. *Mater. Today Energy* 18: 100516. https://doi.org/10.1016/j.mtener.2020.100516.

246 Sun, Q., Wang, J., Zhang, L. et al. (2020). Achieving high energy density and discharge efficiency in multi-layered PVDF-PMMA nanocomposites composed of 0D $BaTiO_3$ and 1D $NaNbO_3$@SiO_2. *J. Mater. Chem. C* 8: 7211–7220. https://doi.org/10.1039/d0tc00838a.

247 Liu, Y., Hou, Y., Ji, Q. et al. (2020). Modulation of individual-layer properties results in excellent discharged energy density of sandwich-structured composite films. *J. Mater. Sci.: Mater. Electron.* 31: 7663–7671. https://doi.org/10.1007/s10854-020-03302-0.

248 Li, L., Zhou, Y., Liu, Y. et al. (2022). Perspective on scalable high-energy-density polymer dielectrics with ultralow loadings of inorganic nanofillers. *Appl. Phys. Lett.* 120: 50502. https://doi.org/10.1063/5.0080825.

249 Zhang, Y., Li, S., Cong, X. et al. (2022). Interesting influence of different inorganic particles on the energy storage performance of a polyethersulfone-based dielectric composite. *ACS Appl. Energy Mater.* 5: 3545–3557. https://doi.org/10.1021/acsaem.1c04075.

250 Su, Y., Chen, C., Pan, H. et al. (2021). Muscle fibers inspired high-performance piezoelectric textiles for wearable physiological monitoring. *Adv. Funct. Mater.* 31: 2010962. https://doi.org/10.1002/adfm.202010962.

251 Bi, K., Bi, M., Hao, Y. et al. (2018). Ultrafine core-shell $BaTiO_3@SiO_2$ structures for nanocomposite capacitors with high energy density. *Nano Energy* 51: 513–523. https://doi.org/10.1016/j.nanoen.2018.07.006.

252 Song, Y., Shen, Y., Liu, H. et al. (2012). Enhanced dielectric and ferroelectric properties induced by dopamine-modified $BaTiO_3$ nanofibers in flexible poly(vinylidene fluoride-trifluoroethylene) nanocomposites. *J. Mater. Chem.* 22: 8063–8068. https://doi.org/10.1039/c2jm30297g.

253 Guo, N., DiBenedetto, S.A., Tewari, P. et al. (2010). Nanoparticle, size, shape, and interfacial effects on leakage current density, permittivity, and breakdown strength of metal oxide-polyolefin nanocomposites: experiment and theory. *Chem. Mater.* 22: 1567–1578. https://doi.org/10.1021/cm902852h.

254 Wang, G., Huang, X., and Jiang, P. (2017). Bio-inspired polydopamine coating as a facile approach to constructing polymer nanocomposites for energy storage. *J. Mater. Chem. C* 5: 3112–3120. https://doi.org/10.1039/c7tc00387k.

255 Tang, H., Lin, Y., and Sodano, H.A. (2013). Synthesis of high aspect ratio $BaTiO_3$ nanowires for high energy density nanocomposite capacitors. *Adv. Energy Mater.* 3: 451–456. https://doi.org/10.1002/aenm.201200808.

256 Pan, Z., Yao, L., Zhai, J. et al. (2017). Ultrafast discharge and enhanced energy density of polymer nanocomposites loaded with $0.5(Ba_{0.7}Ca_{0.3})TiO_3$-$0.5Ba(Zr_{0.2}Ti_{0.8})O_3$ one-dimensional nanofibers. *ACS Appl. Mater. Interfaces* 9: 14337–14346. https://doi.org/10.1021/acsami.7b01381.

257 Pan, Z., Liu, J., Xu, J. et al. (2022). Two-dimensional fillers induced superior electrostatic energy storage performance in trilayered architecture nanocomposites. *ACS Appl. Mater. Interfaces* 14: 8448–8457. https://doi.org/10.1021/acsami.1c23086.

258 Chen, J., Shen, Z., Kang, Q. et al. (2022). Chemical adsorption on 2D dielectric nanosheets for matrix free nanocomposites with ultrahigh electrical energy storage. *Sci. Bull.* 67: 609–618. https://doi.org/10.1016/j.scib.2021.10.011.

259 Wen, F., Lou, H., Ye, J. et al. (2019). Preparation and energy storage performance of transparent dielectric films with two-dimensional platelets. *Compos. Sci. Technol.* 182: 107759. https://doi.org/10.1016/j.compscitech.2019.107759.

260 Zhu, Y., Zhu, Y., Huang, X. et al. (2019). High energy density polymer dielectrics interlayered by assembled boron nitride nanosheets. *Adv. Energy Mater.* 9: 1901826. https://doi.org/10.1002/aenm.201901826.

261 Liu, S., Wang, J., Shen, B. et al. (2017). Poly(vinylidene fluoride) nanocomposites with a small loading of core-shell structured $BaTiO_3@Al_2O_3$ nanofibers exhibiting high discharged energy density and efficiency. *J. Alloys Compd.* 696: 136–142. https://doi.org/10.1016/j.jallcom.2016.11.186.

262 Liu, S., Wang, J.J., Wang, J.J. et al. (2017). Core-shell structured $BaTiO_3@SiO_2$ nanofibers for poly(vinylidene fluoride) nanocomposites with high discharged energy. *Mater. Lett.* 189: 176–179. https://doi.org/10.1016/j.matlet.2016.12.008.

263 Wang, G., Huang, Y., Wang, Y. et al. (2017). Substantial enhancement of energy storage capability in polymer nanocomposites by encapsulation of $BaTiO_3$ NWs with variable shell thickness. *Phys. Chem. Chem. Phys.* 19: 21058–21068. https://doi.org/10.1039/c7cp04096b.

264 Liu, S., Wang, J., Shen, B., and Zhai, J. (2017). Enhanced discharged energy density and efficiency of poly(vinylidene fluoride) nanocomposites through a small loading of core-shell structured $BaTiO_3$@Al_2O_3 nanofibers. *Ceram. Int.* 43: 585–589. https://doi.org/10.1016/j.ceramint.2016.09.198.

265 Pan, Z., Zhai, J., and Shen, B. (2017). Multilayer hierarchical interfaces with high energy density in polymer nanocomposites composed of $BaTiO_3$@TiO_2@Al_2O_3 nanofibers. *J. Mater. Chem. A* 5: 15217–15226. https://doi.org/10.1039/c7ta03846a.

266 Kang, D., Wang, G., Huang, Y. et al. (2018). Decorating TiO_2 nanowires with $BaTiO_3$ nanoparticles: a new approach leading to substantially enhanced energy storage capability of high-k polymer nanocomposites. *ACS Appl. Mater. Interfaces* 10: 4077–4085. https://doi.org/10.1021/acsami.7b16409.

267 Li, Q., Chen, L.L.Q., Gadinski, M.R. et al. (2015). Flexible high-temperature dielectric materials from polymer nanocomposites. *Nature* 523: 576–579. https://doi.org/10.1038/nature14647.

268 Wang, Y., Chen, J., Li, Y. et al. (2019). Multilayered hierarchical polymer composites for high energy density capacitors. *J. Mater. Chem. A* 7: 2965–2980. https://doi.org/10.1039/C8TA11392K.

269 Castro, M., Kumar, B., Feller, J.F. et al. (2011). Novel e-nose for the discrimination of volatile organic biomarkers with an array of carbon nanotubes (CNT) conductive polymer nanocomposites (CPC) sensors. *Sens. Actuators, B* 159: 213–219. https://doi.org/10.1016/j.snb.2011.06.073.

270 Hao, X. (2013). A review on the dielectric materials for high energy-storage application. *J. Adv. Dielectr.* 03: 1330001. https://doi.org/10.1142/s2010135x13300016.

15

Electrocaloric Effect in Ferroelectric Ceramics

Guangzu Zhang, Kailun Zou, and Shenglin Jiang

Huazhong University of Science and Technology, School of Integrated Circuits, Engineering Research Center for Functional Ceramics MOE, Wuhan National Laboratory for Optoelectronics, Luoyu Road, Wuhan 430074, China

15.1 Introduction

Refrigeration technology is widely used in the military, industry, and daily life [1–3]. Currently, refrigeration heavily relies on vapor compression technology, which can't get rid of using environmentally harmful hydrofluorocarbons (HFCs) refrigerants. According to the "United Nations Framework Convention on Climate Change", HFCs, whose greenhouse effect potential (GWP) is greater than 1000, belong to the controlled gas. And the greenhouse effect of HFCs is even three times higher than that of CO_2 gas [4]. It is predicted that by 2050, the direct CO_2 emission equivalent of HFCs gaseous refrigerants will reach 45% of the total CO_2 emissions [5]. The over-reliance on HFCs for compressor refrigeration technology will exacerbate global warming. Furthermore, the low efficiency of the compressor refrigeration causes a large amount of power loss [6]. In addition, the compressor is bulky and heavy, which is unable to satisfy the local cooling such as integrated circuit chips in modern electronics. Therefore, it's urgent to develop new high-efficiency and eco-friendly refrigeration technology with a small size and light weight.

Novel refrigeration technologies based on thermoelectric effect (TE), magnetocaloric effect (MCE), electrocaloric effect (ECE), elastocaloric effect (eCE), and barocaloric effect (BCE) have been developed. Although thermoelectric cooling devices based on the Peltier effect have been commercialized, they suffer from low cooling efficiency [7–9]. Refrigeration technology based on caloric effects (MCE [10], ECE [3], eCE [11], and BCE [12]) utilizes the changes of latent heat during solid-state phase transitions to realize heat transportation and cooling. The process of the phase transition is induced by magnetic field, electric field, uniaxial stress, and isostatic pressure, respectively. Since it's easier to produce and manipulate electric field stimuli than to produce magnetic and mechanical stimuli, much more attention has been paid to the ECE generated in ferroelectrics [2, 13]. Refrigeration based on ECE eliminates the use of a compressor and fluorinated refrigerant

Piezoelectric Materials: From Fundamentals to Emerging Applications, First Edition. Edited by Jiagang Wu.

and features low energy consumption, small size, and lightweight, becoming a promising supplement to conventional vapor-compression refrigeration and being a candidate for on-chip cooling.

ECE refers to the reversible adiabatic entropy change (ΔS) and isothermal temperature change (ΔT) arising from domain switching of ferroelectric materials under the application or removal of an external electric field. The first discovery of the ECE can be traced back to 1930 when Kurtschatov and Kobeko discovered it in Rochelle salt [14]. However, no specific ΔS and ΔT were reported. It was not until 1963 that Wiseman et al. measured the ECE in Rochelle salt, obtaining a ΔT of 0.0036 K under 1.4 kV/cm and 295 K [15]. The research interest in ECE of ferroelectric ceramics has risen in 1960s, when the ECE in bulk ceramics such as $BaTiO_3$ (BT) [16] and $Pb(Zr_{0.65}Ti_{0.35})O_3$ [17] has been mainly focused on. However, due to the restricted preparation technology at that time, the electric breakdown strength (E_b) of the bulk ceramics is generally lower than 20 kV/cm, making the ΔT of the bulk ceramics hard to exceed 1 K [3]. The low ΔT makes the ECE in ferroelectrics not receive much attention. It was not until the early twenty-first century that the high ΔT of 12 K was achieved in the $PbZr_{0.95}Ti_{0.05}O_3$ ferroelectric ceramic films and ferroelectric poly(vinylidene fluoride-trifluoroethylene) (P(VDF-TrFE)) copolymer [18, 19], igniting the enthusiasm of researchers. To make EC cooling toward practical application, tremendous efforts have been devoted to developing EC materials with high ECE. Nowadays, the research into the ECE of inorganic ferroelectrics (bulk ceramics, single crystals, ceramic films, and multilayer ceramics) is blossoming, and the EC cooling devices based on these EC materials are under active development. This chapter mainly introduces the research progress of the EC ceramics and EC cooling devices and summarizes the methods to improve their respective ECE.

15.2 ECE in Ferroelectric Ceramics

15.2.1 ECE in Ceramic Films

There are two kinds of test methods for the measurement of ECE, namely by direct method and indirect method. Direct method, i.e. using temperature sensor (thermocouple [20] and infrared temperature sensor [21, 22]), heat flow sensor [23], differential scanning calorimeter (DSC) [24, 25], thermometer (Th) [26–28] and other sensors or instruments to directly measure the ΔS or ΔT of EC materials under the stimulation of the electric fields. These kinds of methods can directly provide reliable measurement of ECE. However, it's still challenging to detect the ΔT of the thin films with substrates by direct method. The indirect method, which has been expounded exhaustively in the previous reviews [29], is used in the ECE measurements for ferroelectric thin films. The formulas for calculating ΔT and ΔS derived from the Maxwell relation are as follows:

$$\Delta T = -\frac{1}{\rho}\int_{E_1}^{E_2} \frac{T}{C_E}\left(\frac{\partial P}{\partial T}\right)_E dE \tag{15.1}$$

$$\Delta S = -\frac{1}{\rho}\int_{E_1}^{E_2}\left(\frac{\partial P}{\partial T}\right)_E dE \tag{15.2}$$

where E_1 and E_2 are the initial electric field and final electric field respectively, T is the test temperature, C_E is the specific heat capacity, and $\frac{\partial P}{\partial T}$ is the pyroelectric coefficient under a constant electric field. In 2006, the researchers used the indirect method to calculate the ECE of the $PbZr_{0.95}Ti_{0.05}O_3$ (PZT) thin film and obtained a ΔT of 12 K [18]. As indicated by formulas (15.1) and (15.2), the high electric breakdown strength (E_b) is one of the keys to improving the ECE. The relationship between the E_b and the thickness (h) of the EC materials can be expressed as [30]:

$$E_b \propto h^{-0.39} \tag{15.3}$$

Since the thickness of ceramic thin films is generally less than 1.0 μm, the ceramic thin films can withstand higher electric fields and thus larger ECE can be induced.

At present, the most effective way to improve ECE of the ferroelectric thin films is to tune the phase structure by changing the compositions. For instance, in $(1-x)Pb(Mg_{1/3}Nb_{2/3})O_3$–$x$$PbTiO_3$ (PMN–xPT) ceramic thin films, by adjusting the content of $PbTiO_3$ (PT), two-phase and three-phase coexistence can be constructed under specific electric fields and temperatures [31]. Figure 15.1 shows an E–T–xPT three-dimensional phase diagram with an invariant critical point (ICP), where E represents the electric field, T represents the temperature, and xPT represents the content of PT in PMN–xPT. The orange area except the B-ICP line segment in Figure 15.1 belongs to the two-phase coexistence area, where B is the three-phase coexistence point under zero electric field. The B-ICP line segment in Figure 15.1 is a three-phase coexistence region. Compared with other regions, the three-phase coexistence region has more types of phase structures and the transition energy between different phases is lower, which is conducive to the generation of a larger ECE. Based on this phase diagram, the phase transition behavior and the ECE of 0.67PMN–0.33PT [32], 0.9PMN–0.1PT [33], and 0.93PMN–0.07PT [34] were

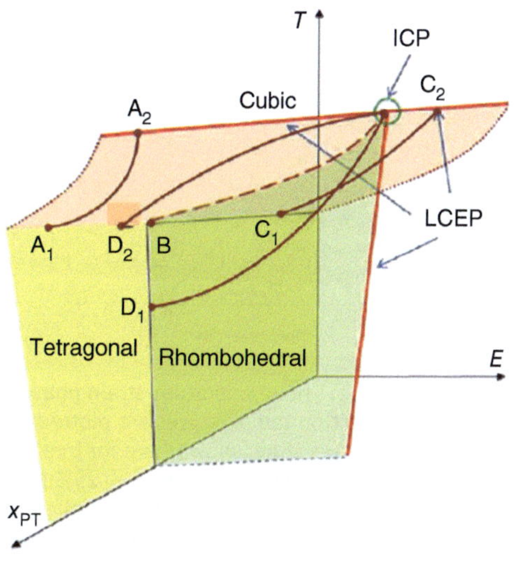

Figure 15.1 Schematic of the E–T–x_{PT} three-dimensional phase diagram with an ICP point (E represents an electric field, T represents temperature, x_{PT} represents the content of $PbTiO_3$ in PMN–xPT). Source: Liu et al. [31]. Reproduced with permission of American Institute of Physics.

studied. In addition, adjusting the phase structure by components to improve ECE has also been applied in $PbZr_xTi_{1-x}O_3$ (PZT) and $BaTiO_3$ (BT)-based thin films [35–39].

In addition to the phase structure, the substrate is another factor that significantly affects the ECE of the thin films. The following aspects need to be considered in selection of the substrate of the EC thin films: (i) The stress caused by the lattice mismatch between the substrate and the thin films; (ii) The influence of the substrate buffer layer on the orientation degree and the quality of the thin films; (iii) The flexible substrate can be used to endow the ceramic thin films with flexibility. Through thermodynamic simulation, it was found that different mismatch stress (compressive stress, stress-free, and tensile stress) was generated between the substrates with different unit cell parameters and the BT thin films. When the stress evolved from compressive stress to tensile stress, the temperature corresponding to the maximum ΔT (ΔT_{max}) of the BT film shifted to a lower temperature [40]. A similar phenomenon was found in the simulation of $PbTiO_3$ films [41]. Figure 15.2a illustrates the phase diagram of BT thin films under different mismatch stresses through molecular dynamics and first-principle calculations, where *c*, *r*, and *aa* are three types of ferroelectric phases with different polarization orientations and *p* is the paraelectric phase [42]. As shown in Figure 15.2b, the ferroelectric-paraelectric phase transition point moves to a higher temperature with the increase of mismatch compressive stress, making the ΔT_{max} shift to a higher temperature. The phase diagram in Figure 15.2a reveals the movement law of BT thin films' phase boundary under mismatch stress.

In addition to the stress caused by the lattice mismatch, laying a suitable buffer layer on the substrate can be utilized to improve the ECE of the films. For instance, by adding a buffer layer $La_{0.5}Sr_{0.5}CoO_3$ on the (001) oriented $Ir/SrTiO_3/TiN/Si$ substrate, the 0.67PMN–0.33PT thin films exhibited a higher degree of orientation than that without buffer layer [43]. Moreover, the pyrochlore phase is eliminated in 0.67PMN–0.33PT thin films with a buffer layer so that the phase structure

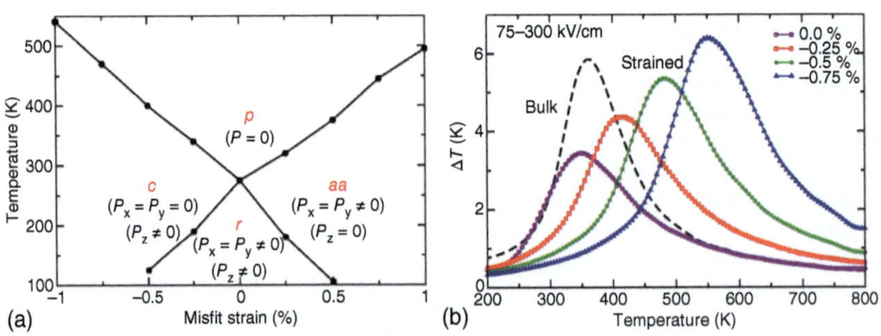

Figure 15.2 (a) The temperature-strain phase diagram for epitaxially strained BT thin films: the transition temperatures are plotted as a function of misfit strain. (b) The ΔT as a function of temperature: shown here for free (dashed line) and strained bulk $BaTiO_3$ with misfit strain equal to 0.0% (circles), −0.25% (squares), −0.5% (stars), and −0.75% (triangles). ΔT is calculated for a variation of the applied field from 75 to 300 kV/cm. Source: Marathe and Ederer [42]. Reproduced with permission of American Institute of Physics.

is purer. The ΔT of the 0.67PMN–0.33PT thin film with a buffer layer is nearly double than that of the 0.67PMN–0.33PT thin film without a buffer layer. This enhancement in ΔT was attributed to the reduction of the residual tensile stress at the substrate after adding the buffer layer and the improvement of the crystallinity and degree of orientation. Similarly, the introduction of $Pb(Zr_{0.3}Ti_{0.7})O_3/PbO_x$ as a buffer layer on the substrate of 0.68PMN–0.32PT thin films played a significant role in improving the ECE [44]. Therefore, introducing a suitable buffer layer can enhance the orientation degree and the quality of the thin films, which is beneficial to improving the ECE. For the fabrication of inorganic thin films, hard substrates such as Si or SiO_2 are generally adopted, making the EC thin films lose flexibility and difficult to be applied to flexible electronic devices. The fluorphlogopite sheets are flexible single-crystal insulating materials that could impart flexibility to thin films [45, 46]. In recent years, researchers have prepared flexible $Pb_{0.82}Ba_{0.08}La_{0.1}Zr_{0.9}Ti_{0.1}O_3$ [47], $Pb_{0.91}La_{0.09}(Zr_{0.65}Ti_{0.35})_{0.9775}O_3$ [48], and Mn-modified $0.65(0.94Na_{0.5}Bi_{0.5}TiO_3$–$0.06BaTiO_3)$–$0.35\ SrTiO_3$ thin films [49] using the fluorphlogopite sheets as substrates. All these flexible thin films obtained a ΔT greater than 12.0 K, which expands the application range of inorganic thin films in the field of EC cooling.

In addition to the substrate's role in regulating the ECE of films, the thickness, defects, and composition gradients of the films are also significant factors to optimize the ECE. Based on the phenomenological Landau-Devonshire thermodynamic theory, the researchers found that the ECE of epitaxial $Ba_{0.6}Sr_{0.4}TiO_3$ thin films strongly depends on the film thickness and a large ECE can be generated near the optimal thickness and room temperature [50]. Then the researchers extended the study on the film thickness to the bilayer $BaTiO_3/SrTiO_3$ films [51]. It was calculated that the orientation of the bilayer films and the thickness ratio between the bilayer films jointly determine the maximum ECE of the bilayer films. The ECE reaches a maximum value at the critical relative thickness of the $SrTiO_3$ layer of 50%, 23%, and 12% for (001), (110), and (111) oriented $BaTiO_3/SrTiO_3$ bilayer thin films, respectively. Moreover, the (110) oriented $BaTiO_3/SrTiO_3$ bilayer thin films have the largest ECE. For the defects in the thin films, the volatilization of oxides (such as PbO) and the ion substitution are often used to construct defect dipoles in the thin film to adjust the ECE. For instance, the researchers deposited the $PbZrO_3$ thin films on pretreated Pt/Si substrates by the sol-gel method. The treated substrates have a porous electrode surface, which can enhance the interface effect between the thin film and the Pt electrode. The porous electrode surface makes the lead oxide at the bottom interface of the $PbZrO_3$ thin films more volatile to form lead vacancies during the high-temperature preparation process and form a defect dipole with the oxygen vacancies [52]. Compared with ordinary dipoles, a larger electric field is required to drive the flip of defect dipoles. So the introduction of defect dipoles in the $PbZrO_3$ films increases the antiferroelectric-ferroelectric phase transition electric fields by nearly two times, which greatly enhances the negative ECE. A giant negative ECE up to −18.5 K was obtained by the indirect method. Moreover, in the $Na_{0.5}Bi_{0.5}(Ti_{0.97}W_{0.01}Fe_{0.02})O_3$ thin films, researchers adopted Fe^{3+} to partially replace Ti^{4+} to construct negatively charged Fe'_{Ti} defects, which formed a defect dipole with the positively charged oxygen vacancies under the electrostatic forces [53]. Due to the decoupling characteristics of defect dipoles during the heating process, the polarization clamped by the defect dipoles gradually increases with the increase of temperature and a favorable negative ECE was achieved near room temperature. For the regulation of composition gradient, researchers designed the $PbZr_xTi_{1-x}O_3$ stacked films with an increasing gradient ($x_1 = 0.52$, $x_2 = 0.80$, $x_3 = 0.95$) and decreasing gradient ($x_1 = 0.95$, $x_2 = 0.80$, $x_3 = 0.52$), where x_i represents the Zr content of the i-th layer PZT film [36]. Different stacking orders of the stacked films will produce different mismatch stresses. Similar to the mismatch stress between the substrate and the film, the stress caused by the composition gradient could change the phase boundary temperature of the films, which in turn affects the temperature corresponding to the ΔT_{max}.

Ferroelectric thin films generally possess large ΔT_{max} due to their high electric breakdown strength. In recent years, the ECE of the thin films has been greatly improved by adjusting the compositions, substrates, and structures. The ΔT_{max} of the thin films is generally greater than 10 K, which has demonstrated great potential in the field of EC refrigeration.

15.2.2 ECE in Bulk Ceramics and Single Crystals

Compared with ferroelectric thin films, bulk ceramics are several orders of magnitude thicker and difficult to withstand higher operating electric fields, which limits the improvement of ECE. However, the higher thickness enables bulk ceramics to absorb more heat during the EC refrigeration process. Moreover, the preparation process and the composition control of the bulk ceramics are relatively easy. Many efforts have been made to optimize the ECE of bulk ceramics through composition control. According to the formula (15.1) and (15.2), improving the pyroelectric coefficient $\frac{\partial P}{\partial T}$ and the electric breakdown strength of the EC materials can effectively enhance the ECE for ergodic ferroelectrics [54]. From the perspective of thermodynamics, ferroelectric phase transitions can be divided into first-order and second-order phase transitions according to the relationship between spontaneous polarization and temperature. The value of $\frac{\partial P}{\partial T}$ is generally higher at the first-order phase transition than that at second-order phase transition, leading to a larger ECE. The second-order phase transition is more diffuse so that the ΔT is smaller but the temperature window is wider than that at first-order phase transition [55–57]. Ferroelectric materials possess ferroelectric-paraelectric and ferroelectric-ferroelectric phase transitions from the perspective of unit cell structure. Among them, the ferroelectric-paraelectric phase transition occurs at the Curie temperature and this phase transition changes sharply, resulting in a large ΔT. Nevertheless, the working temperature range of ECE is narrow, and the temperature corresponding to the maximum ECE is generally much higher than room temperature [58]. Ferroelectric-ferroelectric phase transition refers to the structural change between different ferroelectric lattices. Through the simultaneous adjustment of the composition and temperature, a region where multiple phases coexist, i.e. morphotropic phase boundary (MPB), polycrystalline phase boundary

(PPT), or invariant critical point (ICP), can be constructed. Reasonable regulation of ferroelectric-ferroelectric phase transition enables the EC materials to produce an ECE that is comparable to the ferroelectric-paraelectric phase transition while maintaining high ECE in a wider temperature range around room temperature. Therefore, regulating the phase structures of the bulk ceramics by changing the composition can effectively optimize the ECE and the operating temperature range, which is the key to designing bulk ceramics with favorable EC performance. What's more, improving the insulating properties of the bulk ceramics to increase the electric breakdown strength is also an effective way to enhance the ECE. The improvement of the ceramics' resistivity can not only enhance the ECE but also reduce the joule heat generated by the bulk ceramics with high electric fields, which is beneficial for the bulk ceramics to absorb more heat. The methods used to improve the insulating properties of the bulk ceramics are as follows: (i) doping the ions to inhibit the growth of grains [58, 59]; (ii) introducing dielectrics with high electric breakdown strength at grain boundaries [26, 60, 61]; (iii) adding appropriate amounts of sintering aids to reduce defects and pores at grain boundaries [62, 63]; (iv) using two-step sintering, spark plasma sintering (SPS) and hot pressing sintering to reduce the sintering temperature of the material and inhibit grain growth and reduce porosity [64–68]. These methods can significantly improve the insulation properties of the bulk ceramics so that the EC materials can work under higher electric fields and the ECE is enhanced. In this section, the EC bulk ceramics are divided into lead-free ceramics (BT-based, BNT-based, KNN-based) and lead-containing ceramics (PZT, PMN–xPT, PST), and the regulation of the phase structures and insulation properties of these ceramics are discussed in detail.

15.2.2.1 ECE in BT-based Bulk Ceramics

Barium titanate (BT) ceramics, which undergo a first-order phase transition from a tetragonal ferroelectric phase to a cubic paraelectric phase at 400 K, possess a strong EC strength. The ΔT of BT single crystal and BT bulk ceramics can reach 4.8 and 1.4 K near the Curie point, respectively [69, 70]. However, the application of BT ceramics in EC cooling is restricted since the first-order phase transition temperature region of BT is narrow and the phase transition temperature is far away from room temperature [69]. The methods of A-site doping, B-site doping, and A/B-site co-doping have been adopted to reduce the phase transition temperature of BT. For instance, $SrTiO_3$ with a paraelectric phase was added into BT ceramic to synthesize binary solid solution $(Ba_{1-x}Sr_x)TiO_3$, and its ECE was studied [58]. When the content of Sr^{2+} is 35%, the temperature corresponding to the ΔT_{max} moves to room temperature. Moreover, the EC temperature span is broadened and the ΔT reaches 0.42 K at 20 kV/cm. In addition to the substitution of Sr^{2+}, researchers also investigated the ECE of $Ba_{0.94}R_{0.04}TiO_3$ bulk ceramics, where R represents the rare earth elements La, Ce, Nd, Sm, Eu, Gd, Dy, and Er [57]. It was found that the latent heat of the $Ba_{0.94}R_{0.04}TiO_3$ decreased with the enhancement of the ionic radius of the rare earth doping at the A site, and the phase transition became more diffuse. When Dy^{3+} and Er^{3+} with low radius were introduced into the A site, higher polarization and higher Curie temperature can be obtained, which was mainly attributed to the enhancement of lattice tetragonality caused by the substitution of Ba^{2+} with the rare earth elements. The samples show a favorable EC strength of ~0.35 K m/MV over a wide temperature range from room temperature to 140 °C. Noted that element Ce with adjustable valence can take place both A-site and B-site (Ce^{3+} for A-site and Ce^{4+} for B-site), whereas only A-site doping successfully extends large ECE to a wider temperature range ($\Delta T > 1$ K for 37–70 °C) [71]. In addition to the A-site substitution, more attention has been paid to the B-site substitution for BT-based ceramics, such as $Ba(Zr_xTi_{1-x})O_3$ (BZT), $Ba(Sn_xTi_{1-x})O_3$ (BSn_xT) and $Ba(Hf_xTi_{1-x})O_3$ (BHf_xT) [72–74]. The high EC performance of these B-site substituted ceramics originates from the construction of the ICP, where the BT-based ceramics simultaneously possess four types of phase structures (cubic, tetragonal, orthorhombic, and trigonal) and corresponding 26 polar states. The enhanced local structure heterogeneity and nearly vanished flipping barrier at ICP avails to a low field-induced large ECE [72]. For example, the researchers designed and prepared $Ba(Zr_{0.2}Ti_{0.8})O_3$ ceramics near the ICP point, which displayed a giant ΔT of 4.5 K at 312 K and 145 kV/cm [26]. In the process of preparing $Ba(Zr_{0.2}Ti_{0.8})O_3$ ceramics, a small amount of boron-lead glass was added, which not only improved the EC strength but greatly improved the working electric fields. Similarly, $Ba(Sn_{0.105}Ti_{0.895})O_3$ and $Ba(Hf_{0.11}Ti_{0.89})O_3$ bulk ceramics were designed around the ICP. The inset of Figure 15.3a shows the phase diagram of $Ba(Sn_xTi_{1-x})O_3$. As shown in Figure 15.3a, the cubic, tetragonal, orthorhombic, and trigonal phases coexist when $x = 0.105$, and there is a sharp dielectric constant-temperature peak at room temperature, which contributed to a large ΔT of 0.61 K at 20 kV/cm (Figure 15.3b) [73]. In addition, the partial substitution of Ti^{4+} by Hf^{4+} can also construct multiphase coexistence [74].

In addition to the A-site and B-site doping, A/B-site co-doping can also be used to tune the ICP point and introduce abundant phase transitions to improve the ECE of bulk ceramics. Typical A/B-site co-doped BT-based EC ceramics include

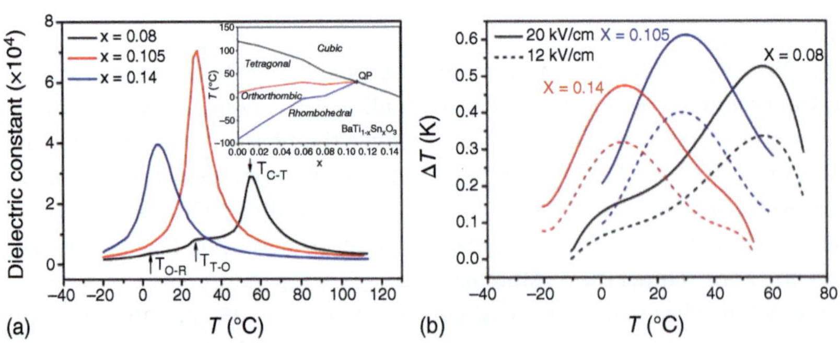

Figure 15.3 (a) The temperature dependence of the dielectric constant of all the BTSn ceramics during a cooling process, was recorded by applying an electric stimulus of 1 V at 100 Hz on the sample. The arrows indicate the phase transition sequence of BTSn with $x = 0.08$. The inset shows the phase diagram of BTSn ceramics with the Sn doping level ranging from 0% to 15%. (b) EC effect in BTSn ceramics with different compositions. Source: Luo et al. [73]. Reproduced with permission of American Institute of Physics.

$(Ba_xCa_{1-x})Ti_{1-y}Hf_yO_3$ [75], $(Ba_xCa_{1-x})Ti_{1-y}Sn_yO_3$ [76], $(Ba_xCa_{1-x})Ti_{1-y}Zr_yO_3$ [77], and $(Ba_xSr_{1-x})Ti_{1-y}Zr_yO_3$ [28], to name but a few. Among them, the $(Ba_{0.8}Sr_{0.15})(Zr_{0.15}Ti_{0.85})O_3$ bulk ceramic not only maintained the high EC strength of $BaZr_{0.15}Ti_{0.85}O_3$ ceramics but the introduction of Sr^{2+} shifted the Curie temperature to room temperature. A ΔT of 2.43 K was obtained at 50 kV/cm and 315 K. In the $0.9Ba(Ti_{0.89}Sn_{0.11})O_3$–$0.1(Ba_{0.7}Ca_{0.3})TiO_3$ ceramics separated ferroelectric-ferroelectric transition and ferroelectric-paraelectric transition in a wide temperature range was achieved, replacing the single ferroelectric-paraelectric transition in relaxor ferroelectric $Ba(Ti_{0.89}Sn_{0.11})O_3$. The separated phase transitions increased the ΔT from 0.40 to 0.50 K at 20 kV/cm and widened the EC temperature window from ~ 45 to ~ 62 K [78]. In a single ferroelectric-paraelectric transition, the gap of free energy between the phases is large and the inversion barrier between the phases is high, which results in a small ECE. Differently, in the separated ferroelectric-ferroelectric and ferroelectric-paraelectric phase transition, the inversion barriers between the phases are relatively lower, leading to a larger ECE. Therefore, the reasonable doping of cations to replace the Ba^{2+} in the A-site and Ti^{4+} in the B-site is an effective way to tune the phase boundary and the ECE in BT-based ferroelectric ceramics. Besides the ion substitution, increasing the ceramics' resistivity was also utilized to enhance the electric breakdown strength and thus improve the ECE of BT-based ceramics. For example, in the $Ba_{0.7}Sr_{0.3}TiO_3$-based ceramics, the SPS was utilized to improve the resistivity and the electric breakdown strength. Compared with the ceramics sintered by conventional solid-state method, the electric breakdown strength of SPS-BST ceramics was increased from 40 to 90 kV/cm, which was attributed to the fully dense ceramics with fine grains. Moreover, the electric breakdown strength of SPS-BST ceramics can be further increased to 120 kV/cm by substituting titanium ions with 0.3 mol% manganese ions. Finally, a high ΔT of 2.53 K was obtained under 120 kV/cm, which was significantly higher than that of conventionally sintered ceramics [65]. What's more, a two-step sintering process was adopted to avoid the abnormal growth of grains. The dense $Ba_{0.65}Sr_{0.35}TiO_3$ ceramics with uniform grains have been successfully prepared and can withstand significantly higher electric fields. Under an electric field of 80 kV/cm, a ΔT of 1.82 K was obtained [66]. Based on the above discussion, in BT-based ceramics, by adjusting the composition and fabrication process, the ECE can be greatly improved.

15.2.2.2 ECE in BNT-based Bulk Ceramics

Sodium bismuth titanate (BNT) is a type of ferroelectrics with a composite-ionic perovskite structure, in which the A site is co-occupied by Na^+ and Bi^{3+}. The phase structure of BNT changes with temperature as the following processes: Trigonal phase *R3c* (523 K) ↔ Trigonal *R3c* + tetragonal *P4bm* (670 K) ↔ Quartet Phase *P4bm* (813 K) ↔ Cubic phase *Pm3m* [79]. It can be seen that the phase transition temperatures of BNT are all greater than 523 K, which is far from room temperature. Therefore, introducing suitable elements to reduce the phase transition temperature is the key point to enhance the ECE of BNT-based bulk ceramics near room temperature. At present, it's still difficult to adjust the Curie temperature of BNT-based bulk ceramics to room temperature and maintain a large ECE. The research on BNT-based EC ceramics is mainly to shift the *R3c-P4bm* transition temperature to room temperature and construct the MPB of the trigonal ferroelectric phase and the tetragonal antiferroelectric phase. The ferroelectric-antiferroelectric phase transition of BNT-based ceramics could induce a drastic change in spontaneous polarization, which was expected to obtain a large EC strength. The widely studied BNT-based solid solutions at this phase transition include $(1-x)Na_{0.5}Bi_{0.5}TiO_3$–$xBaTiO_3$ [80–82], $(1-x)Na_{0.5}Bi_{0.5}TiO_3$–$xKNbO_3$ [83], $(1-x)Na_{0.5}Bi_{0.5}TiO_3$–$x(K_{0.5}Bi_{0.5})TiO_3$ [84], $(1-x)Na_{0.5}Bi_{0.5}TiO_3$–$xSrTiO_3$ [85, 86], and $(0.94-x)Bi_{0.5}Na_{0.5}TiO_3$–$0.06BaTiO_3$–$x(Sr_{0.7}Bi_{0.2})TiO_3$, and so on [87–89]. According to the phase diagram of $(1-x)Na_{0.5}Bi_{0.5}TiO_3$–$xBaTiO_3$, researchers designed a composition $0.94Na_{0.5}Bi_{0.5}TiO_3$–$0.06BaTiO_3$ (0.94BNT–0.06BT), which has a trigonal-tetragonal phase transition around room temperature. A negative ECE with $\Delta T_{max} = -1.0$ K was obtained at 323 K and 60 kV/cm according to the indirect method [80]. Later, the researchers used modified DSC to directly test the ΔT of 0.94BNT–0.06BT again. The result showed a negative ECE with ΔT_{max} of −1.1 K, which was consistent with the previous study [83]. In addition to the BNT-BT, studies have shown that the introduction of $SrTiO_3$ can also shift the trigonal-tetragonal phase transition temperature of BNT-based ceramics to a lower temperature [90] and construct a large ECE near room temperature. For instance, $0.75Na_{0.5}Bi_{0.5}TiO_3$–$0.25SrTiO_3$ and $0.74Na_{0.5}Bi_{0.5}TiO_3$–$0.26SrTiO_3$ bulk ceramics exhibited favorable ECE near room temperature. The former obtained a ΔT_{max} of 1.64 K and a ΔS_{max} of 2.52 J/(kg K) at 50 kV/cm and 333 K, while the latter had a wide temperature range (ΔT greater than 1.0 K in the range of 303–343 K) due to the increase in relaxation degree [85]. In the $0.75Na_{0.5}Bi_{0.5}TiO_3$–$0.25SrTiO_3$ (0.75BNT–0.25ST) bulk ceramics, ergodic relaxation (ER) phase and non-ergodic relaxation phase (NER) exist in the two-phase coexistence region [86]. The low connection strength between the dipoles in these two relaxation phases results in a large ECE with a ΔT_{max} of 0.51 K at 60 kV/cm (Figure 15.4). Similarly, with the coexistence of ER and NER, the $0.86Bi_{0.5}Na_{0.5}TiO_3$–$0.06BaTiO_3$–$0.08(Sr_{0.7}Bi_{0.2})TiO_3$

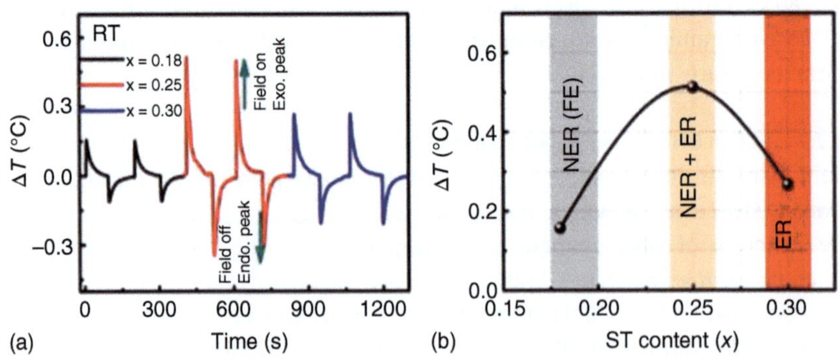

Figure 15.4 (a) The directly measured temperature versus time with $E = 6$ kV/mm and (b) ΔT value for BNT–xST ceramics ($x = 0.18$, 0.25, and 0.30). Source: Li et al. [86]. Reproduced with permission of American Institute of Physics.

bulk ceramic was designed, which exhibited a high ΔT_{max} of 0.4 K at 60 kV/cm and near room temperature [87]. The addition of Sr^{2+} in $[Bi_{0.5}(Na_{0.72}K_{0.18}Li_{0.1})_{0.5}]TiO_3$ ceramics even achieves a giant ΔT of 2.51 K and $\Delta T/\Delta E$ of 0.386 K mm/kV [88]. In summary, forming solid solution in BNT-based ceramics to shift the MPB of the trigonal-tetragonal phase to room temperature is conducive to the enhancement of ECE near room temperature.

15.2.2.3 ECE in KNN-based Bulk Ceramics

$K_{0.5}Na_{0.5}NbO_3$ (KNN)-based ferroelectric ceramics have attracted extensive attention due to their excellent piezoelectric and electromechanical properties [91, 92]. However, there are few studies on the ECE of KNN-based bulk ceramics compared with other lead-free ceramics such as BT and BNT-based ceramics. KNN ceramics undergo a phase transition from trigonal to orthorhombic at about 150 K, orthorhombic to tetragonal phase transition at approximately 483 K, and tetragonal to cubic phase transition at ~683 K. Since each phase transition temperature is far away from room temperature, adjusting the phase boundary temperature is also needed in investigating the ECE of the KNN-based ceramic systems. Numerous studies have shown that reasonable doping can shift the trigonal-orthogonal phase transition to a higher temperature and the orthorhombic-tetragonal phase transition to a lower temperature, constructing a three-phase coexistence at room temperature. Since it is easier to produce a large ECE when the material is in the multi-phase coexistence point, the research on the ECE of KNN-based ceramics is still mainly based on the construction of multi-phase coexistence at room temperature. For example, the $(1-x)(Na_{0.52}K_{0.48-x})(Nb_{0.92-x}Sb_{0.08})O_3$–$x$LiTaO$_3$ bulk ceramics were designed to improve the ECE. In this composition, with the increase of $LiTaO_3$ content, the representative dielectric peak of the orthorhombic-tetragonal phase transition gradually diffuses and shifts to a lower temperature, which is consistent with the phase diagram trend of this material [93]. When $x = 0.0375$, the orthorhombic and tetragonal phases coexist near room temperature, and the ceramics obtain a favorable ΔT of 0.30 K at 20 kV/cm [94]. Moreover, due to the dispersion of the phase transition, the ceramics maintain a favorable ECE within the temperature range from 303 to 323 K. In addition, a trigonal-orthogonal-tetragonal polymorphic phase transition at room temperature was constructed in $(1-x)K_{0.5}Na_{0.5}NbO_3$–$xBi_{0.5}Na_{0.5}ZrO_3$ bulk ceramics. As a result, a ΔT of 0.16 K was obtained at 35 kV/cm in the composition $x = 0.05$ [95]. Similarly, researchers designed solid solutions such as $0.95(K_{0.49}Na_{0.49}Li_{0.02})(Nb_{0.8}Ta_{0.2})O_3$–$0.05CaZrO_3$ [96], $0.95(K_{0.48}Na_{0.52})(Nb_{0.95}Sb_{0.05})O_3$–$0.05Bi_{0.5}(Na_{0.82}K_{0.18})_{0.5}ZrO_3$ [97], $0.975K_{0.5}Na_{0.5}NbO_3$–$0.025(Bi_{0.99}Dy_{0.01})_{0.5}Na_{0.5}TiO_3$ [98], and $0.97(Na_{0.5}K_{0.5})Nb_{0.87}Sb_{0.13})NbO_3$–$0.03CaZrO_3$–1.0 mol% CuO [99] and other components, and the ECE of the KNN-based ceramics were optimized.

15.2.2.4 ECE in PZT-based Bulk Ceramics

PZT is a solid solution of the orthorhombic antiferroelectric $PbZrO_3$ and the tetragonal ferroelectric $PbTiO_3$. In the $(1-x)PbZrO_3$–xPbTiO$_3$ binary solid solution, with the increasing proportion of $PbTiO_3$, the solid solution will successively pass through the orthorhombic antiferroelectric phase, the trigonal ferroelectric phase, and the tetragonal ferroelectric phase. The Curie temperature of PZT ceramics is higher than 503 K, and the ΔT_{max} of the PZT ceramics occurs at high temperatures, leading to a small ECE at room temperature [100]. Moving the phase transition point to room temperature through A-site doping (La^{3+}, Ba^{2+}, and Sr^{2+}) and B-site doping (Sn^{4+} and Nb^{5+}) is one of the main strategies to optimize the ECE of the PZT-based bulk ceramics. From the phase diagram of La^{3+}-doped PZT, it can be seen that an orthorhombic antiferroelectric-trigonal ferroelectric phase transition or an orthorhombic antiferroelectric-cubic paraelectric phase transition at room temperature can be constructed in PLZT ceramics with low content of lanthanum (La < 11 mol%) and high content of zirconium (Zr > 60 mol%) [101]. The EC properties of these PLZT ceramics have been intensively investigated. For example, relaxor ferroelectric $Pb_{0.89}La_{0.11}(Zr_{0.7}Ti_{0.3})_{0.9725}O_3$ and orthogonal antiferroelectric $Pb_{0.93}La_{0.07}(Zr_{0.82}Ti_{0.18})_{0.9825}O_3$ ceramics were designed at the phase boundary. Under an electric field of 30 kV/cm, the ΔT of 1.4 and 1.3 K were measured by a direct method, respectively [102]. The $(Pb_{0.98}La_{0.02})(Zr_{0.95}Ti_{0.05})_{0.995}O_3$ bulk ceramics, which exhibits an orthorhombic antiferroelectric-trigonal ferroelectric phase boundary, obtained a negative ECE with a ΔT of 5 K at 60 kV/cm [103]. The electric breakdown strength of the above-mentioned PLZT ceramics is lower than 60 kV/cm, which limits the further improvement of ECE. To enhance the resistivity and the electric breakdown strength, researchers prepared $Pb_{0.85}La_{0.1}(Zr_{0.65}Ti_{0.35})O_3$ ceramics by hot pressing sintering technology. The electric breakdown strength of the ceramics can reach 200 kV/cm and the ΔT is 3.1 K, which is nearly double that of $Pb_{0.85}La_{0.1}(Zr_{0.65}Ti_{0.35})O_3$ ceramics prepared by conventional sintering method [67]. In addition to reducing the phase transition temperature of PZT by doping La^{3+} at the A site, the researchers also adjusted the atomic mole ratio of Sn, Zr and Ti in the $Pb_{0.97}La_{0.02}(Zr, Sn, Ti)O_3$ and $Pb_{0.99}Nb_{0.02}(Zr, Sn, Ti)_{0.98}O_3$ ceramics and their phase diagrams at 298 K were obtained [104]. Around the phase boundary in the phase diagrams, the researchers designed a large number of components with high ECE near the room temperature, such as $Pb_{0.99}Nb_{0.02}[(Zr_{0.57}Sn_{0.43})_{0.92}Ti_{0.08}]_{0.98}O_3$ [105], $Pb_{0.99}Nb_{0.02}[(Zr_{0.58}Sn_{0.43})_{0.92}Ti_{0.08}]_{0.98}O_3$ [106], $Pb_{0.97}La_{0.02}(Zr_{0.80}Sn_{0.14}Ti_{0.06})O_3$ [107], $Pb_{0.97}La_{0.02}(Zr_{0.75}Sn_{0.19}Ti_{0.06})O_3$ [108], and $Pb_{0.97}La_{0.02}(Zr_{0.73}Sn_{0.22}Ti_{0.05})O_3$ [109]. In addition, a two-step sintering technique was used to prepare dense PLZST ceramics, which exhibited a peak $\Delta T > 2$ K at an electric field of 50 kV/cm and a $\Delta T > 1.78$ K in the temperature range of 293–423 K [110].

15.2.2.5 ECE in PMN–*x*PT-based Bulk Ceramics

Lead magnesium niobate (PMN) is a typical relaxor ferroelectric [111]. The binary solid solution PMN–*x*PT exhibits favorable ECE. The Curie temperature of PMN ceramics is around 250 K, and the maximum ΔT of the PMN ceramics measured by a thermocouple at 226 K is 0.15 K (15 kV/cm) [112], indicating that the pure PMN ceramics are not suitable for EC refrigeration at room temperature. The introduction of PT into PMN to form PMN–*x*PT was used to shift the phase transition temperature to room temperature. According to the phase diagram of PMN–*x*PT ceramics [113], when the value of x is in the range of 0.05–0.2, the trigonal ferroelectric phase is formed and the phase transition of the trigonal ferroelectric phase to the cubic paraelectric phase occurs at room temperature [114]. In the 0.9PMN–0.1PT ceramics, a first-order phase transition at room temperature was constructed and a ΔT of 1.35 K (@15 kV/cm) was obtained at 301 K [20]. In addition, in the depolarized 0.85PMN–0.15PT sample, the researchers obtained a ΔT of 1.71 K at 291 K [115]. The phase transition (ferroelectric-paraelectric) temperature of PMN–xPT ceramics moves toward higher temperatures when excess $PbTiO_3$ is introduced. The 0.75PMN–0.25PT ceramic was fabricated and a favorable ΔT of 2.10 K was achieved at 398 K and 25 kV/cm [116]. Therefore, in the PMN–xPT system, the ferroelectric-paraelectric phase transition temperature along with the ΔT can be effectively regulated by changing the content of PT.

15.2.2.6 ECE in PST-based Bulk Ceramics

Scandium lead tantalite (PST) displays a high ECE and excellent cooling potential at room temperature [27, 117]. The B sites of PST ceramics are occupied by Sc^{3+} and Ta^{5+}, forming different sublattices. When the B-site ions in the same sublattice inside the PST are ordered along the {111} plane, a highly ordered PST lattice structure can be formed [118]. However, the ions' thermal vibration in the lattice has a strong hindering effect on the ordered arrangement of the lattice, which makes the two lattice states of ordered arrangement and disordered arrangement mixed in the PST ceramics. For PST, the degree of sublattice ordering directly determines the type of phase transition and the strength of the ECE. The degree of order in PST ceramics can be intuitively represented by the intensity of the (111) XRD diffraction peak. The research on the ECE of PST is mainly based on the degree of order (Ω), which is related to the annealing time. For example, after being annealed at 1273 K for 15 minutes, PST ceramics with a Ω of 0.46 can be obtained. While annealed at 1273 K for 103 hours, the Ω is greatly increased to 0.86 [118]. For the PST ceramics with a high Ω, the permittivity-temperature peak is sharper, which shows a first-order ferroelectric-paraelectric phase transition with a large latent heat. By contrast, for the PST ceramics with a low Ω the permittivity-temperature peak is diffused. The phase transition becomes a diffused second-order phase transition with low latent heat. Therefore, the degree of order directly affects the types of phase transition of the PST ceramics, which also determines the values of ΔT. For instance, a high ΔT of 1.8 K can be obtained in PST ceramics with a high Ω (0.85), which is significantly larger than that of PST ceramics with a low Ω (0.2) (0.1 K) [119]. The Ω of the PST ceramics can be tuned by regulating the solid-phase sintering process. Moreover, the ions (Mn, Ni, Fe, Co, Sb, Nb, Ti, In, and Y) were incorporated into the B-sites of the PST ceramics with high Ω to replace Sc^{3+} or Ta^{5+}. The ΔT of these doped samples was obtained by the indirect method [120]. According to the test results, it was found that: (i) The doped samples with a high Ω still possess high ΔT; (ii) The introduction of ions with large radii (such as In and Sb) will reduce the tolerance factor of the PST ceramics, causing the weakened stability of the perovskite structure, the precipitation of pyrochlore phase and thus the reduced ΔT; (iii) Doping ions with high polarizability (such as Y) into the PST with high Ω is beneficial to improving ΔT. To sum up, changing the Ω of the PST ceramics by sintering process and doping ions in the B-sites are effective ways to tune the EC properties of the PST ceramics.

15.2.3 ECE in Multi-Layer Ceramics (MLCs)

Compared with the bulk ceramics, the ceramic thin films have a thinner thickness, which can withstand a higher electric field, and thus can produce a larger ECE. However, the effective volume of the thin films is very small, leading to a low heat absorption capacity. In addition, the thin films have to be deposited on the substrates, which are not free-standing, and therefore, the heat exchange between the thin films and the substrates will greatly reduce the cooling efficiency. For the bulk ceramics, it is difficult to generate a ΔT larger than 5 K because of the low electric breakdown strength. To overcome these problems, attention has been paid to ceramic thick films with multi-layer structures (MLCs). On the one hand, the thickness of the single-layer thick film in MLCs is generally within 30.0 μm, which can withstand an electric field of hundreds of kV/cm [121]. Consequently, a significantly larger ΔT can be obtained. Moreover, the effective volume of MLCs is much higher than that of thin films, which have a heat absorption capacity being orders of magnitude higher than that of thin films. The MLCs EC materials show great potential in practical EC refrigeration [27, 117]. In this section, we introduce the investigation of the ECE of MLCs based on lead-free thick films and lead-containing thick films.

Through stacking the ceramic thick films and printing with inner and outer electrodes, the MLCs are formed (Figure 15.5). The interdigitated electrodes combine the layers of thick films in parallel connection, which enables MLCs not only to have a volume similar to that of the bulk ceramic but also to withstand higher electric fields close to thin films. In the study of the ECE of the MLCs, the selection of materials (the composition of the thick films and the selection of internal electrodes) and the structure of the MLCs (the thickness of a single-layer film, the proportion of active parts in a single-layer film, and the numbers of active layers, etc.) are the key factors affecting the EC properties. At present, the adopted compositions of the thick films in MLCs are similar to that of bulk ceramics, which possess a phase transition near room temperature, such as $BaZr_{0.2}Ti_{0.8}O_3$ [123, 124], $Ba(Zr_{0.20}Ti_{0.80})O_3–Ba(Sn_{0.11}Ti_{0.89})O_3$ [125], $PbSc_{0.5}Ta_{0.5}O_3$ [22, 119], $0.9Pb(Mg_{1/3}Nb_{2/3})O_3–0.1PbTiO_3$ [126] and so on. Since the ceramic thick films need to be co-fired with the internal electrodes, the sintering temperature of the dielectrics and the metal electrodes must be matched, which directly affects the selection of the thick films. Several electrodes and their corresponding sintering atmosphere and sintering upper limit temperature are as follows: Pt (air, 1673 K), Ag/Pd (60/40) (air, 1473 K), Ag/Pd (70/30) (air, 1423 K),

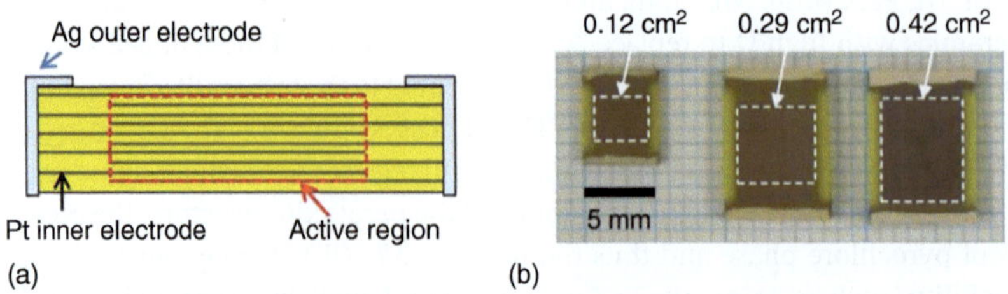

Figure 15.5 (a) Schematic cross-section of MLC structure. (b) Photograph of three MLCs with an active area per layer of 0.12, 0.29 and 0.42 cm^2 within dashed lines. Source: Usui et al. [122]. Reproduced with permission of IOP Publishing.

Ag/Pd (80/20) (air, 1373 K), Ni/Cu (reducing atmosphere, 1573 K) [127]. The sintering temperature of most BT-based EC ceramics is higher than 1573 K. To reduce the cost, it is necessary to introduce sintering aids (such as PbO/B_2O_3, and $LiBiO_2$) into the ceramics to reduce the sintering temperature to match the annealing requirement of the Ag/Pd electrodes and the Ni electrodes [125]. Moreover, the thermal conductivity needs to be considered in the selection of the electrodes. Using metals with high thermal conductivity as the inner electrodes is conducive to improving the EC cooling power of the MLCs. The study has shown that replacing Ni electrodes (thermal conductivity: 94 W/(m K)) with Ag electrodes (thermal conductivity: 428 W/(m K)) as the inner electrodes of the MLCs can improve the cooling power of the MLCs [128].

In addition to the compositions and internal electrodes, the thickness of the single layer of the MLCs also has a great influence on the EC performance. The MLCs with thinner single-layer thickness can withstand higher electric fields and thus produce larger ECE. For example, in 2002, the PST-based MLCs with a single-layer thickness of 128 μm were fabricated, which can withstand an electric field of 138 kV/cm and generate a ΔT of 3.50 K [119]. In 2019, the single-layer thickness of the PST-based MLCs was reduced to 37.9 μm, leading to a higher electric breakdown strength of 290 kV/cm and a higher ΔT of 5.50 K [22]. Moreover, the $BaTiO_3$ MLCs with an ultra-thin single-layer thickness of 3.0 μm can endure a high electric field of 800 kV/cm and produce a large ΔT of 7.10 K [129]. The proportions of the active areas and the number of active layers of the MLCs also have great influences on the ECE. The active areas refer to the dielectric regions that lie in the overlapping areas of the interdigitated electrodes. To study the effect of the active areas on the ΔT of MLCs, the researchers carried out the ECE measurement on $0.9Pb(Mg_{1/3}Nb_{2/3})O_3$–0.1 $PbTiO_3$ MLCs [122]. As shown in Figure 15.6, there is a temperature difference between the active areas and the inactive areas. Particularly, the ΔT in the center of the active areas is the highest. The weak ΔT of the inactive area is mainly due to the heat exchange with the active area. Therefore, the larger the proportion of the active areas, the ΔT measured by the thermocouple is closer to the adiabatic ΔT of the EC materials. Moreover, the researchers investigated the variation of EC response crossing $BaTiO_3$-based MLCs using an infrared camera [130]. It was found that a significant variation of ECE up to 0.285 K existed between the central regions and the regions near the terminals. It was observed that the heat flux perpendicular to the terminal-terminal direction accounted for only

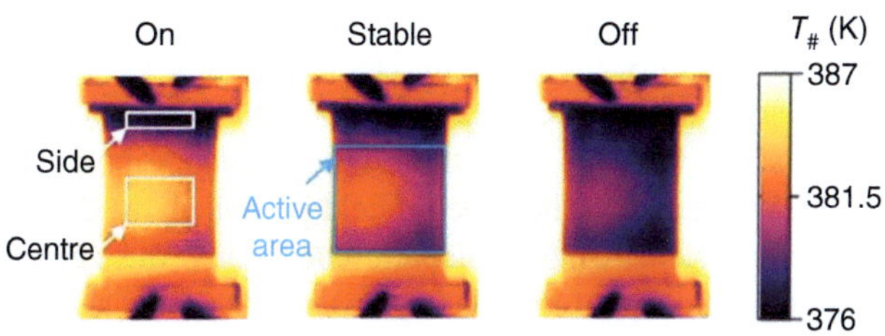

Figure 15.6 Surface ΔT of $0.9Pb(Mg_{1/3}Nb_{2/3})O_3-0.1PbTiO_3$ MLCs measured through an infrared camera. Source: Usui et al. [122] Reproduced with permission of IOP Publishing.

20% of the total heat flux, which indicates that the EC cooling devices should be reasonably designed to favor heat exchange through the MLCs' terminals. The number of active layers also has a great influence on the ECE. The researchers studied the ECE of 19-layer and 49-layer $0.9Pb(Mg_{1/3}Nb_{2/3})O_3$–$0.1PbTiO_3$ MLCs with uniform monolayer thickness, respectively. It was found that the 49-layer $0.9Pb(Mg_{1/3}Nb_{2/3})O_3$–$0.1PbTiO_3$ MLCs have a large internal electrical strain and are easy to breakdown, resulting in a lower working electric field and lower ΔT (103 kV/cm, 1.2 K) than that of the 19-layer MLCs (240 kV/cm, 2.2 K) [131]. In summary, the rational design of the composition and the thickness of a single-layer film, the proportion of active parts, and the number of active layers would help to improve the ECE of MLCs.

The EC MLCs have high requirements for the preparation process and structural design. At present, the research on EC MLCs is mainly aimed at some lead-containing systems with relatively mature processes, such as the mentioned PST and PMN-PT MLCs. For the research on the ECE of lead-free MLCs, the main focus is on BT-based and NBT-based MLCs. For instance, the researchers purchased the BT-based commercial Y5V MLCs (1210ZG226ZAT2A, AVX). The samples show a ΔT ~0.50 K at 300 kV/cm and the ΔT maintains stability in a wide temperature range around room temperature [132]. The researchers also compared the ΔT to that measured with thermal resistance Pt-100 thermometry and infrared imaging and gave the scope of application of these two test methods [133]. Commercial BT-based MLCs have been widely researched, but the EC strength of Y5V-type MLCs at room temperature is lower than 0.1 K m/MV. The size of Y5V-type MLCs is limited, which makes it difficult to meet the requirements of the EC device for size diversification. Then the researchers developed the $0.9(K_{0.49}Na_{0.49}Li_{0.02})(Nb_{0.8}Ta_{0.2})O_3$–$0.1CaZrO_3$ MLCs, which can achieve a ΔT of 2.45 K at 300 kV/cm and 353 K [134]. Meanwhile, the researchers developed the $(1-x)Na_{0.5}Bi_{0.5}TiO_3$–$xSr_{0.85}Bi_{0.10}TiO_3$ MLCs, which can produce a ΔT of 2.59 K and possess favorable temperature stability between the 308 and 368 K according to the indirect method [135]. To sum up, there are few lead-free MLCs with large ECE and EC strength at room temperature. There is still an EC performance gap between the lead-free MLCs and the lead-containing MLCs, which needs to be further studied by the researchers.

15.3 Designs and Demonstrations of Ceramics-based Cooling Devices

To realize the refrigeration applications, the EC materials used in the EC devices need to undergo thermodynamic cycle processes. Taking the Brayton cycle for example [136], in the initial stage, the dipoles in the EC materials are randomly oriented at zero electric field. When an external electric field is applied, the dipoles are orientated along the electric field. Under adiabatic conditions, the entropy remains constant while the temperature of the EC materials increases. Then, the EC materials are in contact with the heat sink while keeping the applied electric field constant. When the electric field is removed, the dipoles change from a state of order to disorder. The temperature of the EC materials decreases while the entropy remains the same at this stage. Finally, the EC materials are in contact with a heat source and absorb the heat from the heat source. The temperature and the entropy of the system increase to the initial state to complete a cycle. The heat would be transferred continuously from the heat source to the heat sink by repeating this cycle.

The first EC device prototype dates back to 1989 when Sinyavsky et al. designed an active EC regenerator (AER) based on the PST ceramics [137, 138]. The AER was proposed from the magnetocaloric regenerator which was invented by Brown in 1975 [139]. Later in 1995, Sinyavsky et al. investigated an EC prototype with a length of 30 cm and made of 400 g PST and it was claimed that a large temperature span (ΔT_{span}) of 15 K could be obtained [140]. Since the great breakthrough in ΔT in PZT ferroelectric ceramic films and P(VDF-TrFE) polymers, many prototypes of EC devices have been designed [141–145], which follow the principle of regeneration. These developed EC devices are generally classified as fluid-based and solid-based, depending on the mediums that transfer the heat generated by the EC materials [146]. For instance, Plaznik et al. developed an EC cooling device utilizing the $[Pb(Mg_{1/3}Nb_{2/3})O_3]_{0.9}[PbTiO_3]_{0.1}$ (PMN–10PT) relaxor ceramics as active EC elements and dielectric silicon oil as the medium to transfer heat [143]. Figure 15.7

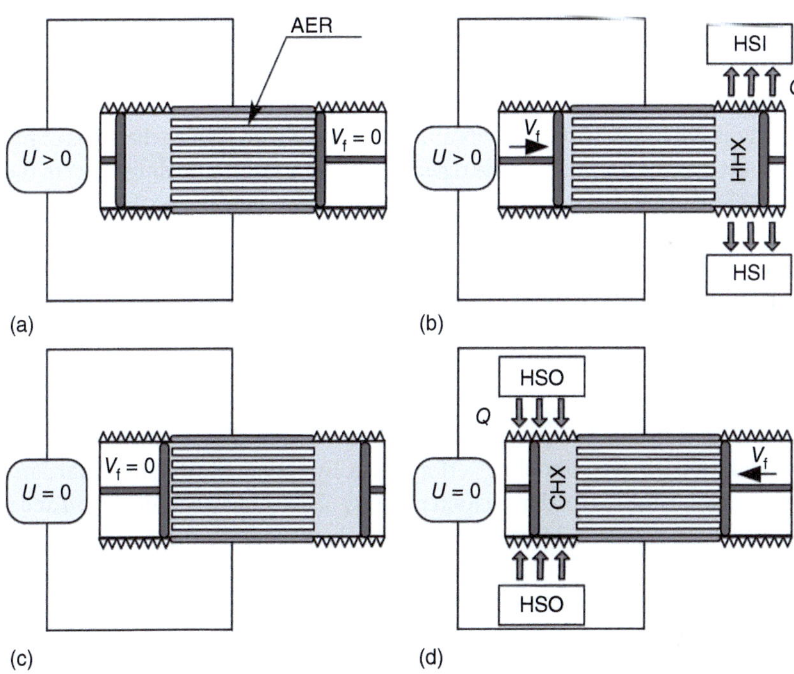

Figure 15.7 Depicts the schematic illustration of the four operating steps of the EC system with an AER. Firstly, the EC material is polarized and heated up with an applied electric field (a). Then, the silicon oil is pumped from the heat sink to the heat source and the heat is transferred from the EC material to the silicon oil. Subsequently, the silicone oil reaches the hot end and transfers the heat to the heat sink (b). Thirdly, the EC material is depolarized and the temperature of the EC material decreases (c). Meanwhile, the silicon oil flows from the heat source to the heat sink and is cooled down. The cooled silicone oil enters the cold heat exchanger and absorbs the heat from the heat source (d).

depicts the schematic illustration of the four operating steps of the EC system with an AER. Firstly, the EC material is polarized and heated up with an applied electric field (Figure 15.7a). Then, the silicon oil is pumped from the heat sink to the heat source and the heat is transferred from the EC material to the silicon oil. Subsequently, the silicone oil reaches the hot end and transfers the heat to the heat sink (Figure 15.7b). Thirdly, the EC material is depolarized and the temperature of the EC material decreases (Figure 15.7c). Meanwhile, the silicon oil flows from the heat source to the heat sink and is cooled down. The cooled silicone oil enters the cold heat exchanger and absorbs the heat from the heat source (Figure 15.7d). Repeating these steps can achieve a stable temperature gradient along the fluid direction in the AER and a ΔT_{span} several times higher than the ΔT_{ECE} of the EC material. The ratio of ΔT_{span} to ΔT_{ECE} is defined as the regeneration factor. With an operating frequency of 0.75 Hz and an electric field of 50 kV/cm, the maximum ΔT_{span} of 3.3 K and a maximum regeneration factor of 3.7 were obtained in the EC cooling device based on the PMN–10PT ceramics. A simulation based on a 2D dynamic numerical model was carried out to improve the EC performance of the device. The simulation results show that the length of AER, the types of working fluids, and the operating frequency can be regulated to further elevate the ΔT_{span} of the devices. The highest ΔT_{span} of 14 K and a regeneration factor of 9.6 could be obtained under the electric field of 118 kV/cm by choosing the water as the working fluid and extending the regenerator length to 0.2 m [143]. Similarly, Blumenthal et al developed a fluid-based cooler with PMN–8PT MLCs, which were submerged in silicone oil [145].

Besides the choice of working fluids and the operating frequency, the types of the EC elements and the geometry of the regenerator also affect the cooling effect of the EC devices [147, 148]. Defay et al. designed a fluid-based cooler using the PST MLCs as the EC elements because of PST ceramics' sharp first-order phase transition and favorable ΔT_{ECE} [117]. Their first EC regenerator, i.e. formed with 15 0.9-mm-thick PST-MLCs distributed in a matrix of three columns by five rows, exhibited a maximum ΔT_{span} of 1 K and a low regeneration factor of 0.45. Then, they used finite element numerical modeling to investigate different configurations of the EC prototypes to optimize the ΔT_{span}. It was found that the inactive structural parts in the first device acted as heat traps and absorbed a non-negligible amount of heat, which weakened the overall cooling performance. By optimizing the structural parts of the regenerator, the thermal insulation property of the system was improved and a ΔT_{span} of 2 K was achieved. Moreover, the researchers modified the thickness of the MLCs since thinner MLCs could increase the heat exchange area. A higher ΔT_{span} of 4 K was obtained by decreasing the PST MLCs' thickness from 0.9 to 0.5 mm, which was realized by decreasing the number of PST layers from 19 to 9. Increasing the length of the regenerator can further widen the temperature gradient and a fourfold increase in the overall length of the MLCs leads to a ΔT_{span} of 9 K. The influence of the types of oscillating fluid on the EC performance has also been considered. Using water instead of dielectric fluid further increased the ΔT_{span} by 20%. The reason is that water has six times larger thermal conductivity. The final optimized EC device consisted of 128 PST MLCs that exhibited a ΔT_{span} of 13.0 K and a regenerator factor of 5.9 after 1500 seconds of operation.

Although the large ΔT_{span} has been achieved in the fluid-based coolers, they still have some disadvantages: (i) the use of working fluid and the pump to drive the oscillatory liquid flow make it difficult to miniaturize the EC devices for chip-scale applications; (ii) mixing of the fluid along the flow direction will weaken the regeneration-induced temperature gradient and lower the cooling power of the device [139]. The solid-state coolers are promising candidates for achieving high ΔT_{span} and high cooling power with a compact size [149]. Zhang and coworkers developed a chip-scale cooling device [141]. As shown in Figure 15.8, the cooling cycle of this device consists of four steps, in which the EC module moves cyclically between the left and right sides of a solid-state regenerator coupled with applying/removing electric fields. The heat is transported from the left to right of the regenerator, which generates a temperature span across the regenerator. The solid-state cooler was accurately driven and controlled by miniature actuators, which eliminated the bulky pumping units and reduced the size of the whole device. Moreover, the solid-state cooler has no mixing problem existing in the liquid-based devices, leading to a high cooling power. Adopting the regenerator with an anisotropic thermal conductivity (higher orthogonal heat exchange with the heat transfer media but lower thermal conductivity along the direction of the temperature gradient) is conducive to improving the ΔT_{span}, which can be practically realized in the solid-state regenerator [149]. Zhang and coworkers also designed a type of rotary AER cooling device [150, 151]. As shown in Figure 15.9a,

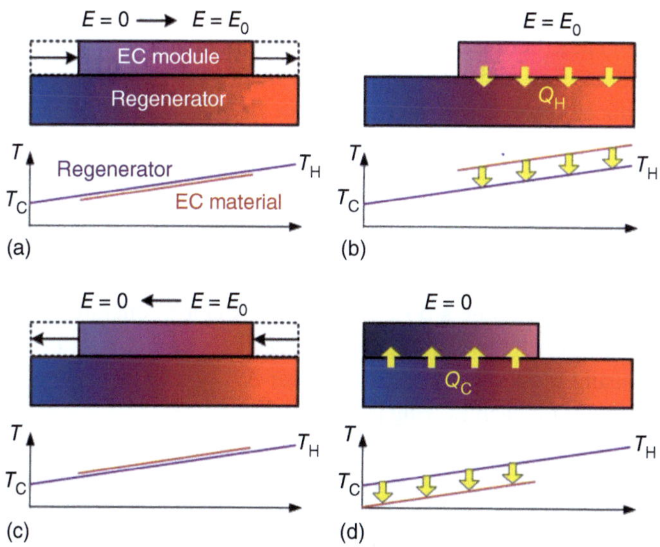

Figure 15.8 The illustration of the operation principle of the EC cooling cycle: (a) The EC module moves to the right side of the regenerator, and the electric field increases from 0 to E_0; (b) The EC module remains fixed on the right side for a period of time, during which heat is ejected from the EC module to the regenerator; (c) The EC module then moves to the left side of the regenerator, and the electric field decreases from E_0 to 0; (d) The EC module stays on the left side for a period of time and absorbs heat from the regenerator at the cold end. Source: Gu et al. [141]. Reproduced with permission of American Institute of Physics.

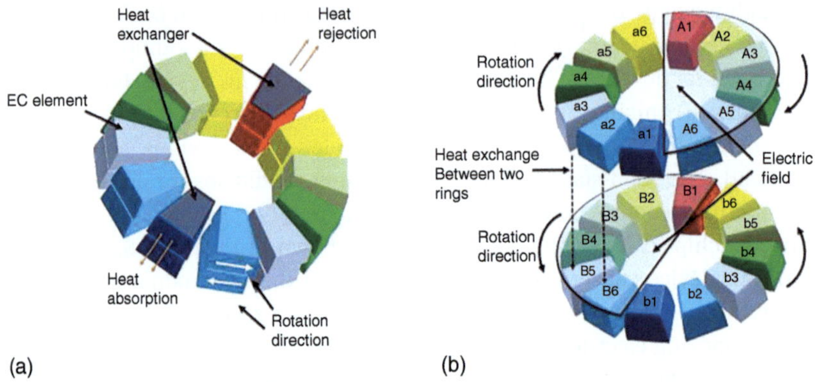

Figure 15.9 (a) The illustration of the EC cooling device with EC rings. (b) Schema of the operation principle as the two EC rings rotating. Source: Zhang et al. [151]. Reproduced with permission of American Institute of Physics.

the device consisted of two or more EC rings and each EC ring contained multiple EC elements (commercial Y5V MLCs). The adjacent EC rings rotate at the same speed, but in opposite directions, as illustrated by the arrows in Figure 15.9b. The electric field can be applied to each EC element of the EC ring. For each EC ring, half of the EC elements are applied with an electric field while the others are grounded. These two EC rings rotate in a step-wise rotation manner, in which a fraction of time is spent on rotating while the remaining fraction of time is spent on staying for heat exchange between the EC elements in the two rings. When an EC element in the half EC ring under an electric field rotates toward the cold end, the electric field is reduced to zero, and the EC element absorbs heat. When an EC element in the other half EC ring which is grounded rotates toward the hot end, an electric field is applied to the EC element, and the EC element ejects heat. The half EC ring under an electric field faces the zero-field region of the adjacent EC ring. Thus, the direct heat exchange between the two adjacent EC rings generates a direct solid-to-solid regeneration process. An EC device based on two adjacent EC rings was developed, which could generate a ΔT_{span} of ~2.58 K and a regenerator factor of 3 [151]. Moreover, a finite element simulation indicates that the ΔT_{span} can be further improved by using the EC elements with larger ΔT_{ECE} and increasing the numbers of the EC element for each EC ring [150].

To further increase the ΔT_{span} of the EC devices, the cascade cooling system similar to the multistage Peltier stack structure was designed [152]. The cascaded EC systems directly transfer the heat between every adjacent two isolated bulk EC modules instead of transferring heat through an oscillating media. Each EC element undergoes a unique and non-overlapping thermodynamic cycle, so the ΔT_{span} of individual EC elements can be superimposed to obtain larger ΔT_{span} in the cascaded EC devices. Wang et al. developed a scalable, high-performance system based on the PST MLCs (a ΔT_{ECE} of 2.5 K at 10.8 MV/m and room temperature) in a modular cascaded self-regenerating architecture with low thermal loss [27]. The PST MLCs-based cascaded EC system consists of a pair of stacked EC modules, each of which contains EC

MLCs distanced with thermal insulating materials. The inclusion of the insulators mitigates the thermal shunting along the temperature gradient (a significant source of heat loss), being a vital improvement compared to the continuous active regenerators. When the polarizing electric fields are switched synchronously, the EC modules are moved laterally relative to one another. In this way, a ΔT_{span} between the cold end and the hot end of the device significantly higher than the ΔT_{ECE} of MLCs can be generated. Moreover, anisotropically thermally conductive plates, designed to have high through-plane and low in-plane thermal conductivity, are adopted to enhance the heat exchange between layers while maintaining low lateral thermal leakage. Finally, under an electric field of 10.8 MV/m and a switching frequency of 0.2 Hz, the cascaded system achieves a ΔT_{span} of 5.2 K and a heat flux of 135 mW/cm^2.

To conclude, the active electrocaloric regeneration method has been widely used to design EC cooling devices and obtain ΔT_{span} much higher than the ΔT_{ECE}. The fluid-based cooler has achieved a ΔT_{span} higher than 10 K, but they face the challenges of miniaturization due to the necessity of liquid pumping systems. Differently, in the solid-state regenerators, the size of the whole device can be reduced by using the miniature actuators. Moreover, the solid-state cascade structure further stimulates the cooling potential of the EC materials by superimposing the ECE, providing an important path for the development of EC devices with a high-temperature span and large cooling capacity.

15.4 Conclusions and Prospects

In this chapter, we list the researches on the ECE in different ferroelectric ceramic materials (films and bulks), compare their EC performance, and summarize the methods of improving the ECE of various inorganic ferroelectrics. The methods to tune the EC properties of these materials mainly include three aspects: the electric breakdown strength, the phase transition, and the working temperature range of the materials. According to the previous discussions, the high working electric fields, the sharp internal phase transition, and the phase transition temperature close to room temperature are helpful for the ferroelectrics to achieve a large ECE near room temperature. The $Ba_{0.8}Zr_{0.2}TiO_3$, 0.94BNT–0.06BT, $PbSc_{0.5}Ta_{0.5}O_3$, $0.9Pb(Mg_{1/3}Nb_{2/3})O_3$–$0.1PbTiO_3$ and many other ferroelectric ceramic materials with favorable ECE have been developed. Moreover, the MLCs with the optimal compositions have been successfully fabricated, which further improves the EC performance. Many EC devices based on the MLCs have been reported. However, since the working electric fields of the EC devices can only be about 30% of the breakdown electric fields of the EC materials in the actual operation process, it's still difficult to put the EC devices into practical application. The limited working electric fields make the ΔT of the EC materials and EC devices undesirable. Therefore, further improving the ECE of the EC materials and greatly widening the temperature difference between the cold ends and hot ends of the devices are the major problems currently faced in the field of EC refrigeration. At present, the MLCs are widely investigated because the multi-layer EC ceramics can superimpose heat absorption and the multi-layer structure is beneficial to reducing the working voltages and miniaturizing the EC devices. Unfortunately, expensive Pt electrodes are usually required during fabricating the MLCs with high EC performance. The PST MLCs, which have exhibited desirable ECE, are not suitable for large-scale applications due to the expensive Sc. It's urgent to develop MLCs with high ECE, which can be fabricated with cheap raw materials and cheap electrodes (Cu, Ni). In addition to the EC materials, it faces great challenges in the design of the EC devices, such as the heat storage problem during the thermal transport, the thermal conduction problem when the EC materials are in contact with the cold and hot ends, and the thermal insulation problem when the material is in the thermodynamic cycle. The rational design of the EC devices and the reduced heat loss during the cooling process are the keys to achieving $\Delta T_{\mathrm{device}}$ much greater than ΔT_{ECE}. So far, the maximum temperature difference between the cold and hot ends of the designed EC devices can reach 13 K. In the future, the further improvement of the EC performance of the EC devices will still depend on the breakthroughs in the ECE of EC materials and the improved structural design of the EC devices. The EC cooling, as a green, efficient, and easy-to-integrate refrigeration method, will promote the development of the refrigeration field in the direction of environmental protection and low energy consumption.

References

1 Defay, E., Crossley, S., Kar-Narayan, S. et al. (2013). The electrocaloric efficiency of ceramic and polymer films. *Adv. Mater.* 25: 3337–3342.

2 Moya, X., Kar-Narayan, S., and Mathur, N.D. (2014). Caloric materials near ferroic phase transitions. *Nat. Mater.* 13: 439–450.

3 Valant, M. (2012). Electrocaloric materials for future solid-state refrigeration technologies. *Prog. Mater Sci.* 57: 980–1009.

4 Li, G. (2021). Performance evaluation of low global warming potential working fluids as R134a alternatives for two-stage centrifugal chiller applications. *Korean J. Chem. Eng.* 38: 1438–1451.

5 Velders, G.J.M., Fahey, D.W., Daniel, J.S. et al. (2009). The large contribution of projected HFC emissions to future climate forcing. *Proc. Natl. Acad. Sci.* 106: 10949–10954.

6 Shi, J., Han, D., Li, Z. et al. (2019). Electrocaloric cooling materials and devices for zero-global-warming potential, high-efficiency refrigeration. *Joule* 3: 1200–1225.

7 Hsu, K.F., Loo, S., Guo, F. et al. (2004). Cubic $AgPb_mSbTe_{2+m}$: bulk thermoelectric materials with high figure of merit. *Science* 303: 818–821.

8 Poudel, B., Hao, Q., Ma, Y. et al. (2008). High-thermoelectric performance of nanostructured bismuth antimony telluride bulk alloys. *Science* 320: 634–638.

9 Russ, B., Glaudell, A., Urban, J.J. et al. (2016). Organic thermoelectric materials for energy harvesting and temperature control. *Nat. Rev. Mater.* 1: 1–14.

10 Taguchi, Y., Sakai, H., and Choudhury, D. (2017). Magnetocaloric materials with multiple instabilities. *Adv. Mater.* 29: 1606144.

11 Liu, Y., Infante, I.C., Lou, X. et al. (2014). Giant room-temperature elastocaloric effect in ferroelectric ultrathin films. *Adv. Mater.* 26: 6132–6137.

12 Li, B., Kawakita, Y., Kawamura, S.O. et al. (2019). Colossal barocaloric effects in plastic crystals. *Nature* 567: 506–510.

13 Li, W., Jafri, H.M., Zhang, C. et al. (2020). The strong electrocaloric effect in molecular ferroelectric $ImClO_4$ with ultrahigh electrocaloric strength. *J. Mater. Chem. A* 8: 16189–16194.

14 Kobeko, P. and Kurtschatov, J. (1930). Dielektrische Eigenschaften der Seignettesalzkristalle. *Z. Angew. Phys.* 66: 192–205.

15 Wiseman, G.G. and Kuebler, J.K. (1963). Electrocaloric effect in ferroelectric Rochelle salt. *Phys. Rev.* 131: 2023–2027.

16 Karchevskii, A.J. (1962). Electrocaloric effect in polycrystalline barium titanate. *Soviet Phys. Solid State* 3: 2249–2254.

17 Thacher, P.D. (1968). Electrocaloric effects in some ferroelectric and antiferroelectric $Pb(Zr, Ti)O_3$ compounds. *J. Appl. Phys.* 39: 1996–2002.

18 Mischenko, A.S., Zhang, Q., Scott, J.F. et al. (2006). Giant electrocaloric effect in thin-film $PbZr_{0.95}Ti_{0.05}O_3$. *Science* 311: 1270–1271.

19 Neese, B., Chu, B., Lu, S.G. et al. (2008). Large electrocaloric effect in ferroelectric polymers near room temperature. *Science* 321: 821–823.

20 Xiao, D.Q., Wang, Y.C., Zhang, R.L. et al. (1998). Electrocaloric properties of $(1-x)Pb(Mg_{1/3}Nb_{2/3})O_3-xPbTiO_3$ ferroelectric ceramics near room temperature. *Mater. Chem. Phys.* 57: 182–185.

21 Lu, S.G., Rožič, B., Zhang, Q.M. et al. (2010). Organic and inorganic relaxor ferroelectrics with giant electrocaloric effect. *Appl. Phys. Lett.* 97: 162904.

22 Nair, B., Usui, T., Crossley, S. et al. (2019). Large electrocaloric effects in oxide multilayer capacitors over a wide temperature range. *Nature* 575: 468–472.

23 Zhang, G., Li, Q., Gu, H. et al. (2015). Ferroelectric polymer nanocomposites for room-temperature electrocaloric refrigeration. *Adv. Mater.* 27: 1450–1454.

24 Sebald, G., Seveyrat, L., Guyomar, D. et al. (2006). Electrocaloric and pyroelectric properties of $0.75Pb(Mg_{1/3}Nb_{2/3})O_3-0.25PbTiO_3$ single crystals. *J. Appl. Phys.* 100: 124112.

25 Goupil, F.L., Berenov, A., Axelsson, A.K. et al. (2012). Direct and indirect electrocaloric measurements on <001>-$PbMg_{1/3}Nb_{2/3}O_3-30PbTiO_3$ single crystals. *J. Appl. Phys.* 111: 124109.

26 Qian, X.S., Ye, H.J., Zhang, Y.T. et al. (2014). Giant electrocaloric response over a broad temperature range in modified $BaTiO_3$ ceramics. *Adv. Funct. Mater.* 24: 1300–1305.

27 Wang, Y., Zhang, Z., Usui, T. et al. (2020). A high-performance solid-state electrocaloric cooling system. *Science* 370: 129–133.

28 Jian, X.D., Lu, B., Li, D.D. et al. (2019). Enhanced electrocaloric effect in Sr^{2+}-modified Lead-free $BaZr_xTi_{1-x}O_3$ ceramics. *ACS Appl. Mater. Interfaces* 11: 20167–20173.

29 Liu, Y., Scott, J.F., and Dkhil, B. (2016). Direct and indirect measurements on electrocaloric effect: recent developments and perspectives. *Appl. Phys. Rev.* 3: 031102.

30 Gerson, R. and Marshall, T.C. (1959). Dielectric breakdown of porous ceramics. *J. Appl. Phys.* 30: 1650–1653.

31 Liu, Z.K., Li, X., and Zhang, Q.M. (2012). Maximizing the number of coexisting phases near invariant critical points for giant electrocaloric and electromechanical responses in ferroelectrics. *Appl. Phys. Lett.* 101: 954–959.

32 Feng, Z., Shi, D., and Dou, S. (2011). Large electrocaloric effect in highly (001)-oriented $0.67PbMg_{1/3}Nb_{2/3}O_3$–$0.33PbTiO_3$ thin films. *Solid State Commun.* 151: 123–126.

33 Mischenko, A.S., Zhang, Q., Whatmore, R.W., and Mathur, N.D. (2006). Giant electrocaloric effect in the thin film relaxor ferroelectric $0.9PbMg_{1/3}Nb_{2/3}O_3$–$0.1PbTiO_3$ near room temperature. *Appl. Phys. Lett.* 89: 242912.

34 Correia, T.M., Young, J.S., Whatmore, R.W. et al. (2009). Electrocaloric refrigeration: thermodynamics, state of the art and future perspectives. *Appl. Phys. Lett.* 95: 182904–182904.

35 Li, B., Wang, J.B., Zhong, X.L. et al. (2011). Enhancing the electrocaloric effect of $PbZr_{0.4}Ti_{0.6}O_3/PbTiO_3$ superlattices via composition tuning. *Europhys. Lett.* 95: 67004.

36 Chen, C., Wang, S., Zhang, T. et al. (2020). Designing coexisting multi-phases in PZT multilayer thin films: an effective way to induce large electrocaloric effect. *RSC Adv.* 10: 6603–6608.

37 Shirsath, S.E., Cazorla, C., Lu, T. et al. (2020). Interface-charge induced giant electrocaloric effect in lead free ferroelectric thin-film bilayers. *Nano Lett.* 20: 1262–1271.

38 Guo, F., Wu, X., Lu, Q., and Zhao, S. (2018). Near room temperature giant negative and positive electrocaloric effects coexisting in lead-free $BaZr_{0.2}Ti_{0.8}O_3$ relaxor ferroelectric films. *Ceram. Int.* 44: 2803–2808.

39 Engelhardt, S., Molin, C., Gebhardt, S. et al. (2018). $BaZr_xTi_{1-x}O_3$ epitaxial thin films for electrocaloric investigations. *Energy Technol.* 6: 1526–1534.

40 Akcay, G., Alpay, S.P., Rossetti, G.A., and Scott, J.F. (2008). Influence of mechanical boundary conditions on the electrocaloric properties of ferroelectric thin films. *J. Appl. Phys.* 103: 024104.

41 Li, B., Zhang, X., Wang, J.B. et al. (2014). Giant electrocaloric effect of $PbTiO_3$ thin film tuned in a wide temperature range by the anisotropic misfit strain. *Mech. Res. Commun.* 55: 40–44.

42 Marathe, M. and Ederer, C. (2014). Electrocaloric effect in $BaTiO_3$: a first-principles-based study on the effect of misfit strain. *Appl. Phys. Lett.* 104: 212902.

43 He, Y., Li, X.M., Gao, X.D. et al. (2011). Enhanced electrocaloric properties of PMN-PT thin films with LSCO buffer layers. *Funct. Mater. Lett.* 4: 45–48.

44 Feng, Z., Shi, D., Zeng, R., and Dou, S. (2011). Large electrocaloric effect of highly (100)-oriented $0.68PbMg_{1/3}Nb_{2/3}O_3$–$0.32PbTiO_3$ thin films with a $Pb(Zr_{0.3}Ti_{0.7})O_3/PbO_x$ buffer layer. *Thin Solid Films* 519: 5433–5436.

45 Wang, D., Yuan, G., Hao, G., and Wang, Y. (2018). All-inorganic flexible piezoelectric energy harvester enabled by two-dimensional mica. *Nano Energy* 43: 351–358.

46 Xu, X., Liu, W., Li, Y. et al. (2018). Flexible mica films for high-temperature energy storage. *J. Materiomics* 4: 173–178.

47 Wang, D., Chen, X., Yuan, G. et al. (2019). Toward artificial intelligent self-cooling electronic skins: large electrocaloric effect in all-inorganic flexible thin films at room temperature. *J. Materiomics* 5: 66–72.

48 Shen, B.Z., Li, Y., and Hao, X. (2019). Multifunctional all-inorganic flexible capacitor for energy storage and electrocaloric refrigeration over a broad temperature range based on PLZT 9/65/35 thick films. *ACS Appl. Mater. Interfaces* 11: 34117–34127.

49 Yang, C., Han, Y., Feng, C. et al. (2020). Toward multifunctional electronics: flexible NBT-based film with a large electrocaloric effect and high energy storage property. *ACS Appl. Mater. Interfaces* 12: 6082–6089.

50 Qiu, J.H. and Jiang, Q. (2008). Misfit strain dependence of electrocaloric effect in epitaxial $Pb(Zr_{1-x}Ti_x)O_3$ thin films. *J. Appl. Phys.* 103: 084105.

51 Qiu, J.H. and Jiang, Q. (2009). Orientation dependence of the electrocaloric effect of ferroelectric bilayer thin films. *Solid State Commun.* 149: 1549–1552.

52 Wu, M., Song, D., Guo, M. et al. (2019). Remarkably enhanced negative electrocaloric effect in $PbZrO_3$ thin film by interface engineering. *ACS Appl. Mater. Interfaces* 11: 36863–36870.

53 Yang, C., Feng, C., Lv, P. et al. (2021). Coexistence of giant positive and large negative electrocaloric effects in lead-free ferroelectric thin film for continuous solid-state refrigeration. *Nano Energy* 88: 106222.

54 Lu, S.G., Rožič, B., Zhang, Q.M. et al. (2010). Comparison of directly and indirectly measured electrocaloric effect in relaxor ferroelectric polymers. *Appl. Phys. Lett.* 97: 202901.

55 Sun, X., Huang, H., Wang, J. et al. (2019). Strain-induced broadening temperature range of electrocaloric effects in ferroelectric superlattices. *J. Alloys Compd.* 777: 821–827.

56 Qian, J., Hu, P., Liu, C. et al. (2018). High electrocaloric cooling power of relaxor ferroelectric $BaZr_xTi_{1-x}O_3$ ceramics within broad temperature range. *Sci. Bull.* 63: 356–361.

57 Han, F., Bai, Y., Qiao, L.J., and Guo, D. (2016). A systematic modification of the large electrocaloric effect within a broad temperature range in rare-earth doped $BaTiO_3$ ceramics. *J. Mater. Chem. C* 4: 1842–1849.

58 Bai, Y., Han, X., Ding, K., and Qiao, L.J. (2013). Combined effects of diffuse phase transition and microstructure on the electrocaloric effect in $Ba_{1-x}Sr_xTiO_3$ ceramics. *Appl. Phys. Lett.* 103: 162902.

59 Zhang, C., Du, Q., Li, W. et al. (2020). High electrocaloric effect in barium titanate-sodium niobate ceramics with core-shell grain assembly. *J. Materiomics* 6: 618–627.

60 Srikanth, K.S., Patel, S., and Vaish, R. (2017). Enhanced electrocaloric effect in glass-added $0.94Bi_{0.5}Na_{0.5}TiO_3$–$0.06BaTiO_3$ ceramics. *J. Aust. Ceram. Soc.* 53: 523–529.

61 Wang, S.B., Dai, G.Z., Yao, Y.B. et al. (2021). Direct and indirect measurement of large electrocaloric effect in B_2O_3-ZnO glass modified $Ba_{0.65}Sr_{0.35}TiO_3$ bulk ceramics. *Scr. Mater.* 193: 59–63.

62 Wang, Y., Li, J., Yuan, R. et al. (2021). Enhanced electrocaloric effect in $BaSn/TiO_3$ ceramics by addition of CuO. *J. Alloys Compd.* 851: 156772.

63 Kumar, R. and Singh, S. (2017). Enhanced electrocaloric effect in lead-free $0.9(K_{0.5}Na_{0.5})NbO_3$–$0.1Sr(Sc_{0.5}Nb_{0.5})O_3$ ferroelectric nanocrystalline ceramics. *J. Alloys Compd.* 723: 589–594.

64 Vrabelj, M., Uršič, H., Kutnjak, Z. et al. (2016). Large electrocaloric effect in grain-size-engineered $0.9Pb(Mg_{1/3}Nb_{2/3})O_3$–$0.1PbTiO_3$. *J. Eur. Ceram. Soc.* 36: 75–80.

65 Liu, X.Q., Chen, T.T., Fu, M.S. et al. (2014). Electrocaloric effects in spark plasma sintered $Ba_{0.7}Sr_{0.3}TiO_3$-based ceramics: effects of domain sizes and phase constitution. *Ceram. Int.* 40: 11269–11276.

66 Dai, G., Wang, S., Huang, G. et al. (2019). Direct and indirect measurement of large electrocaloric effect in barium strontium titanate ceramics. *Int. J. Appl. Ceram. Technol.* 17: 1354–1361.

67 Zhang, G., Chen, Z., Fan, B. et al. (2016). Large enhancement of the electrocaloric effect in PLZT ceramics prepared by hot-pressing. *APL Mater.* 4: 064103.

68 Liu, X.Q., Chen, T.T., Wu, Y.J. et al. (2013). Enhanced electrocaloric effects in spark plasma-sintered $Ba_{0.65}Sr_{0.35}TiO_3$-based ceramics at room temperature. *J. Am. Ceram. Soc.* 96: 1021–1023.

69 Moya, X., Stern-Taulats, E., Crossley, S. et al. (2013). Giant electrocaloric strength in single-crystal $BaTiO_3$. *Adv. Mater.* 25: 1360–1365.

70 Bai, Y., Han, X., Zheng, X.C., and Qiao, L. (2013). Both high reliability and giant electrocaloric strength in $BaTiO_3$ ceramics. *Sci. Rep.* 3: 2895.

71 Huang, Y.L., Zhao, C.L., Zhong, S.Z., and Wu, J.G. (2021). Highly tunable multifunctional $BaTiO_3$-based ferroelectrics via site selective doping strategy. *Acta Mater.* 209: 116792.

72 Wei, X.W., Zhao, C.L., Zheng, T. et al. (2022). Understanding the enhanced electrocaloric effect in $BaTiO_3$-based ferroelectrics at critical state. *Acta Mater.* 227: 117735.

73 Luo, Z., Zhang, D., Liu, Y. et al. (2014). Enhanced electrocaloric effect in lead-free $BaTi_{1-x}Sn_xO_3$ ceramics near room temperature. *Appl. Phys. Lett.* 105: 102904.

74 Li, J., Zhang, D., Qin, S. et al. (2016). Large room-temperature electrocaloric effect in lead-free $BaHf_xTi_{1-x}O_3$ ceramics under low electric field. *Acta Mater.* 115: 58–67.

75 Wang, X., Wu, J., Dkhil, B. et al. (2017). Large electrocaloric strength and broad electrocaloric temperature span in lead-free $Ba_{0.85}Ca_{0.15}Ti_{1-x}Hf_xO_3$ ceramics. *RSC Adv.* 7: 5813–5820.

76 Wang, X., Tian, F., Zhao, C. et al. (2015). Giant electrocaloric effect in lead-free $Ba_{0.94}Ca_{0.06}Ti_{1-x}Sn_xO_3$ ceramics with tunable Curie temperature. *Appl. Phys. Lett.* 107: 252905.

77 Singh, G., Tiwari, V.S., and Gupta, P.K. (2013). Thermal stability of piezoelectric coefficients in $(Ba_{1-x}Ca_x)(Zr_{0.05}Ti_{0.95})O_3$: a lead-free piezoelectric ceramic. *Appl. Phys. Lett.* 102: 162905.

78 Zhao, C., Yang, J., Huang, Y. et al. (2019). Broad-temperature-span and large electrocaloric effect in lead-free ceramics utilizing successive and metastable phase transitions. *J. Mater. Chem. A* 7: 25526–25536.

79 Jones, G.O. and Thomas, P.A. (2010). Evidence for a non-rhombohedral average structure in the lead-free piezoelectric material $Na_{0.5}Bi_{0.5}TiO_3$. *Acta Crystallogr.* 56: 426–430.

80 Zheng, X.C., Zheng, G.P., Lin, Z., and Jiang, Z.Y. (2012). Electro-caloric behaviors of lead-free $Bi_{0.5}Na_{0.5}TiO_3$–$BaTiO_3$ ceramics. *J. Electroceram.* 28: 20–26.

81 Cao, W.P., Li, W.L., Xu, D. et al. (2014). Enhanced electrocaloric effect in lead-free NBT-based ceramics. *Ceram. Int.* 40: 9273–9278.

82 Turki, O., Slimani, A., Seveyrat, L. et al. (2016). Structural, dielectric, ferroelectric, and electrocaloric properties of 2% Gd_2O_3 doping $(Na_{0.5}Bi_{0.5})_{0.94}Ba_{0.06}TiO_3$ ceramics. *J. Appl. Phys.* 120: 054102.

83 Jiang, X., Luo, L., Wang, B. et al. (2014). Electrocaloric effect based on the depolarization transition in $(1-x)Bi_{0.5}Na_{0.5}TiO_3$–$xKNbO_3$ lead-free ceramics. *Ceram. Int.* 40: 2627–2634.

84 Goupil, F.L., Bennett, J., Axelsson, A.K. et al. (2015). Electrocaloric enhancement near the morphotropic phase boundary in lead-free NBT-KBT ceramics. *Appl. Phys. Lett.* 107: 172903.

85 Cao, W.P., Li, W.L., Dai, X.F. et al. (2016). Large electrocaloric response and high energy-storage properties over a broad temperature range in lead-free NBT-ST ceramics. *J. Eur. Ceram. Soc.* 36: 593–600.

86 Li, F., Li, J., Zhai, J. et al. (2018). Influence of structural evolution on electrocaloric effect in $Bi_{0.5}Na_{0.5}TiO_3$–$SrTiO_3$ ferroelectric ceramics. *J. Appl. Phys.* 124: 164108.

87 Li, F., Chen, G., Liu, X. et al. (2017). Phase-composition and temperature dependence of electrocaloric effect in lead-free $Bi_{0.5}Na_{0.5}TiO_3$–$BaTiO_3$–$(Sr_{0.7}Bi_{0.2})TiO_3$ ceramics. *J. Eur. Ceram. Soc.* 37: 4732–4740.

88 Zhang, L., Zhao, C.L., Zheng, T., and Wu, J.G. (2020). Large electrocaloric effect in $(Bi_{0.5}Na_{0.5})TiO_3$-based relaxor ferroelectrics. *ACS Appl. Mater. Interfaces* 12: 33934–33940.

89 Zhang, L., Zhao, C.L., Zheng, T., and Wu, J.G. (2021). Electrocaloric refrigeration capacity in BNT-based ferroelectrics benefiting from low depolarization temperature and high breakdown electric field. *J. Mater. Chem. A* 9: 12772–12781.

90 Hiruma, Y., Imai, Y., Watanabe, Y. et al. (2008). Large electrostrain near the phase transition temperature of $(Bi_{0.5}Na_{0.5})TiO_3$–$SrTiO_3$ ferroelectric ceramics. *Appl. Phys. Lett.* 92: 262904.

91 Wang, K. and Li, J.F. (2010). Domain engineering of lead-free Li-modified (K,Na)NbO_3 polycrystals with highly enhanced piezoelectricity. *Adv. Funct. Mater.* 20: 1924–1929.

92 Wang, K., Li, J.F., and Liu, N. (2008). Piezoelectric properties of low-temperature sintered Li-modified (Na, K)NbO_3 lead-free ceramics. *Appl. Phys. Lett.* 93: 092904.

93 Zuo, R., Fu, J., and Lv, D. (2009). Phase transformation and tunable piezoelectric properties of lead-free $(Na_{0.52}K_{0.48-x}Li_x)(Nb_{1-x-y}Sb_yTa_x)O_3$ system. *J. Am. Ceram. Soc.* 92: 283–285.

94 Li, J., Bai, Y., Qin, S. et al. (2016). Direct and indirect characterization of electrocaloric effect in (Na,K)NbO_3 based lead-free ceramics. *Appl. Phys. Lett.* 109: 162902.

95 Yang, J. and Hao, X. (2019). Electrocaloric effect and pyroelectric performance in (K,Na)NbO_3-based lead-free ceramics. *J. Am. Ceram. Soc.* 102: 6817–6826.

96 Yang, J., Zhao, Y., Lou, X. et al. (2020). Synergistically optimizing electrocaloric effects and temperature span in KNN-based ceramics utilizing a relaxor multiphase boundary. *J. Mater. Chem. C* 8: 4030–4039.

97 Wang, X., Wu, J., Dkhil, B. et al. (2017). Enhanced electrocaloric effect near polymorphic phase boundary in lead-free potassium sodium niobate ceramics. *Appl. Phys. Lett.* 110: 063904.

98 Aissa, M., Zannen, M., Benyoussef, M. et al. (2021). Large direct and inverse electrocaloric effects in lead-free Dy doped 0.975KNN–0.025NBT ceramics. *Ceram. Int.* 47: 31286–31293.

99 Zhang, N., Zheng, T., Zhao, C. et al. (2021). Enhanced electrocaloric effect in compositional driven potassium sodium niobate-based relaxor ferroelectrics. *J. Mater. Res.* 36: 1142–1152.

100 Sawaguchi, E. (1953). Ferroelectricity versus antiferroelectricity in the solid solutions of $PbZrO_3$ and $PbTiO_3$. *J. Phys. Soc. Jpn.* 8: 615–629.

101 Haertling, G.H. (1987). PLZT electrooptic materials and applications-a review. *Ferroelectrics* 75: 25–55.

102 Lu, B., Li, P., Tang, Z. et al. (2017). Large electrocaloric effect in relaxor ferroelectric and antiferroelectric lanthanum doped lead zirconate titanate ceramics. *Sci. Rep.* 7: 45335.

103 Mendez-González, Y., Peláiz-Barranco, A., Yang, T., and Guerra, J.D.S. (2018). Enhanced electrocaloric effect in La-based PZT antiferroelectric ceramics. *Appl. Phys. Lett.* 112: 122904.

104 Berlincourt, D. (1966). Transducers using forced transitions between ferroelectric and antiferroelectric states. *IEEE Trans. Sonics Ultrason.* 13: 116–124.

105 Xu, Z., Fan, Z., Liu, X., and Tan, X. (2017). Impact of phase transition sequence on the electrocaloric effect in $Pb(Nb,Zr,Sn,Ti)O_3$ ceramics. *Appl. Phys. Lett.* 110: 082901.

106 Novak, N., Weyland, F., Patel, S. et al. (2018). Interplay of conventional with inverse electrocaloric response in (Pb,Nb)(Zr,Sn,Ti)O_3 antiferroelectric materials. *Phys. Rev. B* 97: 094113.

107 Zhuo, F., Li, Q., Gao, J. et al. (2018). Giant negative electrocaloric effect in (Pb,La)(Zr,Sn,Ti)O_3 antiferroelectrics near room temperature. *ACS Appl. Mater. Interfaces* 10: 11747–11755.

108 Zhao, Y., Hao, X., and Zhang, Q. (2015). A giant electrocaloric effect of a $Pb_{0.97}La_{0.02}(Zr_{0.75}\ Sn_{0.18}Ti_{0.07})O_3$ antiferroelectric thick film at room temperature. *J. Mater. Chem. C* 3: 1694–1699.

109 Zhao, Y., Gao, H., Hao, X., and Zhang, Q. (2016). Orientation-dependent energy-storage performance and electrocaloric effect in PLZST antiferroelectric thick films. *Mater. Res. Bull.* 84: 177–184.

110 Lu, B., Yao, Y., Jian, X. et al. (2019). Enhancement of the electrocaloric effect over a wide temperature range in PLZT ceramics by doping with Gd^{3+} and Sn^{4+} ions. *J. Eur. Ceram. Soc.* 39: 1093–1102.

111 Park, S.E., Mulvihill, M.L., and RISCH, G. (1997). Ultrahigh strain and piezoelectric behavior in relaxor based ferroelectric single crystals. *Jpn. J. Appl. Phys.* 82: 1804–1811.

112 Shebanov, L.A., Kapostin, P.P., and Birks, E.H. (1986, 1986). Some peculiarities in the rearrangement of the crystal-structure and electrocaloric effect of single lead magnoniobate crystals in the region of a diffused phase-transition. *Kristallografiya* 31 (2): 317–320.

113 Kiat, J.M. and Dkhil, B. (2008). *Handbook of Advanced Dielectric, Piezoelectric and Ferroelectric Materials*. Cambridge: Woodhead Publishing Limited, CRC Press.

114 Ye, Z.G. and Dong, M. (2000). Morphotropic domain structures and phase transitions in relaxor-based piezo-/ferroelectric $(1-x)Pb(Mg_{1/3}Nb_{2/3})O_3-xPbTiO_3$ single crystals. *J. Appl. Phys.* 87: 2312–2319.

115 Liu, S.B. and Li, Y.Q. (2004). Research on the electrocaloric effect of PMN/PT solid solution for ferroelectrics MEMS microcooler. *Mater. Sci. Eng., B* 113: 46–49.

116 Kriaa, I., Abdelmoula, N., Maalej, A., and Khemakhem, H. (2015). *J. Electron. Mater.* 44: 4852–4856.

117 Torelló, A., Lheritie, P., Usui, T. et al. (2020). Giant temperature span in electrocaloric regenerator. *Science* 370: 125–129.

118 Setter, N. and Cross, L.E. (1980). The role of B-site cation disorder in diffuse phase transition behavior of perovskite ferroelectrics. *J. Appl. Phys.* 51: 4356–4360.

119 Shebanov, L., Borman, K., Lawless, W.N., and Kalvane, A. (2002). Electrocaloric effect in some perovskite ferroelectric ceramics and multilayer capacitors. *Ferroelectrics* 273: 137–142.

120 Shebanov, L. and Borman, K. (1992). On lead-scandium tantalate solid solutions with high electrocaloric effect. *Ferroelectrics* 127: 143–148.

121 Li, J., Shen, Z., Chen, X. et al. (2020). Grain-orientation-engineered multilayer ceramic capacitors for energy storage applications. *Nat. Mater.* 19: 999–1005.

122 Usui, T., Hirose, S., Ando, A. et al. (2017). Effect of inactive volume on thermocouple measurements of electrocaloric temperature change in multilayer capacitors of $0.9Pb(Mg_{1/3}Nb_{2/3})O_3$–$0.1PbTiO_3$. *J. Phys. D: Appl. Phys.* 50: 424002.

123 Hou, Y., Yang, L., Qian, X. et al. (2016). Enhanced electrocaloric effect in composition gradient bilayer thick films. *Appl. Phys. Lett.* 108: 133501.

124 Hou, Y., Yang, L., Zhao, X., and Zhang, Q.M. (2017). Sintering aids modified electrocaloric response in $BaZr_{0.2}Ti_{0.8}O_3$ bilayer films. *J. Alloys Compd.* 724: 8–13.

125 Li, J., Chang, Y., Yang, S. et al. (2019). Lead-free bilayer thick films with giant electrocaloric effect near room temperature. *ACS Appl. Mater. Interfaces* 11: 23346–23352.

126 Crossley, S., Usui, T., Nair, B. et al. (2016). Direct electrocaloric measurement of $0.9Pb(Mg_{1/3}Nb_{2/3})O_3$–$0.1PbTiO_3$ films using scanning thermal microscopy. *Appl. Phys. Lett.* 108: 032902.

127 Fan, P., Liu, K., Ma, W. et al. (2021). Progress and perspective of high strain NBT-based lead-free piezoceramics and multilayer actuators. *J. Materiomics* 7: 508–544.

128 Kar-Narayan, S. and Mathur, N.D. (2009). Predicted cooling powers for multilayer capacitors based on various electrocaloric and electrode materials. *Appl. Phys. Lett.* 95: 242903.

129 Bai, Y., Zheng, G., Ding, K. et al. (2011). The giant electrocaloric effect and high effective cooling power near room temperature for $BaTiO_3$ thick film. *J. Appl. Phys.* 110: 094103.

130 Liu, Y., Dkhil, B., and Defay, E. (2016). Spatially resolved imaging of electrocaloric effect and the resultant heat flux in multilayer capacitors. *ACS Energy Lett.* 1: 521–528.

131 Hirose, S., Usui, T., Crossley, S. et al. (2016). Progress on electrocaloric multilayer ceramic capacitor development. *APL Mater.* 4: 064105.

132 Kar-Narayan, S. and Mathur, N.D. (2010). Direct and indirect electrocaloric measurements using multilayer capacitors. *J. Phys. D: Appl. Phys.* 43: 032002.

133 Kar-Narayan, S., Crossley, S., Moya, X. et al. (2013). Direct electrocaloric measurements of a multilayer capacitor using scanning thermal microscopy and infra-red imaging. *Appl. Phys. Lett.* 102: 032903.

134 Yang, J., Zhao, Y., Zhu, L., and Hao, X. (2021). Enhanced electrocaloric effect of relaxor potassium sodium niobate lead-free ceramic via multilayer structure. *Scr. Mater.* 193: 97–102.

135 Zhu, L., Meng, X., Zhu, J. et al. (2021). Enhanced room temperature electrocaloric effect in lead-free relaxor ferroelectric NBT ceramics with excellent temperature stability. *J. Alloys Compd.* 892: 162241.

136 Zhang, Q.M. and Zhang, T. (2017). The refrigerant is also the pump. *Science* 357: 1094–1095.

137 Sinyavsky, Y.V., Pashkov, N.D., Gorovoy, Y.M. et al. (1989). The optical ferroelectric ceramic as working body for electrocaloric refrigeration. *Ferroelectrics* 90: 213–217.

138 Sinyavsky, Y.V. and Brodyansky, V.M. (1992). Experimental testing of electrocaloric cooling with transparent ferroelectric ceramic as a working body. *Ferroelectrics* 131: 321–325.

139 Brown, G.V. (1976). Magnetic heat pumping near room temperature. *J. Appl. Phys.* 47: 3673–3680.

140 Sinyavskii, Y.V. (1995). Analysis of the efficiency of an electrocaloric cryorefrigerator. *Chem. Pet. Eng.* 31: 501–506.

141 Gu, H., Qian, X., Li, X. et al. (2013). A chip scale electrocaloric effect based cooling device. *Appl. Phys. Lett.* 102: 122904.

142 Wang, Y.D., Smullin, S.J., Sheridan, M.J. et al. (2015). A heat-switch-based electrocaloric cooler. *Appl. Phys. Lett.* 107: 134103.

143 Plaznik, U., Kitanovski, A., Rožič, B. et al. (2015). Bulk relaxor ferroelectric ceramics as a working body for an electrocaloric cooling device. *Appl. Phys. Lett.* 106: 043903.

144 Sette, D., Asseman, A., Gérard, M. et al. (2016). Electrocaloric cooler combining ceramic multi-layer capacitors and fluid. *APL Mater.* 4: 091101.

145 Blumenthal, P., Molin, C., Gebhardt, S., and Raatz, A. (2016). Active electrocaloric demonstrator for direct comparison of PMN-PT bulk and multilayer samples. *Ferroelectrics* 497: 1–8.

146 Blumenthal, P. and Raatz, A. (2016). Classification of electrocaloric cooling device types. *Europhys. Lett.* 115: 17004.

147 Tušek, J., Kitanovski, A., Prebil, I., and Poredoš, A. (2011). Dynamic operation of an active magnetic regenerator (AMR): numerical optimization of a packed-bed AMR. *Int. J. Refrig.* 34: 1507–1517.

148 Guo, D., Gao, J., Yu, Y.J. et al. (2014). Design and modeling of a fluid-based micro-scale electrocaloric refrigeration system. *Int. J. Heat Mass Transfer* 72: 559–564.

149 Meng, Y., Pu, J., and Pei, Q. (2021). Electrocaloric cooling over high device temperature span. *Joule* 5: 780–793.

150 Gu, H., Qian, X.S., Ye, H.J., and Zhang, Q.M. (2014). An electrocaloric refrigerator without external regenerator. *Appl. Phys. Lett.* 105: 162905.

151 Zhang, T., Qian, X.S., Gu, H. et al. (2017). An electrocaloric refrigerator with direct solid to solid regeneration. *Appl. Phys. Lett.* 110: 243503.

152 Ma, M. and Yu, J. (2014). An analysis on a two-stage cascade thermoelectric cooler for electronics cooling applications. *Int. J. Refrig.* 38: 352–357.

16

Piezo-bioelectronic Materials and Emerging Applications

Yaojin Wang, Yang Liu, Lisha Liu, and Xiaolong Tang

Nanjing University of Science and Technology, School of Materials Science and Engineering, 200 Xiaolingwei Street, Xuanwu District, Nanjing, Jiangsu 210094, China

16.1 Introduction

Piezoelectric materials, which are organic and inorganic, can convert mechanical forces into electrical energy and vice versa. Inorganic piezoelectric crystals originate from the arrangement of ions in the asymmetric structure of dielectric materials [1]. The internal polarization of the material varies linearly with applied stress, resulting in electric field (EF) changes at the material boundary. The piezoelectric effect of organic piezoelectric polymers originates from molecular structures and orientations [2]. Piezoelectric biomaterials can interact with biological systems to enhance or replace natural functions. The piezoelectric biomaterials should be non-toxic, harmless, and non-immunogenic (also known as biocompatibility); moreover, they are widely used in tissue repair, biosensors, drug delivery, cancer treatment, and other applications. Bio-piezoelectric materials exhibit both piezoelectric and biomaterial properties and have attracted great research interest. Bio-piezoelectric materials are mammalian tissues, hair, keratin, bone, cartilage, ligaments, and tendons [2–4].

Among the numerous bio-piezoelectric materials, inorganic piezoelectric materials are biocompatible after being encapsulated and feature a high piezoelectric coefficient. The inorganic bio-piezoelectric materials include lead zirconate titanate, aluminum nitride, zinc oxide, barium titanate, lithium niobate, and quartz [5–10]. Organic bio-piezoelectric materials include polar polymers such as polyvinylidene fluoride (PVDF) and its copolymers, optically active polymers such as poly(L-lactic acid) and poly(D-lactic acid), and are characterized by biocompatibility, flexibility, and environmentally friendly properties [2, 11]. Devices constructed from piezoelectric polymers are less expensive in terms of material cost and processing; however, organic piezoelectric materials tend to exhibit low piezoelectric coefficients. To meet the application requirements, organic and inorganic bio-piezoelectric materials are combined to leverage their respective advantages [12, 13].

Piezoelectric Materials: From Fundamentals to Emerging Applications, First Edition. Edited by Jiagang Wu.

In this section, we introduce bio-piezoelectric materials and devices, structures and modifications, biosensing, disease treatment, and energy harvesting. We also summarized the future development and application of bio-piezoelectric materials.

16.2 Materials of Piezo-bioelectronic Devices

The development of piezoelectric materials for biomedical applications has experienced rapid progress over the past decade. Currently, bio-piezoelectric materials should possess high piezoelectric properties, and biocompatibility, or even biodegradability. In addition, high flexibility is essential for bio-piezoelectric materials to fit and deform synchronously with the target tissue conformally. From the point of view of chemical structure, piezoelectric materials currently used for biomedical applications can be roughly divided into three categories: inorganic ceramics, polymers, and biomolecular materials. Here, we give a summary of recently studied piezoelectric materials and their typical material properties (Table 16.1).

16.2.1 Inorganic Ceramics

Inorganic piezoelectric materials have received great attention owing to their excellent electrical output performances. Numerous inorganic piezoelectric materials have been developed for biomedicine applications. As a typical inorganic piezoelectric material, $Pb(Zr_{1-x}Ti_x)O_3$ (PZT) has been widely used in biosensing, disease treatment, and energy harvesting owing to its high piezoelectric coefficient, good electromechanical coupling, and inexpensive manufacturing cost. However, the biocompatibility of PZT must be considered owing to the presence of Pb in its material. Biocompatibility of PZT is vital for biomedicine applications. Several approaches, such as surface modification have been used to improve the biocompatibility of PZT, but the presence of Pb still causes environmental and human health issues. To overcome this problem, various lead-free ferroelectric materials have been developed. Lead-free piezoelectric materials, including $BaTiO_3$(BTO), $BiFeO_3$(BFO), $LiNbO_3$(LNO), $K_{0.5}Na_{0.5}NbO_3$ (KNN), $Na_{0.5}Bi_{0.5}TiO_3–BaTiO_3$ (NBT–BT), and Bi_2O_3, have been developed since the 1990s. However, numerous inorganic lead-free piezoelectric materials exhibit weak cytotoxicity, which limits their applications in the biomedical field, while their intrinsic hardness and brittleness still pose a great challenge. In addition, the techniques for fabricating high piezoelectric, flexible, and biocompatible inorganic piezoelectric materials are characterized by several limitations. To overcome these limitations, current research has focused on exploiting unique methods for preparing inorganic piezoelectric materials with outstanding characteristics. The most common method used is the direct preparation of inorganic piezoelectric materials onto flexible substrates. For example, Hong Goo Yeo et al. prepared a PZT film on an ultra-thin Ni steel metal foil substrate via an radio frequency sputtering (RF-sputtering) method for wrist-worn piezoelectric energy harvesters [29]. Ding Wang et al. prepared a high-quality PZT

Table 16.1 Summary of typical properties of some representative piezoelectric materials.

Compounds	Piezoelectric constant	Biocompatibility	Biodegradability	Published reports for biomedical applications	References
$BaTiO_3$	$d_{33} \sim 190$ pC/N	Low toxic	Nondegradable	Extensive	[14]
$Pb(Zr_{1-x}Ti_x)O_3$	$d_{33} \sim 590$ pC/N	Potential toxicity of metal	Nondegradable	Extensive	[15]
$K_{0.5}Na_{0.5}NbO_3$	$d_{33} \sim 570$ pC/N	Low toxic	Nondegradable	Extensive	[16]
$LiNbO_3$	$d_{33} \sim 16$ pC/N	Low toxic	Nondegradable	Extensive	[17]
$BiFeO_3$	$d_{33} \sim 50$ pC/N	Low toxic	Nondegradable	Considerable	[18]
AlN	$d_{33} \sim 6$ pC/N	Low toxic	Nondegradable	Considerable	[19]
$Pb(Mg_{1/3}Nb_{2/3})O_3–PbTiO_3$	$d_{33} \sim 2000$ pC/N	Potential toxicity of metal	Nondegradable	Extensive	[20]
ZnO	$d_{33} \sim 13$ pC/N	Low toxic	Nondegradable	Extensive	[21]
PVDF	$d_{33} \sim -20$ pC/N	Nontoxic	Nondegradable	Extensive	[22]
P(VDF-TrFE)	$d_{33} \sim -65$ pC/N	Nontoxic	Nondegradable	Extensive	[23]
PLLA	$d_{14} \sim 12$ pC/N	Biocompatible	Degradable	Extensive	[24]
Nylon	$d_{31} \sim 12$ pC/N	Nontoxic	Nondegradable	Rare	[25]
β-Glycine	$d_{16} \sim 190$ pm/V	Biocompatible	Degradable	Rare	[26]
Collagen	$d_{14} \sim 0.1$ pm/V	Biocompatible	Degradable	Considerable	[27]
Peptide nanotubes	$d_{15} \sim 60$ pm/V	Biocompatible	Degradable	Considerable	[28]

film on mica substrates for muscle movement energy harvesting [30]. The mica substrate is characterized by high thermal stability but can be mechanically exfoliated like graphene to a few tens of micrometers. Furthermore, inorganic-based laser lift-off (ILLO) has been widely used to transfer inorganic piezoelectric materials onto flexible substrates for biomedicine applications [31, 32]. ILLO is a physical lift-off transfer technique that uses a laser to vaporize the interface between the piezoelectric materials and the rigid substrate, allowing the transfer of the inorganic piezoelectric materials onto flexible substrates. For example, Dae Yong Park et al. transferred the PZT film to a flexible polyethylene terephthalate (PET) substrate through the ILLO process for self-powered real-time arterial pulse monitoring [33]. In addition, chemical etching is a useful technique for obtaining flexible piezoelectric films. For example, Dagdeviren et al. prepared a cutaneous pressure sensor through the chemical etching method [34].

16.2.2 Polymers

Polymers have received a lot of attention among piezoelectric materials due to their superior flexibility, varied designability, low processing temperature, and simplicity of processing over inorganic piezoelectric materials. More importantly, most piezoelectric polymers have low toxicity and are biocompatible, and some of them, such as poly-β-hydroxybutyrate and poly-3-hydroxybutyrate-3-hydroxy valerate, may even degrade naturally in the human body. These benefits make polymers attractive candidates for the development of tiny, implantable, and flexible electronic devices. PVDF is one of the most common piezoelectric polymers composed of $-CH_2CF_2-$ monomers. The strong electric moment is associated with the highest electro-negativity of fluorine atoms, and thus the PVDF can deliver a non-zero dipole moment. In general, PVDF can be divided into five crystalline phases according to the dipole arrangement, with the α-, β-, and γ- phases being the most common. The α-phase of PVDF exhibits the strongest dipole moment and superior piezoelectric properties since all the dipoles are parallel and contribute to the highest dipole moment per unit cell. Moreover, P(VDF-TrFE), as the PVDF copolymers, have been developed to further enhance the performance of PVDF, which leads to P(VDF-TrFE) copolymers being more suitable as a bio-piezoelectric sensor or copolymers piezoelectric energy harvester.

The piezoelectric effect has also been found in other polymers such as poly-L-lactic acid (PLLA) [24], poly-β-hydroxybutyrate (PHB) [35], poly vinyl chloride (PVC) [36], polyamides (PA), polyacrylonitrile (PAN) [37], polyurethanes (PU) [38] and odd-numbered nylons (such as nylon-11) [[25], p. 11].

However, the piezoelectric polymer still suffers from low piezoelectric coefficients compared to the inorganic piezoelectric materials described above, which results in low energy conversion efficiency and sensitivity. To improve the piezoelectric performance of piezoelectric polymers, a variety of strategies have been employed. For instance, electrospinning, spin coating, solution casting, mechanical drawing, polarization under a high electric field, and high-temperature heat treatment have been used to achieve a higher β phase ratio in PVDF. Moreover, Liu et al. reported stereochemically induced behavior in P(VDF-TrFE) copolymers, which are similar to the morphotropic phase boundaries of perovskite [23]. The P(VDF-TrFE) copolymer with the morphotropic composition exhibits a d_{33} of −63.5 pC/N, outperforming state-of-the-art piezoelectric polymers.

Piezoelectric composites, which combine piezoelectric polymers with inorganic ceramic fillers, are another way to improve piezoelectric performance. For example, Zhang et al. developed a piezoelectric composite with a three-dimensional (3D) interconnected structure [39]. The 3D interconnected structure presents a continuous path for load transfer, resulting in the excellent piezoelectric performance of the piezoelectric composite. In conclusion, piezoelectric composites combine the good flexibility and biocompatibility of piezoelectric polymers with the excellent piezoelectric performances of inorganic piezoelectric ceramics to provide materials with good comprehensive properties that can be used in the biomedical field.

16.2.3 Biomolecular Materials

Compared with the piezoelectric materials discussed above, biomolecular materials are attractive for the biomedical field due to their high biocompatibility and stable piezoelectric properties. It has been shown that the piezoelectric effect is also present in many hierarchical biomolecules, such as bones, ligaments, tendons, viruses, peptides, proteins, polypeptides, or amino acids. Biomolecular piezoelectric materials can be used to promote tissue growth and wound healing due to the generation of surface charge polarization or electric field in response to mechanical stimulation. Amino acids are the basic units of biomolecules, consisting of a carboxyl group (–COOH), an amino group ($-NH_2$), and a variable side chain. The differences between various amino acids depend on their side chain structures. For example, Glycine is a polymorphic amino acid that can form three different crystal structures (α, β, and γ structures) depending on crystallization conditions, as shown in Figure 16.1a. α-Glycine is centrosymmetric and therefore non-piezoelectric, while β- and γ-glycine have non-centrosymmetric structures and exhibit shear piezoelectricity. The d_{16} of β-glycine was measured to be around 190 pm/V, which is comparable to traditional organic piezoelectric materials [26]. The high piezoelectricity of glycine, as well as its biocompatibility and controlled biodegradability, lead to this piezoelectric biomolecular material being potentially useful for biomedical applications.

Peptides and proteins with various structures and functions are constructed using amino acids as basic units. Amino acid sequences and spatial configurations determine the biological function of peptides and proteins, resulting in structure-dependent piezoelectric properties. For instance, the self-assembled diphenylalanine nanotubes have non-centrosymmetric hexagonal structures (Figure 16.1b), thus exhibiting piezoelectricity. Similarly, proteins, especially collagen, exhibit piezoelectricity due to their asymmetrical spatial structure. Collagen is a triple helix structure formed by three twisted polypeptide chains. There are abundant polar groups and charged groups in the main chain, as shown in Figure 16.1c.

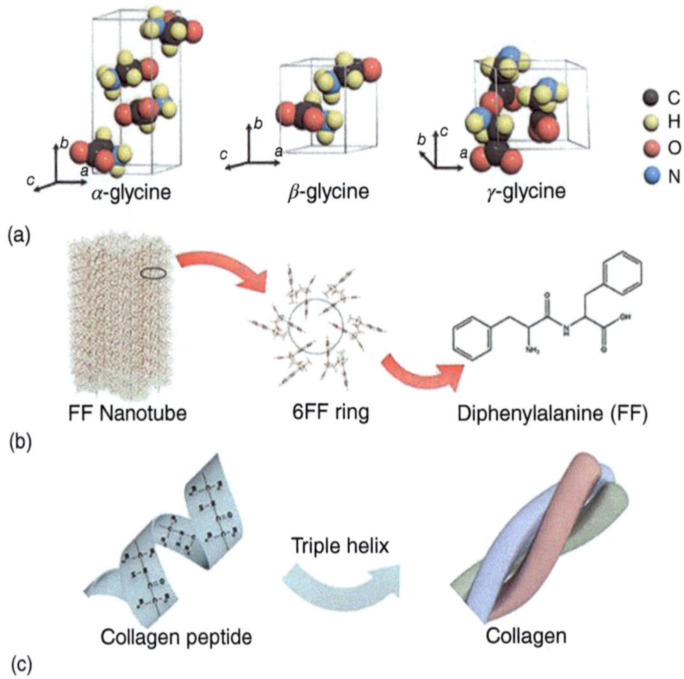

Figure 16.1 The atomic structures of piezo-bioelectronic materials: (a) glycine, (b) diphenylalanine (FF) peptide nanotube, and (c) collagen. Source: Xu et al. [5]/ John Wiley & Sons/CC BY 4.0.

Besides, piezoelectricity has been found in many specific biological tissues. Due to the presence of piezoelectric protein molecules, some biological tissues, including bones and tendons, show the piezoelectric effect, such as bones [40, 41], small intestine [42], and tendons [43]. Similarly, the piezoelectric effect has also been discovered in plant tissues. For example, piezoelectricity in wood is caused by the displacement of crystalline cellulose in reaction to mechanical stress, which results in the creation of piezoelectric charges [44]. Furthermore, the biocompatibility and biodegradability of these biomolecular piezoelectric materials are particularly advantageous for implantable applications, which suggests that biomolecular material-based devices have considerable promise for biomedicine applications.

16.3 Fabrication of Piezo-bioelectronic Device

16.3.1 Structure Design

16.3.1.1 0D

In general, nanoparticles (NPs), nanoclusters, and quantum dots (QDs) are referred to as 0D piezo-bioelectronic nanomaterials. Large surface areas, superior piezoelectric capabilities, and inherent single domains are all hallmarks of 0D piezo-bioelectronic nanomaterials. Based on WS_2 QDs (3 nm), Yong et al. reported

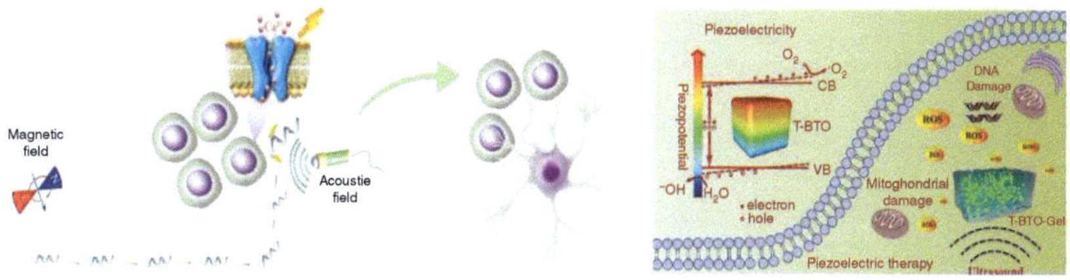

(a) (b)

Figure 16.2 (a) Schematic illustration of a highly controllable micromotor to induce the differentiation of the targeted neural stem-like cell [47]. Source: Reproduced with permission of John Wiley & Sons. (b) BTO-mediated piezoelectric therapy triggered by ultrasonic vibration [48]. Source: Reproduced with permission of John Wiley & Sons.

that multifunctional nanomedicine has a significant effect on X-ray computed tomography (CT)/photoacoustic (PA) imaging signal enhancement and a remarkable effect on photothermal therapy (PTT)/radiotherapy (RT) for treating tumors [45]. With CT/PA imaging and the synergistic effect of PTT and RT, the tumor could be located accurately and completely removed in vivo after WS_2 QDs were injected into the bloodstream. The scavenging of over-accumulated free radicals [46], and the specific differentiation of neural stem cells in vivo (Figure 16.2a) [47], were achieved using 0D piezoelectric bioelectronic materials, respectively.

Sonodynamic therapy (SDT), a promising noninvasive treatment approach, has attracted an increasing amount of interest in recent years. Low-intensity ultrasonic (US) activates its specific chemical agents, known as sonosensitizers, to create deadly reactive oxygen species (ROS) for oncotherapy. SDT offers numerous remarkable prospects and advantages over phototherapeutic approaches, including deeper penetration depth, the absence of phototoxicity, and fewer side effects. Figure 16.2b illustrates a $BaTiO_3$ NPs-mediated piezoelectric therapy triggered by ultrasonic vibrations [48]. Additionally, to gain enhanced piezoelectric capabilities, 0D piezo-bioelectronic NPs can be integrated with other piezo-bioelectronic materials, such as polymer films. The benefits of 0D piezo-bioelectronic nanomaterials enable their use in a variety of flexible biological applications, including disease therapy and bio-catalysis.

16.3.1.2 1D

Nanowires, nanobelts, nanotubes, nanorods (NR), and nanofibers are some of the common forms of 1D piezo-bioelectronic nanomaterials employed today. Due to their wirelike shape, 1D piezo-bioelectronic nanomaterials demonstrate better charge transfer efficiency than 0D piezo-bioelectronic nanomaterials. They can also get beyond the aggregation issues that many 0D piezo-bioelectronic nanomaterials have. Thus, 1D piezo-bioelectronic nanomaterials show excellent piezoelectric properties, high sensitivity, and outstanding flexibility. In piezoelectric biosensing or energy harvesting systems, many 1D piezo-bioelectronic nanomaterials, including PVDF and ZnO nanowires [49], have been employed as electromechanical conversion components. Under their superior mechanical flexibility, PVDF nanowires, for instance, can bear significant strain and have a

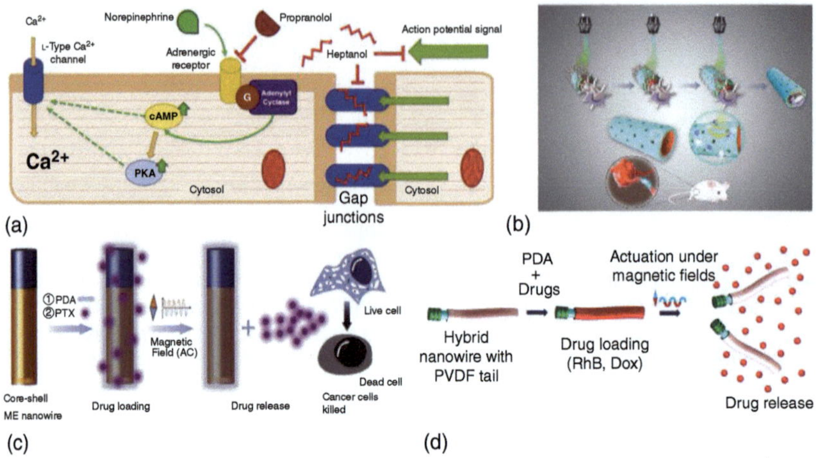

Figure 16.3 (a) Scheme showing the pharmacological action of the drugs tested [49]. Source: Reproduced with permission of Elsevier. (b) Scheme of injectable multilayer biofabrication for the ZnO-loaded polycaprolactone (PCL) conduit [50]. Source: Reproduced with permission of John Wiley and Sons. (c) Scheme showing on-demand drug delivery experiment. The anticancer drug is loaded onto a PDA-treated FeGa@P(VDF-TrFE) core-shell nanowire and then released by applying an alternating magnetic field. PDA, polydopamine; PTX, paclitaxel [51]. Source: Reproduced with permission of John Wiley and Sons. (d) Scheme depicting functionalization of hybrid nanoeels with PDA and drugs, followed by magnetically triggered drug release [52]. Source: Reproduced with permission of John Wiley and Sons.

long operational lifetime. To improve piezoelectric performance, it is very simple to include 1D piezo-bioelectronic nanomaterials in piezoelectric nanowire arrays. It is important to note that the majority of 1D biopiezoelectric nanomaterials have good biocompatibility, enabling widespread use in electronic skins, smart fabrics, and biosensors. As shown in Figure 16.3a,b, promoting drug transport by modulating ion channels [50] and forming microelectronic field stimulation to induce the proliferation of nerve cells [51] have been demonstrated. In Figure 16.3c,d, 1D piezo-bioelectronic nanomaterials have promising applications in drug release under the synergistic effect of multiple physical fields [52]. Targeted drug delivery can also be achieved by surface functionalization treatment [53].

16.3.1.3 2D

Piezoelectric materials, and even typical nonpiezoelectric materials, can lose their centrosymmetry in one direction and exhibit an increased amount of piezoelectricity when thinned to a nanoscale thickness to generate a 2D shape. The produced planar 2D piezo-bioelectronic nanomaterials come in a variety of 2D shapes, including nanoplatelets, nanoplates, nanosheets, and nanoflowers. Examples of 2D piezo-bioelectronic nanomaterials include monolayer transition metal dichalcogenide, boron nitride, carbon nitride, and black phosphorus [49]. Monolayer MoS_2, which is a typical in-plane piezo-bioelectronic material, has a Mo plane sandwiched between two S planes, resulting in a triangle prism structure with Mo atoms at

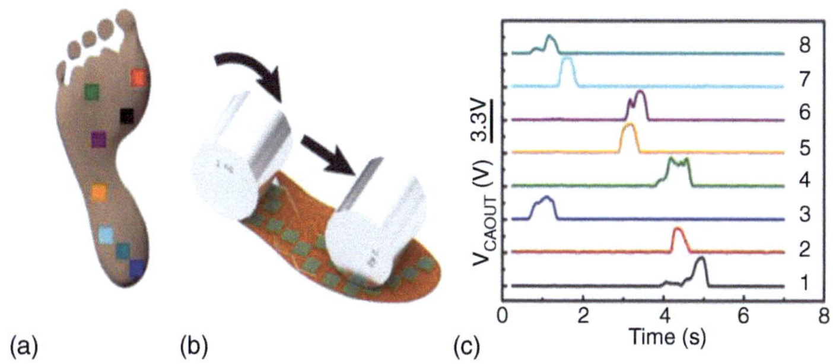

Figure 16.4 (a) Schematic diagram of the placement of selected eight sensor nodes. (b) Schematic of a cylinder rolling along different piezo-pixels. (c) The real-time measurement of the selected eight sensor nodes in the moving trajectory of the cylinder [53].

its core. The displacement of the Mo^{4+} and S_2 molecules, which results in the generation of an electric dipole and polarization charges on the surface, gives it piezoelectricity. Because of their exceptional biocompatibility, high redox activity, photosensitivity, and thermosensitivity, MoS_2 nanosheets have also been widely researched in the field of biomedicine. MoS_2 nanosheets have a lot of potential for biosensor and bioelectronic applications, thanks to all the aforementioned characteristics.

In addition, 2D piezoelectric films are also widely adopted in biosensors and bioelectronics. As shown in Figure 16.4a–c, a self-powered, continuous, real-time pressure distribution monitoring system was developed by combining a hybrid triboelectric-electromagnetic nanogenerator that uses a piezoelectric nanogenerator as a sensor array to acquire pressure signals and a self-designed data acquisition circuit board to process and wirelessly send the signals to a mobile terminal (e.g. smartphone) [54]. The entire system can accurately monitor and visualize real-time pressure distribution during walking, providing a viable solution for sports/exercise biomechanical information acquisition, injury prevention, and foot ulcer prediction. Liu et al. have reported a self-powered photodynamic therapy (s-PDT) system with two different irradiation modes that can be autonomously managed by the patient and meets the requirements of the development of telemedicine and Internet medicine. The system is based on twin-structured piezoelectric nanogenerators for s-PDT, driven by energy obtained from body movements, to achieve effective tumor tissue killing and inhibition [55]. This innovative s-PDT system provides a good example of the development of wearable/implantable and self-controllable long-term photodynamic therapy devices.

16.3.1.4 3D

As shown in Figure 16.5, common 3D piezo-bioelectronic materials are mainly embodied in bio-scaffolds and core-shell structures. The use of biocompatible materials (hydrogel [56], polylactic acid [57], etc.) as scaffolds and composite functional materials has become one of the mainstream approaches for current practical applications. 3D scaffolds provide adequate mechanical support and are commonly

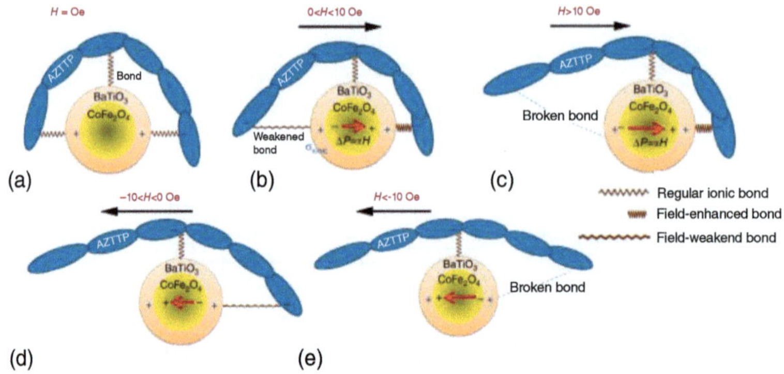

Figure 16.5 Underlying physics of the triggered on-demand drug (AZTTP) release by magnetoelectric NPs (MEN) stimulated by a magnetic field. (a) At zero field, only the ionic charge is present in the MEN shell. (b) An additional dipole moment (proportional to the magnetic field) breaks the original symmetry of the charge distribution in the shell. (c) As the field is increased above the threshold value, the bond on one side is broken. (d, e) The field is reversed to break the bond on the opposite side of the nanoparticle (NP). The red arrows show the electric dipole due to the magnetoelectric effect (ME) effect. In practice, owing to the random configurations of nanoformulations with respect to the field, the effect is present along with every central bond orientation [55]. Source: Reproduced with permission of Elsevier.

used for tissue repair. On-demand release of azidothymidine 50-triphosphate, an anti-human immunodeficiency virus drug, using $CoFe_2O_4$@BTO core-shell structure, based on magneto-piezo synergy, enables targeted drug delivery across the blood-brain barrier [58]. This approach enriches anti-HIV treatment while providing a reference for other therapeutic modalities that require targeted drug delivery.

16.3.2 Modification and Packaging

To satisfy the stringent criteria of biomedical applications, piezo-bioelectronic devices should have excellent piezoelectric performance as well as adequate surface qualities that are customized to their use, such as hydrophilicity, roughness, and porosity. More importantly, the biocompatibility and biodegradability of the device are critical for biomedical applications to ensure biosafety. Many piezoelectric materials, however, are unable to meet all of the aforementioned specifications. The surface modification and packaging of piezoelectric material are crucial for achieving the application of devices.

Kang et al. designed a core-shell ZnO@microporous organic polymer nanosphere to construct multifunctional energy harvesting devices [56], as shown in Figure 16.6a. The triboelectric microporous organic polymer is uniformly coated with piezoelectric ZnO NR, which results in enhanced energy harvesting performance. Marino et al. prepared a platform based on BTO NPs functionalized with an anti-HER2 antibody for ultrasound-activated anticancer treatment, as shown in Figure 16.6b [59]. The modified BTO NPs can target and remotely stimulate

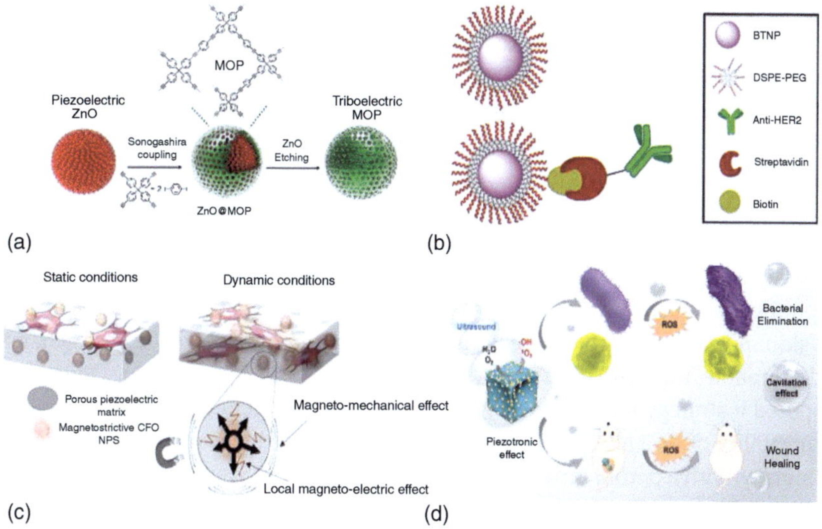

Figure 16.6 (a) Synthesis of ZnO@MOP [56]. Source: Reproduced with permission of John Wiley and Sons. (b) Schematic of the functional modification of BTO NPs with anti-HER_2 antibody through 1,2-distearoyl-*sn*-glycero-3-phosphoethanolamine-*N*-[methoxy (polyethylene glycol)-5000] -biotin coating [59] Springer Nature / CC BY 4.0. (c) Schematic of magneto-mechanical and local magneto-electrical properties of 3D scaffolds under magnetic stimulation [58]. Source: Reproduced with permission of American Chemical Society. (d) Schematic illustration of sonodynamic bacterial elimination and wound healing based on piezoelectric nanocomposites Au@BTO [60]. Source: Reproduced with permission of Elsevier.

human epidermal growth factor type 2 receptor (HER2)-positive breast cancer cells because HER2-positive breast cancers possess higher growth rates and a higher probability of generating metastases and invading other tissues. Wu et al. developed BTO nanocubes with Schottky junctions modified by Au NPs for highly efficient SDT, as shown in Figure 16.6c [60]. The Schottky barrier that forms between Au and BTO facilitates the separation and migration of charge carriers at the piezoelectric/metal interface, which further effectively increases ROS generation via a redox reaction. Fernandes et al. reported a magneto-active 3D porous scaffold based on a piezoelectric polymer of PVDF and magnetostrictive particles of $CoFe_2O_4$ for bone regeneration, as shown in Figure 16.6d [58]. The negatively charged $CoFe_2O_4$ NPs interact with the positively charged CH_2 groups of the bio-piezoelectric PVDF, thereby inducing the nucleation of the electroactive β-phase of the polymer and improving magnetoelectric coupling.

The surface structure of piezoelectric materials is an essential factor for biomedical applications. For example, the wettability of the material surface can control cell attachment and spread during tissue regeneration. A sufficient amount of porosity promotes blood circulation and the transport of physiologically active chemicals. In addition, the proper surface roughness of the material surface is beneficial to cell adhesion and proliferation. For instance, Qian et al. developed a piezoelectric scaffold manufactured by 3D injectable multilayer biofabrication technology [50].

ZnO with nontoxic, ecofriendly, antimicrobial, and proangiogenic characteristics is loaded into PCL to increase the surface roughness of the scaffold and induce polarization of its surface-loaded PCL. The fabricated scaffold exhibits the potential to promote cell adhesion and reproduction and thus demonstrates the potential to repair large nerve defects and restore biomimetic microenvironments for sensory and functional activities.

Biocompatibility and biosafety are the key factors determining the successful application of piezoelectric materials in biomedicine. For example, the cells in the outermost layer of the skin rely entirely on oxygen absorbed through surface respiration for their metabolic activities. Therefore, electronic devices attached to the skin must consider the characteristics of human skin, such as the gas exchange and secretion functions of the body surface. Inspired by human skin, Chen et al. proposed a strategy for fabricating a biocompatible temperature sensor with good permeability of air and high quality that is water-proof by using a semipermeable membrane with a porous structure as the base material, as seen in Figure 16.7a [61]. In this study, the waterproof and breathable properties are mainly achieved by using the porous structure of the encapsulation layer and the substrate. There are many pores with sizes ranging from a few tens of nanometers to several micrometers in the semipermeable film. The pore sizes are larger than the size of air and water vapor molecules, but smaller than that of liquid water droplets and bacteria, allowing the device to be permeable for the perspiration to go out as water vapor to avoid maceration and outside gas (such as O_2) to get in and out for breathing. The water vapor transmittance of the fabricated devices with porous structures is significantly higher than that of the skin, which ensures that the device will not affect the respiration of human skin. The volunteer wears such devices for 24 hours with two times of shower and the in vitro test shows no sign of maceration or stimulation

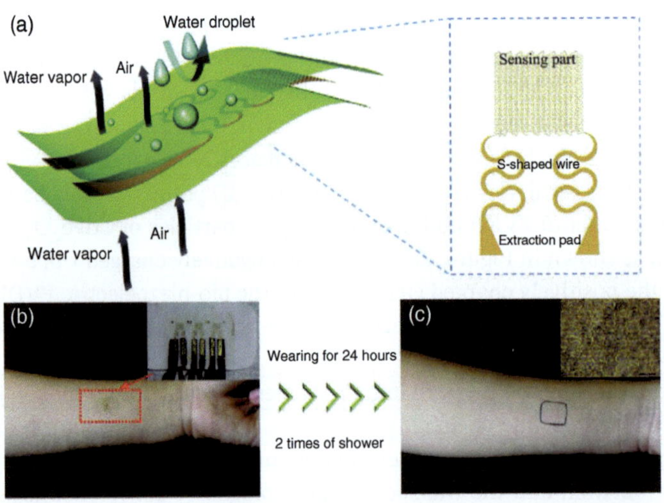

Figure 16.7 (a) The skin-like temperature sensor's illustration. (b) Wearing the device for 24 hours with two times of shower. (c) The skin under the device after 24 hours' wearing [63] / Springer Nature / CC BY 4.0.

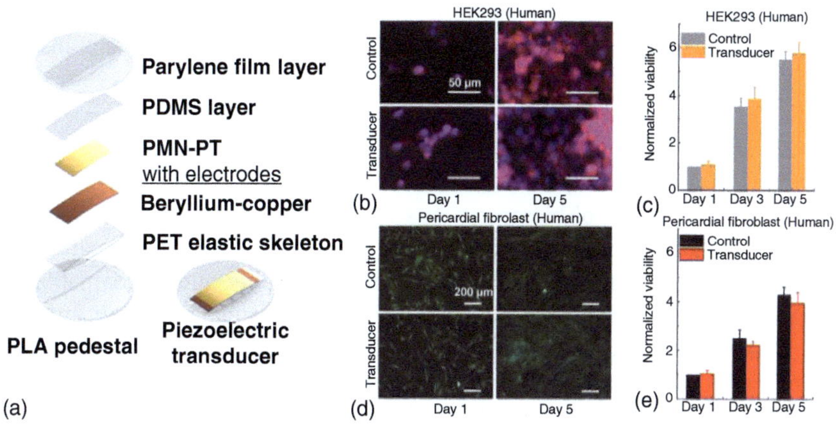

Figure 16.8 (a) A 3D model of piezoelectric device. Fluorescent images of (b) Human HEK293 cells and (d) Human pericardial fibroblasts. Normalized viability of (c) HEK293 and (e) Pericardial fibroblasts after being cultured for one, three, and five days [63]. Source: Reproduced with permission of John Wiley and Sons.

to the skin (Figure 16.7b,c), demonstrating the biocompatibility and safety of the devices.

The toxicity of lead-based piezoelectric materials due to the presence of lead reduces their biocompatibility. As a result, efforts have been made to improve the biocompatibility of lead-based piezoelectrics through modification and packaging. Yi et al. encapsulated $Pb(Mg_{1/3}Nb_{2/3})O_3$–$PbTiO_3$ (PMN–PT) single crystal by polydimethylsiloxane (PDMS) layer and parylene film layer, as seen in Figure 16.8a [63]. Two different cell types, HEK293 cells (Figure 16.8b) and primary pericardial fibroblasts (Figure 16.8d), derived from tissues surrounding the implantation site, were selected as cellular models to evaluate the biocompatibility of the devices. These cells were grown in culture dishes alongside the piezoelectric devices. The cells exhibited stable growth rates, intact cytoarchitecture, and unbroken nuclei, demonstrating that the piezoelectric devices did not influence cellular morphology. The viability of two different cells in the control group did not differ significantly from those of the experimental groups, as shown in Figure 16.8c,e.

16.4 Emerging Biomedical Application of Piezo-bioelectronic Devices

The electromechanical coupling properties of piezoelectric materials enable them to convert external stimuli (pressure, acoustic waves, strain, and motion) into electrical energy. This enables piezoelectric materials not only to overcome the limitation of frequent battery replacement but also to effectively sense weak environmental changes for real-time and self-powered biosensing. Furthermore, the charge produced by poled piezoelectric materials can regulate various physiological behaviors, such as cell growth, migration, differentiation, and apoptosis, to achieve physiological regulation and disease treatment. In general, the biomedical applications of piezoelectric materials fall into three broad categories: biosensing, disease treatment, and energy harvesting.

16.4.1 Biosensing

The first demonstration of biosensors was established by Leland Clark in 1962. Thereafter, biosensors have developed rapidly toward miniaturization, wearability, and portability. However, the structural flexibility, reliability, and biocompatibility of conventional capacitor-based biosensors are still inadequate for their applications. More importantly, these biosensors require an additional power supply, thus requiring periodic battery replacement and increasing the size of the biosensors. To address this challenge, piezoelectric-based biosensors are under intensive investigation.

The output voltage (V_{ij}) generated by a piezoelectric sensor when subjected to stress (σ) can be expressed as follows:

$$V_{ij} = \sigma g_{ij} L_{ij} \tag{16.1}$$

where g_{ij} is the piezoelectric voltage constant and L_{ij} is the distance between the electrodes. As illustrated in Eq. (16.1), the output voltage of a piezoelectric sensor is linearly related to the applied strain or stress. Therefore, piezoelectric sensors can be used to detect and measure stimuli associated with external forces. In addition, no external power source is required to generate output voltage when the mechanical stress is applied, thus eliminating the requirement to include, manage, or replace the battery. On this basis, self-powered piezoelectric sensors can be developed as a highly desirable solution for biosensing applications because of their advantages of flexible structure, high sensitivity, and high accuracy. In recent years, piezoelectric biosensors have been widely developed for various vital sign monitoring and application scenarios, promising to open a new era for the development of biosensors.

16.4.1.1 Physical Health Monitoring

The monitoring of physical health signs can not only accurately assess the physiological state of individuals but also provide a reference for the diagnosis of related diseases. Recently, self-powered piezoelectric-based biosensors have been widely developed to monitor human vital signs, such as pulse wave, respiration rate, heart rate, and blood pressure, as shown in Figure 16.9. In addition, those sensors used for health monitoring need to have excellent flexibility to meet the requirements of attached to the skin and tissues of the human body. Pulse wave measurement is the most prevalent application since continuous monitoring of arterial pulsation is essential for detecting the early start of cardiovascular illness and continuously assessing health status. Based on this, Park et al. proposed a self-powered piezoelectric pulse sensor based on flexible PZT films prepared by the ILLO process [33]. As seen in Figure 16.10a, PZT films are transferred to an ultra-thin PET substrate, thus the piezoelectric pulse sensor can complete conformal attachment on the skin in response to the tiny human pulse wave. The piezoelectric pulse

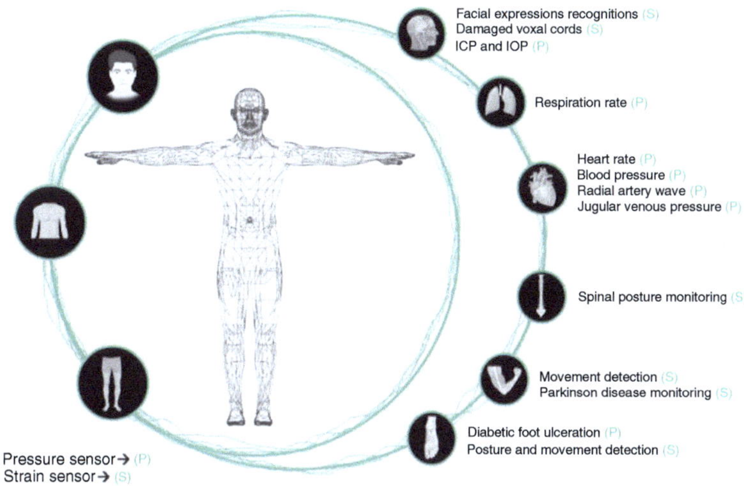

Figure 16.9 Piezoelectric-based biosensors for physical health monitoring applications [64]. Source: Reproduced with permission of Elsevier.

sensor is conformally attached to the wrist using a biocompatible liquid bandage spray, as shown in Figure 16.10b. Piezoelectric pulse sensor signals before and after exercise are shown in Figure 16.10c. The output voltage of the sensors before and after exercise is 65 and 81.5 mV, and the heart rate is ~73 and 100, respectively. In addition, the output signals of the piezoelectric sensor can accurately reveal the characteristic peak of the peripheral artery waveform, as shown in the inset of Figure 16.10c, which contains important biomedical and physiological information, such as arteriosclerosis, coronary artery disease, and myocardial infarction. Piezoelectric sensors can also be used to monitor other human activities, such as carotid pulse and muscle movement, as shown in Figure 16.10d. The output voltage from the carotid artery pulse and saliva swallowing action are 0.4 and 1 V, respectively, as shown in Figure 16.10e.

Due to the good dynamic response and high fidelity inherent in piezoelectric sensors, the arterial pulse wave velocity (PMV) extracted from the pulse waveforms can be used for continuous blood pressure assessment. Petritz et al. developed a P(VDF-TrFE)-based piezoelectric sensor integrated on an ultrathin substrate (1 μm), which greatly improves the sensitivity of the sensors [65]. The PMV can be determined by measuring the pulse wave with two sensors placed a certain distance from each other. As suggested by Ma et al. [66], the human blood pressure P_B (in the range of 5–20 kPa) can be estimated from the PWV as:

$$P_B = \alpha \cdot \text{PMV}^2 + \beta; \alpha = 0.18\,\text{kPa} \cdot S^2; \beta = 2.7\,\text{kPa} \tag{16.2}$$

yielding a blood pressure P_B of 130 mmHg for the 34-year-old man. Thus, the increased sensitivity of ultra-flexible sensors to strain and pressure (>10^3 pC/N), rapid response (<20 ms/N), and excellent mechanical stability result in a wireless electronic health patch for accurate pulse and blood pressure monitoring. However, blood pressure evaluation based on multiple piezoelectric sensors is inherently

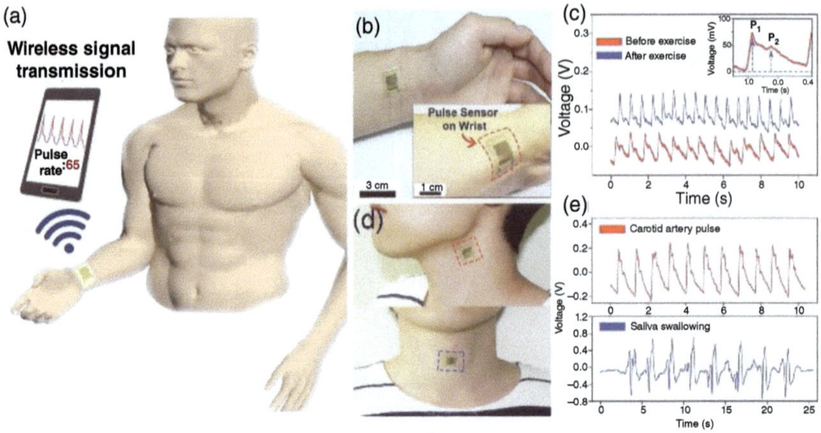

Figure 16.10 (a) Conformal contact of a sensor on a human wrist, and wireless transmission of detected pulse signals to a smartphone. (b) Photograph of a piezoelectric pulse sensor conformally attached to human (b) wrist and (d) neck. (c) Radial artery pulse signals detected by the piezoelectric pulse sensor. (e) The output voltage generated by the carotid artery and swallowing action [33]. Source: Reproduced with permission of John Wiley and Sons.

limited by time synchronization errors and distance errors between different sensors. To end this, Yi et al. developed a single piezoelectric sensor based on arterial-pulse piezoelectric dynamics for real-time blood pressure monitoring [67]. Although arterial pulse waves are directly related to blood pressure waves, the latter are affected by body posture, arterial vessels, and skin characteristics. Therefore, the correlation between piezoelectric pulse waves and blood pressure waves is first elucidated via theoretical analysis of piezoelectric dynamics, as shown in Figure 16.11a. As shown in Figure 16.11b, arterial pulse waves in different postures were obtained using two piezoelectric sensors located near the radial artery of the elbow and the proper palmar digital artery, indicating that arterial pulse waves are sensitive to body postures. Hence, arterial pulse wave monitoring requires extremely precise posture recognition to eliminate motion-derived artifacts when blood pressure is measured. The postural specificity of these eight arterial pulse waves is evaluated by a pattern recognition network, and the model verification results showed that the accuracy of the trained model is up to 96.5%. Figure 16.11c,d show the pulse waves produced by processing the original output voltage obtained at the finger and elbow, respectively. The final pulse waves are consistent with the peak value of blood pressure waves, which proves the feasibility of blood pressure assessment based on arterial pulse piezoelectric dynamics. Based on this, a wearable blood pressure monitoring system consisting of two piezoelectric sensors and a wireless data acquisition module is proposed to monitor the arterial pulse of the lateral epicondyle and digital pulp of the index finger. Continuous blood pressure waves obtained by processing raw pulse data collected using the wireless monitoring system are shown in Figure 16.11e, which shows that the single piezoelectric sensor strategy based on arterial-pulse piezoelectric dynamics enables continuous blood pressure wave monitoring.

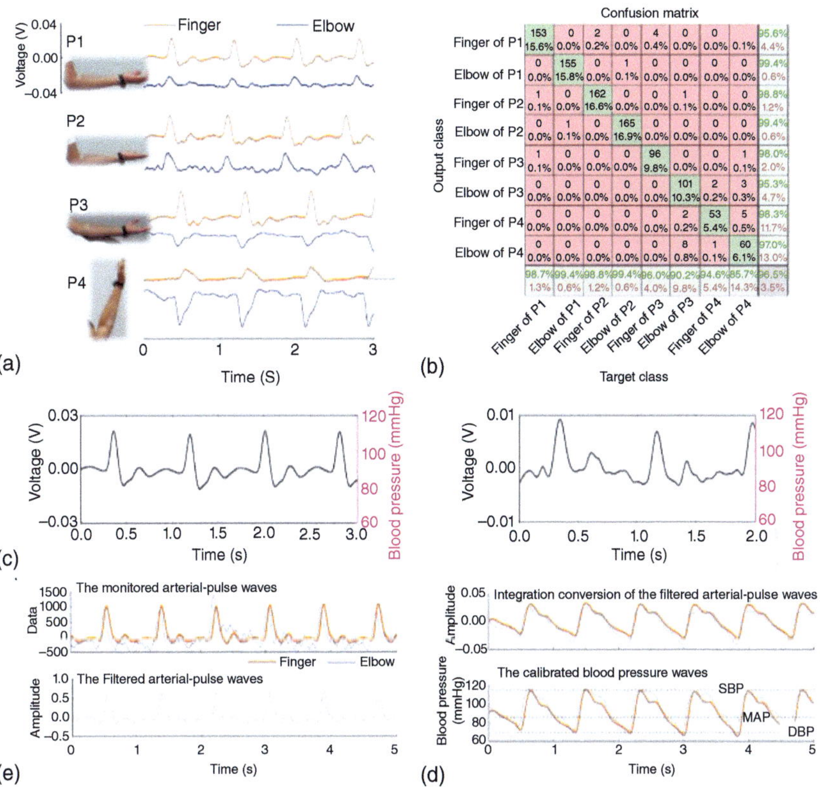

Figure 16.11 (a) Arterial pulse waveforms obtained using piezoelectric sensors on the finger and elbow. (b) Photograph of a piezoelectric pulse sensor conformally attached to a human (b) Confusion matrix of the arterial pulse waves. Processed arterial pulse waves and the calibrated pressure waves from the (c) digital and (d) radial arteries. (e) Blood pressure wave evaluation [67]. Source: Reproduced with permission of John Wiley and Sons.

Polymer-based piezoelectric biosensors have extraordinary flexibility, are responsive to weak deformation and force, and are suitable to sensitively monitor physiological signals. Li et al. developed core-shell PVDF/hydroxylamine hydrochloride (HHE) nanofibers with β-phase-rich and self-orientated nanocrystals for real-time monitoring of cardiovascular walls. PVDF/HHE piezoelectric nanofibers are prepared by electrospinning with HHE as an inducer, which enables the transformation of the nonpolar α-phase to the polar β-phase [68]. As a result, the PVDF/HHE nanofibers possessed excellent stability, high sensitivity and accuracy (1154 V/cm^3 at a pressure of 1 kPa), and good biocompatibility. Based on this, the PVDF/HHE piezoelectric sensor is implanted into the cardiovascular walls of the heart and femoral artery of a living pig to detect the micropressure changes in the distal and proximal parts of the cardiovascular system, as shown in Figure 16.12a. As seen in Figure 16.12b, the PVDF/HHE piezoelectric sensor recorded the subtle changes of blood pressure in the cardiovascular system during different stages of thrombus, which indicated that the sensor can be used as the

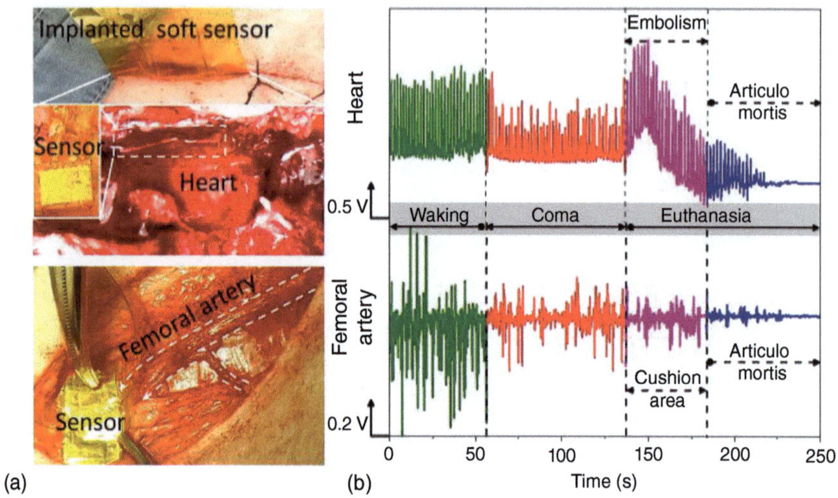

Figure 16.12 (a) Surgical images of the implanted soft sensors attached to the cardiovascular wall of the heart and femoral artery. (b) Output piezoelectric signals of the implanted sensor induced by the cardiovascular elasticity changes of pigs at different physiological states [68]. Source: Reproduced with permission of American Chemical Society.

basis for early diagnosis of acute and chronic cardiovascular diseases and medical prevention, especially for postoperative recurrence of deep thrombus.

So far, many different piezoelectric materials have been reported for pulse measurement. Recently, novel piezoelectric materials have also been developed in this sense: Chen et al. report a 2D layered indium selenide sensor with a large piezoelectric response and long-term mechanical durability that can be used to monitor pulse and respiratory rates in real-time by attaching the device to the wrist and chest [69]. Chen et al. proposed a flexible piezoelectric sensor based on a single-crystalline group III-nitride thin film [70].

Ingestible electronic devices with the ability to sense physiological and pathophysiological states have dramatically improved the standard of patient care. In particular, piezoelectric sensors for sensing the gastrointestinal lumen have expanded our ability to diagnose and treat patients. Dagdeviren et al. reported an ingestible and flexible piezoelectric sensor for monitoring vital signals and ingestion within the gastrointestinal tract [71]. A gastrointestinal lumen sensor consists of 12 groups of PZT ribbons connected in series, where each group consists of ten PZT ribbons connected in parallel and the sensor is then folded into a dissolvable gelatin capsule for implantation into the stomach cavity. The capsule is dissolved by gastric fluid after entering the stomach cavity, and the sensor can be naturally unfolded and placed on the stomach lining in immediate juxtaposition with the mucosa, which allows the sensor to monitor the physiological state of the gastrointestinal tract. The inflation and deflation trials were recorded by the gastrointestinal lumen sensor. When the stomach is filled with air, the gastrointestinal lumen sensor experiences a sudden pressure change, which causes the output voltage to increase significantly from

10 to 60 mV. During the subsequent 40s, the internal pressure stabilized and resulted in a substantially stable output voltage, while when air is released, the voltage drops significantly from 40 mV to the base state. The results of sensing gas changes suggest that the sensor could be used to guide the assessment and treatment of certain diseases that cause gas changes in the stomach cavity, such as aerophilia or significant small intestinal bacterial overgrowth.

The normal breathing rate of a healthy adult is about 12–20 breaths per minute. Tachypnea, that is, more than 20 breaths per minute, is a common clinical symptom of the respiratory system, often due to respiratory insufficiency caused by respiratory diseases or respiratory organ or tissue lesions. If undetected, the disease will further aggravate, leading to the occurrence of respiratory distress, breathing difficulties, and even respiratory failure. Therefore, it is of great significance to real-time monitor human respiratory conditions through wearable piezoelectric sensors. Liu et al. developed a wearable, self-powered respiratory and health monitoring sensor based on electrospinning PVDF piezoelectric nanofibers. The piezoelectric sensor showed great flexibility and deformation-recovery ability, as shown in Figure 16.13a. To monitor human respiratory conditions, the piezoelectric sensor is applied to the human chest, as seen in Figure 16.13b. The respiration rates, recorded by the piezoelectric sensor and Biopac MP150 (commercial physiological signal recording system), are shown in Figure 16.13c,d, and the results indicate that the sensor has excellent sensitivity, reliability, and feasibility as a respiration sensor.

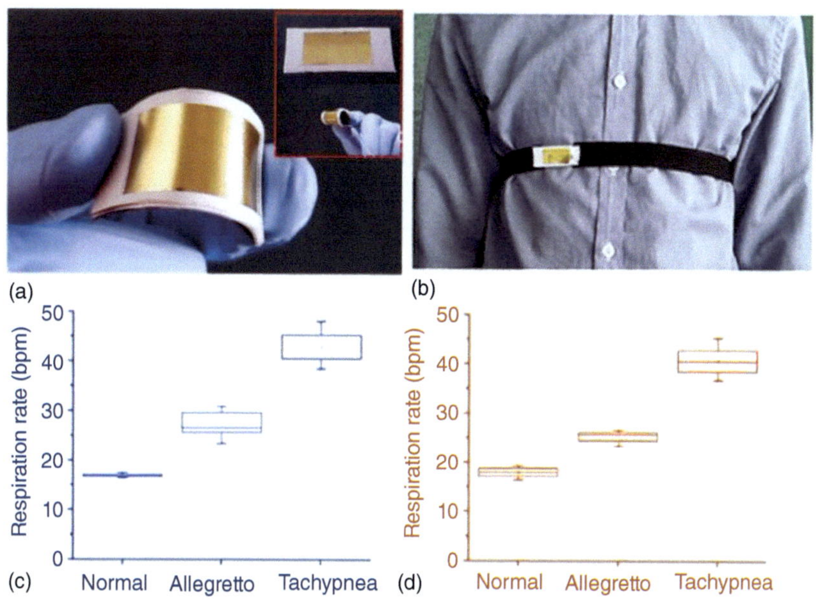

Figure 16.13 (a) Photograph of the piezoelectric sensor. (b) Photograph of the piezoelectric sensor attached to the human chest. Statistical graph of respiration rate for the (c) piezoelectric sensor and (d) Biopac MP150 [72]. Source: Reproduced with permission of IOP Publishing.

16.4.1.2 Biomedical Pressure Sensing

Biomechanics signals associated with the human body can also provide basic physical health information. Persano et al. reported a flexible, self-powered, and light-weight pressure sensor based on electrospinning P(VDF-TrFE) nanofibers [73]. The piezoelectric sensor consists of a soft PDMS post that delivers well-defined levels of pressure to the sensor. The output voltage varies linearly with the applied pressure, indicating that the sensor is capable of measuring small pressures ($\approx$0.1 Pa). Based on this, a piezoelectric sensor can be attached to the finger for biomedical pressure detection. In addition, piezoelectric sensors can also be integrated into solid supports and act as orientation sensors. The results suggest that the capabilities of the sensor could be valuable for a range of biomedical applications.

16.4.1.3 Prosthetic Devices

Bionic prosthesis is an advanced device that mimics the physical and functional properties of a replaced limb, thereby restoring near-natural form and function. For biomimetic prostheses, sensory feedback is highly dependent on the property of the sensor, which requires streamlining the flow of information processing. Piezoelectric sensors are considered to replace traditional sensing elements due to their excellent sensitivity, rapid dynamic response, and convenient signal capture and analysis. Chen et al. proposed an artificial sensory synapse based on a piezoelectric graphene with an ion gel–gated transistor, as shown in Figure 16.14a [74].

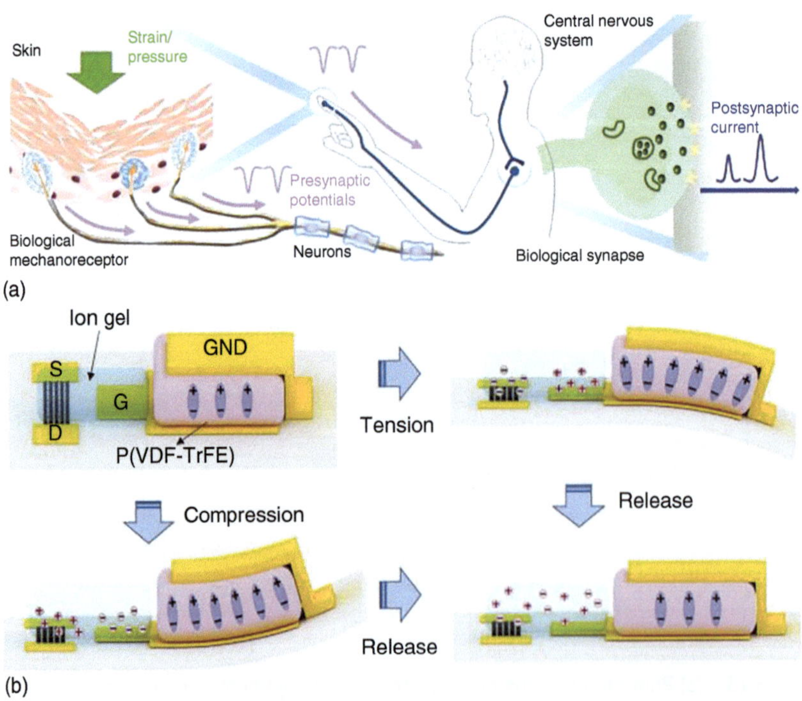

Figure 16.14 (a) The biological sensory nerve system. (b) The working mechanism of the piezotronic graphene transistor [74]. Source: Reproduced with permission of John Wiley and Sons.

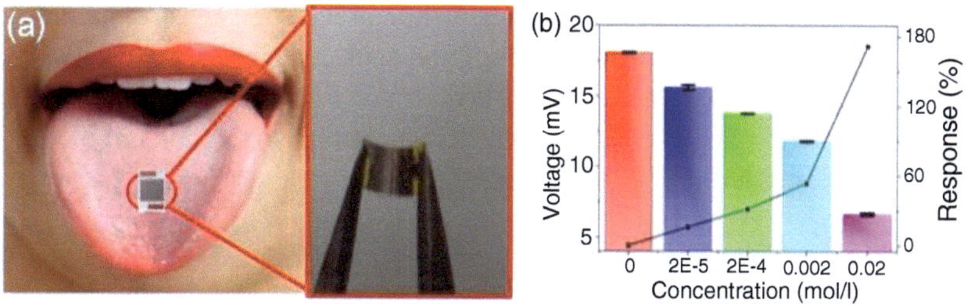

Figure 16.15 (a) The e-skin can mimic taste buds on the tongue. (b) Output piezoelectric voltage and the response [75]. Source: Reproduced with permission of IOP Publishing.

Piezoelectric materials can convert mechanical deformation into electrical energy to power the synaptic device, while the change in postsynaptic current can also reflect the strain amplitude and duration, as shown in Figure 16.14b. In addition, the dynamic modulation based on dual perceptual presynaptic components is also achieved. Therefore, the proposed piezoelectric graphene sensory synapse achieves self-powered sensing and efficient signal processing of external stimuli.

Zhao et al. developed a self-powered gustation electronic skin for mimicking taste buds based on enzyme-modified ZnO nanowire arrays on a patterned-electrode flexible substrate [75]. Self-powered ZnO sensor can act in a similar way as a taste bud on the tongue through the piezoelectric–enzymatic reaction coupling effect, as shown in Figure 16.15a. Different enzymes on the fabricated ZnO sensor catalyze specific enzymatic reactions of taste substances to generate output voltage for taste sensing. Based on this, the output voltage of the sensor decreases with the increase in ascorbic acid content (Figure 16.15b), indicating that the sensor could be used for taste sensing.

Speech recognition is the most intuitive human-computer interaction interface that can help people with hearing impairments communicate and enable smart devices to bilaterally communicate with people. So far, flexible piezoelectric acoustic sensors have garnered a lot of interest as viable candidates in the field of speech recognition due to their high sensitivity, fast response, and wide bandwidth. Han et al. developed a self-powered acoustic sensor based on a PZT membrane mimicking the basilar membrane of the human cochlea [76]. A multi-channel asymmetrical trapezoidal piezoelectric acoustic sensor consisting of the flexible PZT thin films prepared using ILLO can cover the entire range of sound frequency bands due to the capability of resonant sensing and frequency tuning capabilities. The acoustic signals of normalized female speech recorded by the piezoelectric acoustic sensor are similar to the original acoustic signal, which indicates the acoustic sensor successfully reproduces the original human voice. Based on this, the piezoelectric sensor has excellent sensitivity over the whole range of sound frequencies, as shown in the figure. Compared to conventional microphones, the designed piezoelectric sensor delivered a 32 times higher maximum relative sensitivity of −76 dB at 650 Hz, which is ascribed to the tunable multi-resonant frequency bands and low-quality factor at each channel. In addition, based on the above work, Wang et al. reported a flexible piezoelectric mobile acoustic sensor using an integrated acoustic sensor module with a machine learning (ML) algorithm processor [77]. Compared to traditional capacitive microphones, the designed sensor achieves a 90% speaker recognition rate and a 56% reduction in error rate, further demonstrating the commercial mobile applications of piezoelectric acoustic sensors for remote and precise speech recognition.

16.4.2 Disease Treatment

16.4.2.1 Organs and Tissues Regeneration

Regenerative engineering, which was used to replace or repair damaged, non-functional organs and tissues, becomes an option when conventional therapies fail to rehabilitate body parts injured by congenital malformations, trauma, or cancerous diseases. Endogenous electric fields, naturally present in the body, are essential for mammalian growth and tissue regeneration. Piezoelectric materials as exogenous electrical stimuli can cause various phenotypic and genetic alterations in cells, thus accelerating the process of tissue regeneration. To date, piezoelectric bioelectronic materials are widely used for the regenerative engineering of bones, cartilage, tendons, nerves, skin, ligaments, and muscles.

As preset in Figure 16.16, epitaxial piezoelectric bismuth ferrate (BFO+) nanofilms (~10–20 nm thickness), with a constant built-in positive electric field (+75 mV as measured by piezoresponse force microscopy [PFM]), strongly interact with the electronegative potential of the bone (−52 to −87 mV) to stimulate the rat femoral defect [78]. Cortical-like bone deposition of implanted BFO+/strontium titanate (STO) at the rat femoral defect was elicited as well as osseointegration, which promoted migration, attachment, and activity of osteoforming cells. In Figure 16.17, unidirectionally arranged ZnO NR-based piezoelectric patches (PZP) enable potential difference-derived EFs modulation to stimulate skin cells for active wound healing [9]. The piezoelectric potential is generated by the mechanical deformation caused by animal movement and induces EFs at the wound to enhance cell metabolism, migration, and protein synthesis for wound healing. In Figure 16.18, a thermosensitive, stretchable, and piezoelectric substrate (TSPS) based on zinc oxide NR fabricated to mimic the microenvironment of electrical and mechanical stimulation in vivo promotes myogenic differentiation of human umbilical cord blood mesenchymal stem cells (hUCBMSCs), which can effectively promote muscle tissue regeneration [79].

16.4.2.2 Cancer Treatment

As one of the most serious diseases in the world, cancer has become the greatest threat to human health due to its high risk and high fatality rate. At present, surgery, chemotherapy, and radiotherapy are still the first-line treatments for cancer in clinical application, but there are certain adverse reactions, such as drug resistance, serious toxicity to normal tissue, and serious metastasis. Among the new treatment modalities, non-invasive or minimally invasive treatment modalities have become a major development in cancer treatment due to their precise spatial/temporal resolution, higher treatment efficacy, and reduction of unwanted side effects [79].

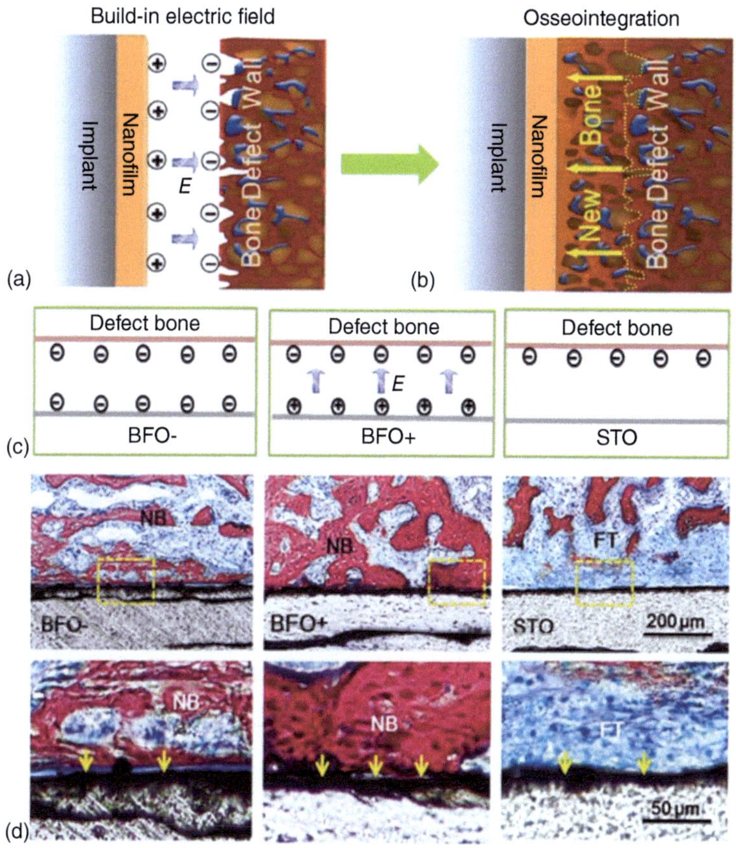

Figure 16.16 Illustration of built-in electric fields promoting implant osseointegration. (a) A built-in electric field is formed between the electropositive nanofilm implant surface and the electronegative endogenous bone defect wall. (b) Rapid and high-quality osseointegration is induced on the implant. Fast and high-quality in vivo osseointegration is triggered by built-in electric fields. (c) Schematic illustration showing that built-in electric fields were only formed between BFO+ nanofilm implants and bone. (d) Histological analysis at two weeks postimplantation demonstrates better osseointegration on BFO+ nanofilm implants than on BFO− and STO implants. Partially magnified images are shown below. Yellow arrows indicate the osseointegration interfaces. NB, nascent bone; FT, fibrous tissue [78]. Source: Reproduced with permission of John Wiley and Sons.

Modern research has shown that reactive oxygen radicals (ROS) have a good eradication effect on cancer cells, based on which numerous treatment options have been developed. In Figure 16.19a,b, tetragonal $BaTiO_3$ (T-BTO) catalyzes the production of reactive oxygen species (ROS) under US excitation, such as toxic hydroxyl (·OH) and superoxide radicals ($\cdot O_2^-$), resulting in in situ tumor eradication [1]. After US irradiation, tumor growth was significantly inhibited by US-driven endoscopic pressure-assisted intratumoral injection of T-BTO-Gel during the first five days after three treatments. In contrast, tumors in the control group, Gel group, T-BTO-Gel group, and Gel + US group all maintained rapid growth. The method

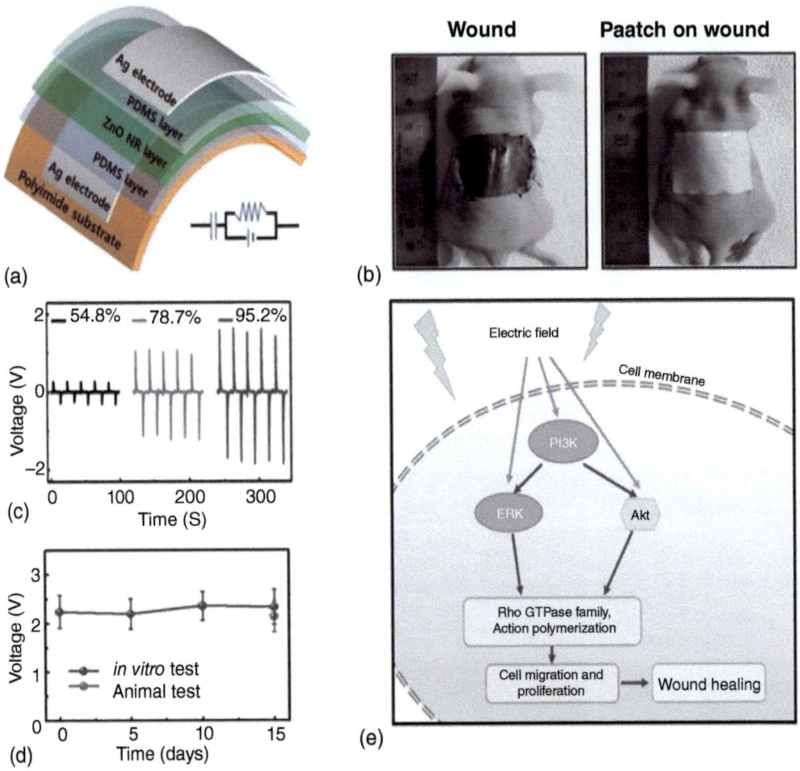

Figure 16.17 (a) Schematic illustration of the PZP modified with Ag electrodes to measure the piezoelectric voltage and current density generated by mechanical bending. (b) In vivo application of a nine-layered PZP. The skin wound was induced on the back of a mouse (left), a nine-layered PZP was placed on the wound, and the PZP was fixed on the wound by dressing a transparent film over the PZP (right). (c) Piezoelectric voltage for different NR filling densities on PDMS (bending radius = 5 mm, the number of bidirectionally grown ZnO NR layer = 1). (d) The piezoelectric voltage differences between the nine-layered PZPs retrieved from 15 days of continuous bending and those detached from the mouse skin wound 15 days after attachment. (e) Schematic illustration showing the in vivo EF-induced intracellular signaling pathways for cell migration and proliferation during wound healing. Source: Reproduced with permission of John Wiley & Sons.

is non-invasive, with low energy attenuation, high tissue penetration, a controlled response, and accompanied by a high therapeutic biosafety in vivo. One of the difficulties of tumor drug therapy is the lack of lymphatic reflux and the increase in tumor interstitial pressure (TIP) caused by tumor interstitial fluid retention, which leads to the unsatisfactory penetration of nanodrugs into the tumor. A "nanolymphatic vessel" based on pyroelectric catalysis achieves photothermal treatment and decomposes tumor interstitial fluid under the irradiation of the NIL-II laser, reducing TIP and enhancing tumor penetration [80]. At the same time, reactive oxygen species produced during pyroelectric catalysis can further damage deep tumor stem cells. The results showed that nanolymatics could relieve 52% of TIP, enhance tumor penetration, and effectively inhibit tumor proliferation (93.75%) and recurrence.

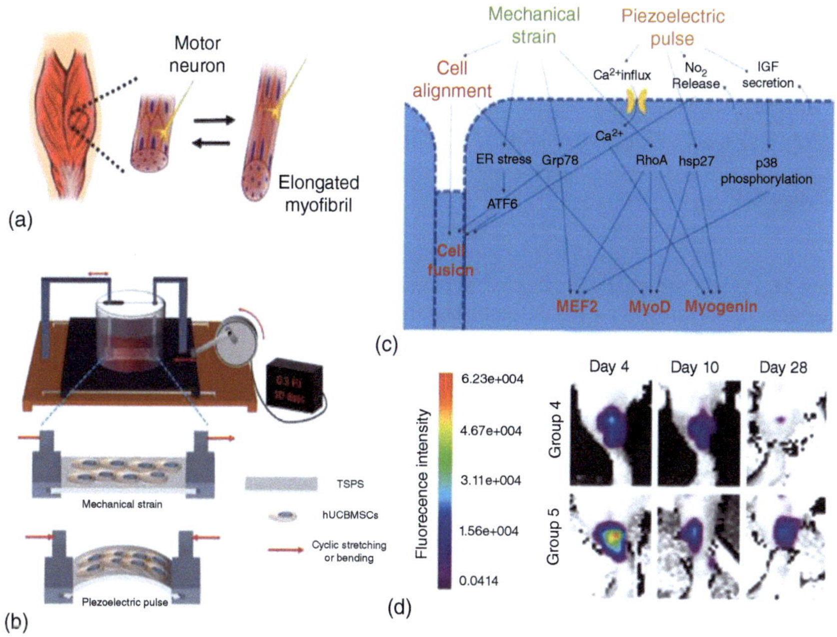

Figure 16.18 (a) Mechanical and electrical stimulation in native skeletal muscle microenvironment, which is mimicked for myogenic differentiation of hUCBMSCs in vitro. (b) A custom-designed cell culture chamber to provide mechanical and electrical stimulation to hUCBMSCs cultured on TSPS for myogenic differentiation. Cyclic bending and stretching of TSPS were repeated at 0.3 Hz for 10 days. (c) A schematic diagram describing the myogenic differentiation-related intracellular signaling induced by mechanical strain (MS) and piezoelectric pulse. (d) In vivo live imaging of DiO-tagged myogenic differentiated hUCBMSCs at 4, 10, and 28 days after implantation [79]. Source: Reproduced with permission of John Wiley and Sons.

16.4.2.3 Oral Care

White and beautiful teeth help boost confidence and show a good appearance during daily communication. However, in our daily lives, we inevitably come into contact with foods and drinks that discolor and stain our teeth. Tooth whitening has become a fast-growing medical beauty field. Compared with traditional teeth-whitening products, new teeth-whitening technologies with non-destructive properties based on piezoelectric biomaterials are expected to serve the public in the future. Based on the piezoelectric effect [81] (Figure 16.20) and pyroelectric effect [82] (Figure 16.21) to produce ROS, it can effectively remove the pigment attached to the teeth and kill the bacteria, thus achieving the effect of teeth whitening. This emerging oral care technology is a highly efficient, non-invasive, low-cost, simple process, and has full potential for application.

16.4.3 Energy Harvesting

Piezoelectric energy harvesting is the process by which mechanical energy, such as vibration, kinetic, and deformation energy, is converted into electrical energy. Biomechanical energy generated by human activities, such as walking, running,

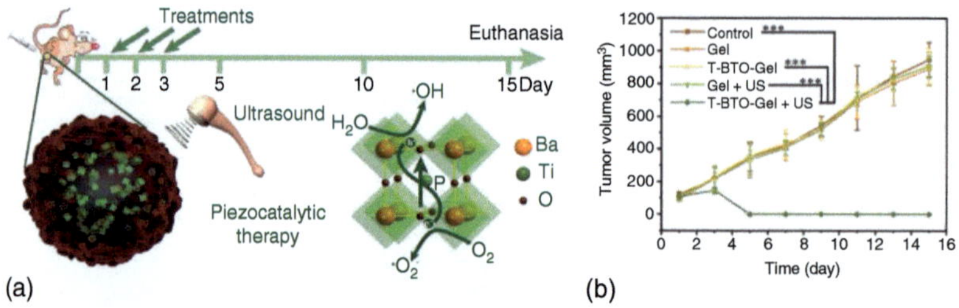

Figure 16.19 (a) Schematic illustration of piezocatalytic therapy in vivo. (b) time-dependent tumor-growth curves of tumor-bearing mice after different treatments, including a control group, Gel group, T-BTO-Gel group, Gel + US group, and T-BTO-Gel + US groups (*$p < 0.05$, **$p < 0.01$, ***$p < 0.001$) [1]. Source: Reproduced with permission of John Wiley and Sons.

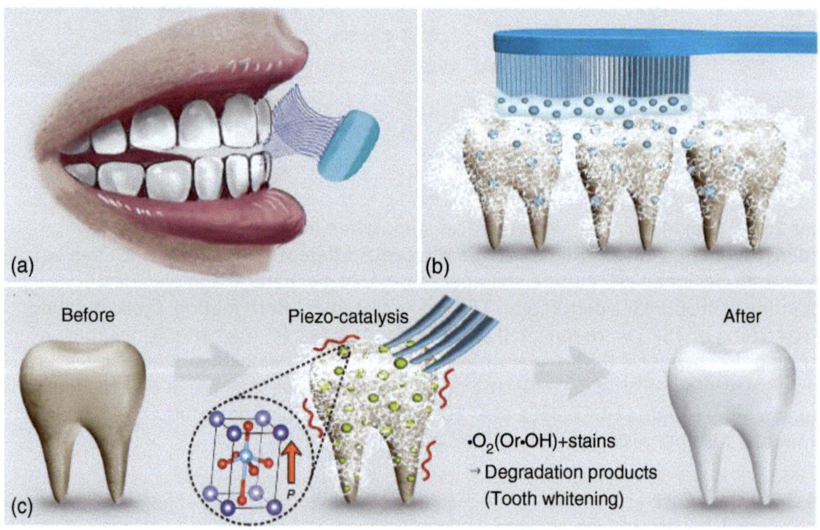

Figure 16.20 Piezo-catalysis for nondestructive tooth whitening: (a) Illustration of mechanical vibration between toothbrush and teeth during daily toothbrushing. (b) The traditional abrasive-based toothpaste method, where tooth whitening is realized by the mechanical friction between teeth and abrasive. (c) The proposed piezo-catalysis effect-based tooth whitening method, wherein piezoelectric particles replace traditional abrasives in toothpaste, generates reactive oxygen species via piezo-catalysis to bleach tooth stains [81]. Source: Reproduced with permission of Springer Nature.

limb movement, and heart beating, can be converted into electrical energy to power implantable medical devices, such as cardiac pacemakers, defibrillators, brain stimulators, and cochlear implants. Self-powered medical devices can avoid the limitations of regular replacement of conventional batteries and the additional risks associated with replacement surgery. Self-powered medical devices break through the limitation that traditional batteries need to be replaced regularly due to their limited operational lifespans, which reduces the additional risk brought by replacement surgery. In the last decade, great progress has been made in converting biomechanical energy into electricity.

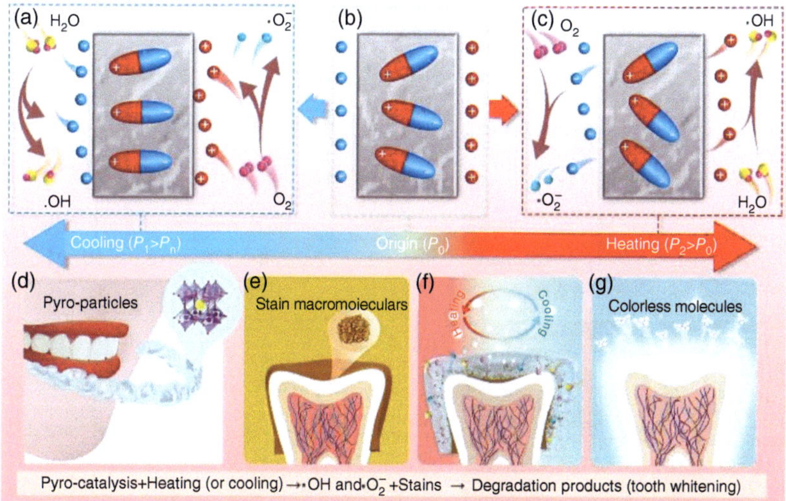

Figure 16.21 Pyro-catalysis for tooth whitening: (a–c) Schematic diagram of pyro-catalysis principle; heating or cooling leads to an alteration of the polarization strength in pyroelectric materials, which further causes the absorption and release of screening charges and the generation of reactive radicals in water. (d) The proposed pyro-catalysis effect-based tooth whitening method, wherein pyroelectric particles are combined with light-curing hydrogel to generate reactive oxygen species via pyro-catalysis to bleach tooth stains. (e–g) Stained teeth can be whitened by wearing braces containing pyroelectric particles that use the pyro-catalytic properties activated by changes in oral temperature brought about by a daily diet to degrade stains on the surface of the tooth [85] Springer Nature / CC BY 4.0.

To power implantable medical devices, the output performance of the piezoelectric energy harvester is critical. Given their high output performance, piezoelectric energy harvesters based on inorganic piezoelectric materials have attracted significant interest. Dagdeviren et al. developed piezoelectric energy harvesting from the motions of the body organs based on high-performance PZT film [83]. As shown in Figure 16.22a, PZT film is transferred to a flexible PI substrate by growth-transfer process and can complete conformal attachment on the bovine hearts in response to the natural contractile and relaxation motions, and the output voltage is shown in Figure 16.22b. The lungs represent another internal organ of interest for mechanical energy harvesting. A photograph of mounting on the lung and diaphragm is shown in Figure 16.22c. As expected, respiratory movement can be converted to output voltage (Figure 16.22d). Overall, the above results indicate that flexible PZT piezoelectric energy harvesters are capable of harvesting mechanical energy from different organs of the body, with output voltages depending on species and organs. In addition, PZT piezoelectric energy harvesters are used to power cardiac pacemakers, as shown in Figure 16.22d,e. To improve the output power, a multi-layer stacked structure is adopted to achieve a time-averaged power density of up to 1.2 $\mu W/cm^2$, (Figure 16.22f) which is sufficient to operate a cardiac pacemaker.

Adequate power supply is a key obstacle for cardiac pacemakers driven by implantable piezoelectric energy harvesters. To further improve the output

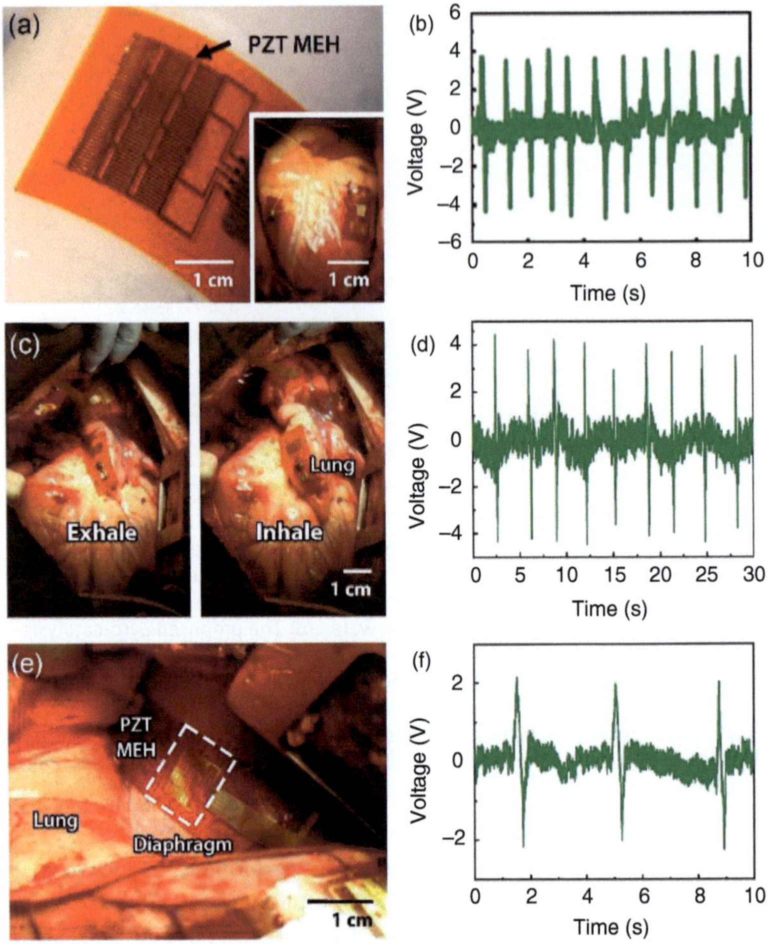

Figure 16.22 (a) A photograph of a flexible PZT piezoelectric energy harvester affixed onto a cow heart. The output voltage is generated from the (b) cow heartbeat, (d) the lung motions, and (f) the diaphragm motions. The photograph of the PZT piezoelectric energy harvester on a cow (c) lung and (e) a cow diaphragm [83]. Source: Reproduced with permission of Proceedings of the National Academy of Sciences.

performance, a piezoelectric single crystal with a high mechanical coupling coefficient is used to harvest heartbeat kinetic energy. Yi et al. developed a leadless heart-worn pacemaker driven by a piezoelectric energy harvester based on PMN-PT single crystal [63]. A mechanical polishing process is used to control the thickness of PMN-PT single crystal to about 50 μm to obtain a flexible piezoelectric film. The commercial cardiac pacemaker is powered by an implantable piezoelectric energy harvester, as shown in Figure 16.23a. The output current generated by the implanted piezoelectric energy harvester in response to the heart beating of an adult pig, which can effectively drive commercial pacemakers. Then, the pulse output terminals are connected to pacemaker probes: one is inserted into the subcutaneous

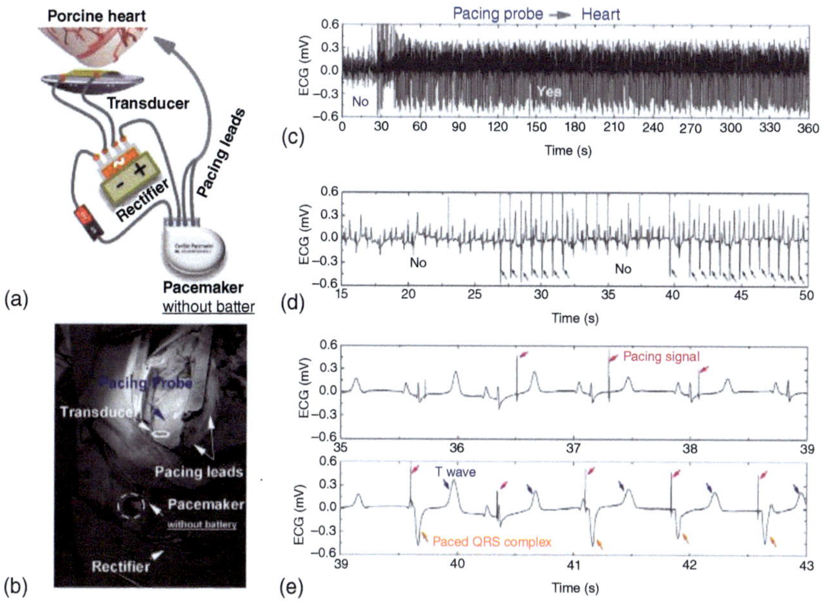

Figure 16.23 (a) Circuit diagram for the exocardial pacing by the batteryless pacemaker. (b) Photograph of the experimental circuit. (c) Electrocardiogram (ECG) signals are regulated by the batteryless pacemaker. (d) The enlarged ECG signals. (e) Invalid and valid pacing ECG signals [63]. Source: Reproduced with permission of John Wiley and Sons.

tissue and the other is inserted into the myocardial tissue from outside the heart, as shown in Figure 16.23b. Figure 16.23c exhibits the electrocardiogram (ECG) signals during this process, indicating that the pig heart is successfully paced by a cardiac pacemaker driven by the piezoelectric energy harvester. In the initial stage of cardiac pacemaker implantation, the paced ECG signals appear unstable (Figure 16.23d,e), which was mainly attributed to the charging process of the pacemaker by the piezoelectric energy harvester in the initial stage. Subsequently, stable-paced ECG signals are obtained.

Neuromodulation, as a non-destructive, adjustable, and reversible therapeutic technique, has attracted extensive attention. It uses neural networks to stimulate or influence neurophysiological signals to manipulate bodily functions to achieve therapeutic goals. Hwang et al. developed a flexible piezoelectric energy harvester based on $Pb(In_{1/2}Nb_{1/2})O_3$–$Pb(Mg_{1/3}Nb_{2/3})O_3$–$PbTiO_3$ (PIN–PMN–PT) single crystal to demonstrate self-powered deep brain stimulation for a live anima [31]. PIN–PMN–PT single crystal is selected due to its remarkable d_{33} of up to 2700 pC/N to generate sufficient output current for deep brain stimulation (30–250 μA) to activate neurons in animals. A stress-controlled nickel exfoliation is used to control the thickness of PIN–PMN–PT single crystal to about 10 μm to obtain a flexible piezoelectric film. The piezoelectric energy harvester generates a short-circuit current of 45 mA when the load resistance is 200 kΩ through mechanical bending motion, which is well matched to the load impedance of the actual deep brain stimulation electrode. These values are sufficient to meet the high standard of brain stimulation. The piezoelectric energy harvester is used to stimulate the M1 cortex in the brain, which generates muscle contractions in the forelimbs. If the stimulation is successfully performed, the muscles of the forelimb should be induced to contract through neural pathways, including the spinal cord. The video-captured and converted binary images are used to track the movement of the red-marked right claw during periodic movements of the piezoelectric energy harvester. The right paw of the mouse was moved about 1.5–2.3 mm by the generated stimulation electricity upon the bending/unbending motions of the piezoelectric energy harvester, indicating that the output current exceeded the threshold current for neuronal activation.

Kim et al. reported a self-powered wireless data transmission device based on flexible single-crystalline $(1-x)Pb(Mg_{1/3}Nb_{2/3})O_3-(x)Pb(Zr,Ti)O_3$ (PMN–PZT) prepared by mechanical polishing [84]. The doping of 0.5 mol% Mn-doping in the PMN-PZT single-crystal yields remarkable enhancements in its piezoelectric performance. Specifically, the crystal exhibits an exceptionally high d_{33} (~1140 pC/N) and electromechanical coupling factor ($k_{33} \approx 0.92$), surpassing the values observed in other PZT materials. The flexible piezoelectric energy harvester was securely attached to the epicardium of an adult male Yorkshire pig (40 kg), demonstrating the feasibility and potential for practical application in physiological environments, as shown in Figure 16.24a. The attachment was accomplished through suturing, ensuring a stable and reliable connection between the harvester and the cardiac tissue. The electrical output signals produced by the flexible energy harvester exhibited remarkable synchronization with the ECG signals obtained from the porcine heart (Figure 16.24b,c), which unequivocally substantiates the origin of the generated electricity as being derived from the intrinsic motion of the cardiac system. Furthermore, owing to its precise synchronization with ECG signals, the flexible piezoelectric energy harvester exhibits promising potential for application in heart monitoring sensors, enabling the accurate detection of vital cardiac information, including normal and abnormal heart rates. The voltage of the capacitor that was charged by the energy harvester was utilized as the drain voltage (V_{DD}) for the transmitting component, which enabled the effective utilization of the harvested energy to power the transmitting part of the system. When the V_{DD} of the capacitor decreases from 3.0 to 2.3 V (the required threshold voltage of the communication module), the estimated energy consumption for wireless transmission is approximately 40.8 μJ (Figure 16.24d). Consequently, following a period of energy harvesting, the capacitor possesses sufficient power to drive the transmitter. A 22 μF capacitor is charged by an energy harvester. As shown in Figure 16.24e, Once the voltage of the capacitor surpasses 2.3 V, it triggers the repetitive on-and-off switching of a distant lamp (about 5 m), which is achieved without any additional power sources.

To remove the external wiring, Jiang et al. developed an ultrasound-induced and self-powered wireless piezoelectric energy harvester via a modified dice-and-fill technique. The core component was an anisotropic one to three piezocomposite comprised of high piezoelectricity lead-free ceramic $(K_{0.48}Na_{0.52})(Nb_{0.95}Sb_{0.05})O_3-(Bi_{0.4}La_{0.1})(Na_{0.4}Li_{0.1})ZrO_3$ pillar array with parallelepipeds shape and the insulating

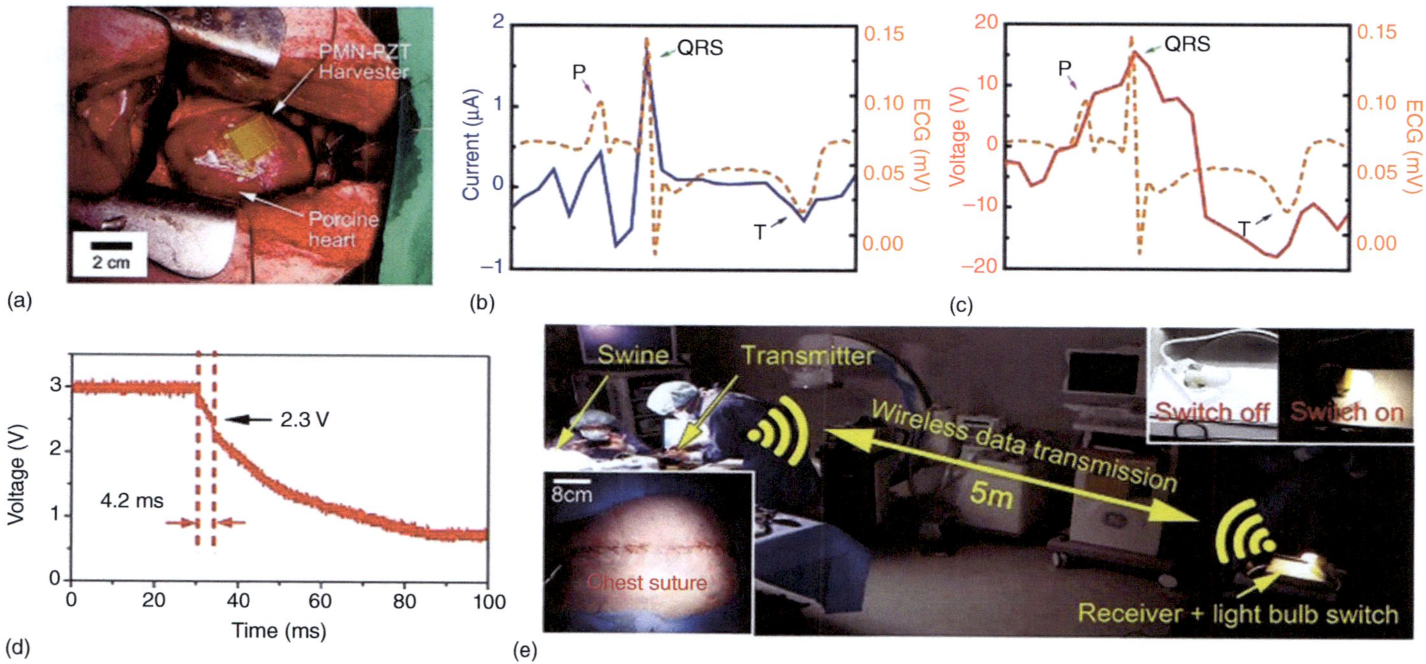

Figure 16.24 (a) Schematic of the device affixed to a porcine heart. Output voltage (b) and short-circuit current (c) of the device were derived from porcine heartbeat, while ECG signals of the pig were recorded. (d) The power consumed by the wireless transmission module. (e) An image capturing the experiment was conducted using a large animal model [84]. Source: Reproduced with permission of John Wiley and Sons.

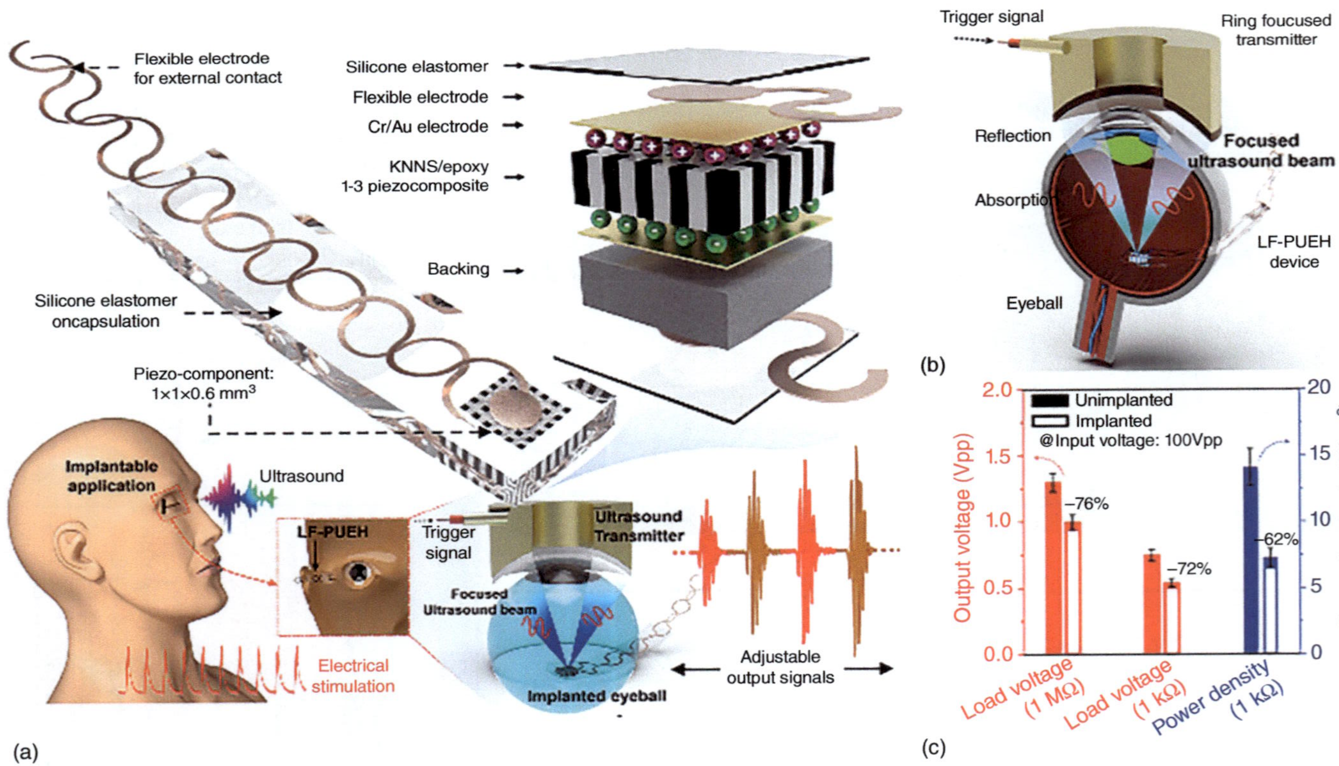

Figure 16.25 Schematic and design of the mm-scale flexible lead-free piezoelectric ultrasonic energy harvester device. (a) Schematics of the device, with key components labeled. The device employs an environmentally friendly lead-free piezocomposite as the core component, wavy-structure-based flexible electrodes for external contact, and a silicone elastomer for encapsulation. The device can be driven by ultrasound to produce adjustable electrical outputs for electrical stimulation applications. (b, c) Schematic profile and optical image of the device implanted into an excised eyeball to mimic an implant situation, with key components labeled [85]. Source: Reproduced with permission of John Wiley and Sons.

polymer filler as shown in Figure 16.25a. The wireless piezoelectric energy harvester can be attached to a complex surface due to its good flexibility and be driven by ultrasound to produce adjustable outputs for retinal electrical stimulation. To demonstrate the feasibility of implanting it inside the human body, the resulting wireless piezoelectric energy harvester was implanted into an excised pig eyeball, and an outline of the working principle is shown in Figure 16.25b. Compared with the in vitro environment, the output voltage and the corresponding power density of the wireless piezoelectric energy harvester in the in vivo environment decreased to 72% and 52%, respectively; this can be seen in Figure 16.25c. Considering the output current of the device exceeds the average threshold required for retinal electrical stimulation, the proposed wireless piezoelectric energy harvester has extensive potential for medical electrical stimulation of implantable biomedical devices.

16.5 The Prospect of Piezo-bioelectronic Materials and Applications

In recent years, bio-piezoelectric materials have attracted considerable research interest owing to their low cost, simple preparation, and stable performance, and have been widely developed and applied in various fields. The electromechanical coupling properties of piezoelectric materials allow them to convert strains generated by biological movements (such as muscle contraction, body movement, breathing, and heartbeat) into electrical energy. Therefore, devices based on piezoelectric materials can not only monitor real time-vital signs such as heart rate, respiration, and blood pressure but also power wearable and implantable medical devices. Piezoelectric nanomaterials with piezoelectric catalytic activities can produce reactive oxygen species for disease treatment and antifouling. In addition, piezoelectric nanomaterials with high piezoelectric performance can effectively generate electrical signals in response to weak strains for regulating biological systems. Despite recent advances in bio-piezoelectric materials and devices, there are still numerous associated challenges and prospects.

Piezoelectric devices are applied in biosensing and energy harvesting owing to their good biocompatibility, excellent flexibility, and high piezoelectric performance. Owing to their biodegradability, immunogenicity, and tissue accumulation, piezoelectric nanomaterials are used for disease treatment. Although numerous researchers have developed bio-piezoelectric materials, the materials available on bio-piezoelectric platforms are still limited, which are investigated for the development of new piezoelectric materials and composites and their modifications and packings. The development of lead-free ferroelectric ceramics, polymers, and composites can overcome this limitation.

For biosensing and energy harvesting applications, piezoelectric devices must adhere tightly to skin or tissue surfaces in response to weak biological pressures or strains. Although numerous studies have demonstrated that piezoelectric devices with excellent flexibility can maintain conformal contact with skin or organs, the fixation of the devices is rarely mentioned. The piezoelectric device connected to a biocompatible tape for fixation may change the distribution of stress and strain, which affects the output signals of the device. To ensure the accurate transfer of deformation, advanced packaging and effective fixing of devices are essential.

The development of new bio-piezoelectric materials with unique properties through a few experiments is vital for accelerating material discovery. ML can be used to search for important information hidden in a large amount of material science data, establish a predictive model of material performance, and thus achieve rapid optimization of material performance. Currently, high-throughput and ML algorithms exhibit great potential for discovering novel bio-piezoelectric materials.

Signal recognition and signal analysis are vital for the practical applications of piezoelectric biosensors. ML algorithms, with the capability of extracting features from a large amount of sensor data, have increased the accuracy of sensor-based measurement, and thus show great potential in data analysis and pattern recognition [77, 86, 87]. Therefore, the integration of ML processors with sensors is vital for piezoelectric biosensing.

The electrical stimulation generated by piezoelectric materials can promote the growth, differentiation, and proliferation of cells. However, the safe range of electrical signals has not been defined, suggesting that safe piezoelectric stimulation for different cells and tissues requires careful evaluation. In addition, because electrical signals are easily affected by various factors, it is difficult to stimulate cells and tissues. Therefore, the precise control of electrical stimulation is vital to achieve the desired therapeutic effects.

The bio-piezoelectric materials and devices are applied in the biomedical field, including materials science, bioengineering, medicine, computer science, physics, and chemistry. Therefore, several researchers from various disciplines are required to work together to optimally address specific thematic areas and ensure development in the biomedical field. Overall, interdisciplinary cooperation is a significant trend in the biomedical field.

References

1 Zhu, P., Chen, Y., and Shi, J. (2020). Piezocatalytic tumor therapy by ultrasound-triggered and $BaTiO_3$-mediated piezoelectricity. *Adv. Mater.* 32 (29): 2001976.

2 Chorsi, M.T., Curry, E.J., Chorsi, H.T. et al. (2019). Piezoelectric biomaterials for sensors and actuators. *Adv. Mater.* 31 (1): 1802084.

3 Ribeiro, C., Sencadas, V., Correia, D.M., and Lanceros-Méndez, S. (2015). Piezoelectric polymers as biomaterials for tissue engineering applications. *Colloids Surf. B* 136: 46–55.

4 Rajabi, A.H., Jaffe, M., and Arinzeh, T.L. (2015). Piezoelectric materials for tissue regeneration: a review. *Acta Biomater.* 24: 12–23.

5 Xu, Q., Gao, X., Zhao, S. et al. (2021). Construction of bio-piezoelectric platforms: from structures and synthesis to applications. *Adv. Mater.* 33 (27): 2008452.

6 Proto, A., Penhaker, M., Bibbo, D. et al. (2016). Measurements of generated energy/electrical quantities from locomotion activities using piezoelectric wearable sensors for body motion energy harvesting. *Sensors* 16 (4): 524.

7 Vilarinho, P.M., Barroca, N., Zlotnik, S. et al. (2014). Are lithium niobate ($LiNbO_3$) and lithium tantalate ($LiTaO_3$) ferroelectrics bioactive? *Mater. Sci. Eng., C* 39: 395–402.

8 Selvarajan, S., Alluri, N.R., Chandrasekhar, A., and Kim, S.-J. (2017). Unconventional active biosensor made of piezoelectric $BaTiO_3$ nanoparticles for biomolecule detection. *Sens. Actuators, B* 253: 1180–1187.

9 Bhang, S.H., Jang, W.S., Han, J. et al. (2017). Zinc oxide nanorod-based piezoelectric dermal patch for wound healing. *Adv. Funct. Mater.* 27 (1): 1603497.

10 Kamel, N.A. (2022). Bio-piezoelectricity: fundamentals and applications in tissue engineering and regenerative medicine. *Biophys. Rev.* 14 (3): 717–733.

11 Beringer, L.T., Xu, X., Shih, W. et al. (2015). An electrospun PVDF-TrFe fiber sensor platform for biological applications. *Sens. Actuators Phys.* 222: 293–300.

12 Genchi, G.G., Ceseracciu, L., Marino, A. et al. (2016). P(VDF-TrFE)/$BaTiO_3$ nanoparticle composite films mediate piezoelectric stimulation and promote differentiation of SH-SY5Y neuroblastoma cells. *Adv. Healthcare Mater.* 5 (14): 1808–1820.

13 Kim, T., Kim, H.J., Choi, W. et al. (2022). Deep brain stimulation by blood–brain-barrier-crossing piezoelectric nanoparticles generating current and nitric oxide under focused ultrasound. *Nat. Biomed. Eng.* 1–15.

14 Bechmann, R. (1956). Elastic, piezoelectric, and dielectric constants of polarized barium titanate ceramics and some applications of the piezoelectric equations. *J. Acoust. Soc. Am.* 28 (3): 347–350.

15 Berlincourt, D. (1971). Piezoelectric crystals and ceramics. In: *Ultrasonic Transducer Materials* (ed. O.E. Mattiat), 63–124. Boston, MA: Springer US.

16 Xu, K., Li, J., Lv, X. et al. (2016). Superior piezoelectric properties in potassium–sodium niobate lead-free ceramics. *Adv. Mater.* 28 (38): 8519–8523.

17 Newnham, R.E. (2005). *Properties of Materials: Anisotropy, Symmetry, Structure.* Oxford University Press on Demand.

18 Zheng, T. and Wu, J. (2015). Enhanced piezoelectric activity in high-temperature $Bi_{1-x-y}Sm_xLa_yFeO_3$ lead-free ceramics. *J. Mater. Chem. C* 3 (15): 3684–3693.

19 Guy, I.L., Muensit, S., and Goldys, E.M. (1999). Extensional piezoelectric coefficients of gallium nitride and aluminum nitride. *Appl. Phys. Lett.* 75 (26): 4133–4135.

20 Hwang, G.-T., Park, H., Lee, J.-H. et al. (2014). Self-powered cardiac pacemaker enabled by flexible single crystalline PMN-PT piezoelectric energy harvester. *Adv. Mater.* 26 (28): 4880–4887.

21 Someya, T. (2012). *Stretchable Electronics.* Wiley.

22 Martins, P., Lopes, A.C., and Lanceros-Mendez, S. (2014). Electroactive phases of poly(vinylidene fluoride): determination, processing and applications. *Prog. Polym. Sci.* 39 (4): 683–706.

23 Liu, Y., Aziguli, H., Zhang, B. et al. (2018). Ferroelectric polymers exhibiting behaviour reminiscent of a morphotropic phase boundary. *Nature* 562 (7725): 96–100.

24 Zhao, C., Zhang, J., Wang, Z.L., and Ren, K. (2017). A poly(L-lactic acid) polymer-based thermally stable cantilever for vibration energy harvesting applications. *Adv. Sustainable Syst.* 1 (9): 1700068.

25 Scheinbeim, J.I. (1981). Piezoelectricity in γ-form nylon 11. *J. Appl. Phys.* 52 (10): 5939–5942.

26 Guerin, S., Stapleton, A., Chovan, D. et al. (2018). Control of piezoelectricity in amino acids by supramolecular packing. *Nat. Mater.* 17 (2): 180–186.

27 Minary-Jolandan, M. and Yu, M.-F. (2009). Nanoscale characterization of isolated individual type I collagen fibrils: polarization and piezoelectricity. *Nanotechnology* 20 (8): 085706.

28 Kholkin, A., Amdursky, N., Bdikin, I. et al. (2010). Strong piezoelectricity in bioinspired peptide nanotubes. *ACS Nano* 4 (2): 610–614.

29 Yeo, H.G., Xue, T., Roundy, S. et al. (2018). Strongly (001) oriented bimorph PZT film on metal foils grown by *rf*-sputtering for wrist-worn piezoelectric energy harvesters. *Adv. Funct. Mater.* 28 (36): 1801327.

30 Wang, D., Yuan, G., Hao, G., and Wang, Y. (2018). All-inorganic flexible piezoelectric energy harvester enabled by two-dimensional mica. *Nano Energy* 43: 351–358.

31 Hwang, G.-T., Kim, Y., Lee, J.-H. et al. (2015). Self-powered deep brain stimulation via a flexible PIMNT energy harvester. *Energy Environ. Sci.* 8 (9): 2677–2684.

32 Lee, H.S., Chung, J., Hwang, G.-T. et al. (2014). Flexible inorganic piezoelectric acoustic nanosensors for biomimetic artificial hair cells. *Adv. Funct. Mater.* 24 (44): 6914–6921.

33 Park, D.Y., Joe, D.J., Kim, D.H. et al. (2017). Self-powered real-time arterial pulse monitoring using ultrathin epidermal piezoelectric sensors. *Adv. Mater.* 29 (37): 1702308.

34 Dagdeviren, C., Su, Y., Joe, P. et al. (2014). Conformable amplified lead zirconate titanate sensors with enhanced piezoelectric response for cutaneous pressure monitoring. *Nat. Commun.* 5 (1): 4496.

35 Ramsay, B.A., Ramsay, J.A., and Cooper, D.G. (1989). Production of poly-β-hydroxyalkanoic acid by *Pseudomonas cepacia*. *Appl. Environ. Microbiol.* 55 (3): 584–589.

36 Ko, H.-U., Kim, H.C., Kim, J.W. et al. (2018). Soft piezoelectric polymer of poly[di(ethylene glycol) adipate] plasticized poly vinyl chloride and its strain sensing. *Mater. Lett.* 227: 276–280.

37 Wang, W., Zheng, Y., Jin, X. et al. (2019). Unexpectedly high piezoelectricity of electrospun polyacrylonitrile nanofiber membranes. *Nano Energy* 56: 588–594.

38 Nayak, S., Sahoo, B., Chaki, T.K., and Khastgir, D. (2013). Development of polyurethane–titania nanocomposites as dielectric and piezoelectric material. *RSC Adv.* 3 (8): 2620–2631.

39 Zhang, G., Zhao, P., Zhang, X. et al. (2018). Flexible three-dimensional interconnected piezoelectric ceramic foam based composites for highly efficient concurrent mechanical and thermal energy harvesting. *Energy Environ. Sci.* 11 (8): 2046–2056.

40 Halperin, C., Mutchnik, S., Agronin, A. et al. (2004). Piezoelectric effect in human bones studied in nanometer scale. *Nano Lett.* 4 (7): 1253–1256.

41 Minary-Jolandan, M. and Yu, M.-F. (2009). Uncovering nanoscale electromechanical heterogeneity in the subfibrillar structure of collagen fibrils responsible for the piezoelectricity of bone. *ACS Nano* 3 (7): 1859–1863.

42 Zhang, Z., Liu, S., Pan, Q. et al. (2022). Van der Waals exfoliation processed biopiezoelectric submucosa ultrathin films. *Adv. Mater.* 34 (26): 2200864.

43 Brown, C.P., Boyd, J.L., Palmer, A.J. et al. (2016). Modulation of mechanical interactions by local piezoelectric effects. *Adv. Funct. Mater.* 26 (42): 7662–7667.

44 Sun, J., Guo, H., Schädli, G.N. et al. (2021). Enhanced mechanical energy conversion with selectively decayed wood. *Sci. Adv.* 7 (11): eabd9138.

45 Yong, Y., Cheng, X., Bao, T. et al. (2015). Tungsten sulfide quantum dots as multifunctional nanotheranostics for in vivo dual-modal image-guided photothermal/radiotherapy synergistic therapy. *ACS Nano* 9 (12): 12451–12463.

46 Zhang, X.-D., Zhang, J., Wang, J. et al. (2016). Highly catalytic nanodots with renal clearance for radiation protection. *ACS Nano* 10 (4): 4511–4519.

47 Liu, L., Chen, B., Liu, K. et al. (2020). Wireless manipulation of magnetic/piezoelectric micromotors for precise neural stem-like cell stimulation. *Adv. Funct. Mater.* 30 (11): 1910108.

48 Liang, S., Deng, X., Ma, P. et al. (2020). Recent advances in nanomaterial-assisted combinational sonodynamic cancer therapy. *Adv. Mater.* 32 (47): 2003214.

49 Gouveia, P.J., Rosa, S., Ricotti, L. et al. (2017). Flexible nanofilms coated with aligned piezoelectric microfibers preserve the contractility of cardiomyocytes. *Biomaterials* 139: 213–228.

50 Qian, Y., Cheng, Y., Song, J. et al. (2020). Mechano-informed biomimetic polymer scaffolds by incorporating self-powered zinc oxide nanogenerators enhance motor recovery and neural function. *Small* 16 (32): 2000796.

51 Chen, X.-Z., Hoop, M., Shamsudhin, N. et al. (2017). Hybrid magnetoelectric nanowires for nanorobotic applications: fabrication, magnetoelectric coupling, and magnetically assisted in vitro targeted drug delivery. *Adv. Mater.* 29 (8): 1605458.

52 Mushtaq, F., Torlakcik, H., Hoop, M. et al. (2019). Motile piezoelectric nanoeels for targeted drug delivery. *Adv. Funct. Mater.* 29 (12): 1808135.

53 Deng, C., Tang, W., Liu, L. et al. (2018). Self-powered insole plantar pressure mapping system. *Adv. Funct. Mater.* 28 (29): 1801606.

54 Liu, Z., Xu, L., Zheng, Q. et al. (2020). Human motion driven self-powered photodynamic system for long-term autonomous cancer therapy. *ACS Nano* 14 (7): 8074–8083.

55 Kolishetti, N., Vashist, A., Arias, A.Y. et al. (2022). Recent advances, status, and opportunities of magneto-electric nanocarriers for biomedical applications. *Mol. Aspects Med.* 83: 101046.

56 Kang, C.W., Lee, D.-M., Park, J. et al. (2022). Core-shell ZnO@microporous organic polymer nanospheres as enhanced piezo-triboelectric energy harvesting materials. *Angew. Chem. Int. Ed.* 61 (46): e202209659.

57 Escrivá-de-Romaní, S., Arumí, M., Bellet, M., and Saura, C. (2018). HER2-positive breast cancer: current and new therapeutic strategies. *Breast* 39: 80–88.

58 Fernandes, M.M., Correia, D.M., Ribeiro, C. et al. (2019). Bioinspired three-dimensional magnetoactive scaffolds for bone tissue engineering. *ACS Appl. Mater. Interfaces* 11 (48): 45265–45275.

59 Marino, A., Battaglini, M., De Pasquale, D. et al. (2018). Ultrasound-activated piezoelectric nanoparticles inhibit proliferation of breast cancer cells. *Sci. Rep.* 8 (1): 6257.

60 Wu, M., Zhang, Z., Liu, Z. et al. (2021). Piezoelectric nanocomposites for sonodynamic bacterial elimination and wound healing. *Nano Today* 37: 101104.

61 Chen, Y., Lu, B., Chen, Y., and Feng, X. (2015). Breathable and stretchable temperature sensors inspired by skin. *Sci. Rep.* 5 (1): 11505.

62 Chen, Y., Chen, Y., Lu, B. et al. (2018). Review of ultra-thin and skin-like solid electronics. *Sci. Sin. Informationis* 48 (6): 605–625.

63 Yi, Z., Xie, F., Tian, Y. et al. (2020). A battery- and leadless heart-worn pacemaker strategy. *Adv. Funct. Mater.* 30 (25): 2000477.

64 Tsikriteas, Z.M., Roscow, J.I., Bowen, C.R., and Khanbareh, H. (2021). Flexible ferroelectric wearable devices for medical applications. *iScience* 24 (1): 101987.

65 Petritz, A., Karner-Petritz, E., Uemura, T. et al. (2021). Imperceptible energy harvesting device and biomedical sensor based on ultraflexible ferroelectric transducers and organic diodes. *Nat. Commun.* 12 (1): 2399.

66 Ma, Y., Choi, J., Hourlier-Fargette, A. et al. (2018). Relation between blood pressure and pulse wave velocity for human arteries. *Proc. Natl. Acad. Sci.* 115 (44): 11144–11149.

67 Yi, Z., Liu, Z., Li, W. et al. (2022). Piezoelectric dynamics of arterial pulse for wearable continuous blood pressure monitoring. *Adv. Mater.* 34 (16): 2110291.

68 Li, T., Feng, Z.-Q., Qu, M. et al. (2019). Core/shell piezoelectric nanofibers with spatial self-orientated β-phase nanocrystals for real-time micropressure monitoring of cardiovascular walls. *ACS Nano* 13 (9): 10062–10073.

69 Chen, X., Li, X., Shao, J. et al. (2017). High-performance piezoelectric nanogenerators with imprinted P(VDF-TrFE)/$BaTiO_3$ nanocomposite micropillars for self-powered flexible sensors. *Small* 13 (23): 1604245.

70 Chen, J., Liu, H., Wang, W. et al. (2019). High durable, biocompatible, and flexible piezoelectric pulse sensor using single-crystalline III-N thin film. *Adv. Funct. Mater.* 29 (37): 1903162.

71 Dagdeviren, C., Javid, F., Joe, P. et al. (2017). Flexible piezoelectric devices for gastrointestinal motility sensing. *Nat. Biomed. Eng.* 1 (10): 807–817.

72 Liu, Z., Zhang, S., Jin, Y.M. et al. (2017). Flexible piezoelectric nanogenerator in wearable self-powered active sensor for respiration and healthcare monitoring. *Semicond. Sci. Technol.* 32 (6): 064004.

73 Persano, L., Dagdeviren, C., Su, Y. et al. (2013). High performance piezoelectric devices based on aligned arrays of nanofibers of poly(vinylidenefluoride-*co*-trifluoroethylene). *Nat. Commun.* 4 (1): 1633.

74 Chen, Y., Gao, G., Zhao, J. et al. (2019). Piezotronic graphene artificial sensory synapse. *Adv. Funct. Mater.* 29 (41): 1900959.

75 Zhao, T., Fu, Y., He, H. et al. (2018). Self-powered gustation electronic skin for mimicking taste buds based on piezoelectric–enzymatic reaction coupling process. *Nanotechnology* 29 (7): 075501.

76 Han, J.H., Kwak, J.-H., Joe, D.J. et al. (2018). Basilar membrane-inspired self-powered acoustic sensor enabled by highly sensitive multi tunable frequency band. *Nano Energy* 53: 198–205.

77 Wang, H.S., Hong, S.K., Han, J.H. et al. (2021). Biomimetic and flexible piezoelectric mobile acoustic sensors with multiresonant ultrathin structures for machine learning biometrics. *Sci. Adv.* 7 (7): eabe5683.

78 Liu, Y., Zhang, X., Cao, C. et al. (2017). Built-in electric fields dramatically induce enhancement of osseointegration. *Adv. Funct. Mater.* 27 (47): 1703771.

79 Yoon, J.-K., Misra, M., Yu, S.J. et al. (2017). Thermosensitive, stretchable, and piezoelectric substrate for generation of myogenic cell sheet fragments from human mesenchymal stem cells for skeletal muscle regeneration. *Adv. Funct. Mater.* 27 (48): 1703853.

80 He, Y., Li, Z., Cong, C. et al. (2021). Pyroelectric catalysis-based "nano-lymphatic" reduces tumor interstitial pressure for enhanced penetration and hydrodynamic therapy. *ACS Nano* 15 (6): 10488–10501.

81 Wang, Y., Wen, X., Jia, Y. et al. (2020). Piezo-catalysis for nondestructive tooth whitening. *Nat. Commun.* 11 (1): 1328.

82 Wang, Y., Wang, S., Meng, Y. et al. (2022). Pyro-catalysis for tooth whitening via oral temperature fluctuation. *Nat. Commun.* 13 (1): 4419.

83 Dagdeviren, C., Yang, B.D., Su, Y. et al. (2014). Conformal piezoelectric energy harvesting and storage from motions of the heart, lung, and diaphragm. *Proc. Natl. Acad. Sci.* 111 (5): 1927–1932.

84 Kim, D.H., Shin, H.J., Lee, H. et al. (2017). In vivo self-powered wireless transmission using biocompatible flexible energy harvesters. *Adv. Funct. Mater.* 27 (25): 1700341.

85 Jiang, L., Yang, Y., Chen, R. et al. (2019). Ultrasound-induced wireless energy harvesting for potential retinal electrical stimulation application. *Adv. Funct. Mater.* 29 (33): 1902522.

86 Jung, Y.H., Pham, T.X., Issa, D. et al. (2022). Deep learning-based noise robust flexible piezoelectric acoustic sensors for speech processing. *Nano Energy* 101: 107610.

87 Zhou, X., Liang, W., Wang, K.I.-K. et al. (2020). Deep-learning-enhanced human activity recognition for internet of healthcare things. *IEEE Internet Things J.* 7 (7): 6429–6438.

17

Porous Piezoelectric Materials for Energy Applications

Yan Zhang

Central South University, Powder Metallurgy Research Institute, State Key Laboratory of Powder Metallurgy, No. 932 Lushannan Road, Changsha 410083, China

17.1 Introduction

As a class of smart materials, piezoelectric materials have wide applications in the fields of sensors, actuators, energy harvesters, transducers, catalysis, nanopositioners, ultrasonic motors, and so on due to the *piezoelectric effect*. The piezoelectric effect was first discovered in 1880 by the Curie brothers [1]. The *direct piezoelectric effect* means that an electric charge or potential can be generated under a mechanical force or strain due to the change of polarization, which is of interest for applications related to sensors, energy harvesters, and catalysis. For the converse piezoelectric effect, a mechanical stress or strain is generated when an external electric field is applied to the piezoelectric materials, which is related to the applications of actuators. Therefore, piezoelectric materials have attracted much attention in recent years.

Piezoelectric materials are usually produced with high relative densities to obtain high piezoelectric properties. The existence of porosities in piezoelectric materials is often considered to be defects, since the mechanical strength can be significantly reduced [2]. Moreover, piezoelectric materials can be easily broken down during the poling process, because the electric field can be concentrated at the pores due to their low relative permittivity, leading to the dielectric breakdown or fracture [3]. Despite these disadvantages, it was found that there are many advantages of porous piezoelectric materials. The introduction of porosities can significantly reduce the relative permittivity, density, and Young's modulus. The polarization and piezoelectric charge coefficient can be maintained if the microstructure is designed, which is beneficial for some specific applications.

Piezoelectric materials can be used as sensors to detect force, strain, and acceleration. The ability of piezoelectric materials for piezoelectric sensing applications can be evaluated by the following parameters and figures of merit [4].

$$g_{33} = \frac{d_{33}}{\varepsilon_{33}^{T}\varepsilon_0} \tag{17.1}$$

Piezoelectric Materials: From Fundamentals to Emerging Applications, First Edition. Edited by Jiagang Wu.

$$g_h = \frac{d_{33} + 2d_{31}}{\varepsilon_{33}^T \varepsilon_0} \tag{17.2}$$

where g_{33} is the piezoelectric voltage coefficient, which is a measure of the voltage, generated from a stress in the poling direction. g_h is the *hydrostatic* voltage coefficient and is a measure of the electric field per unit of hydrostatic stress. d_{33} and d_{31} are the longitudinal and transverse piezoelectric charge coefficients, ε_{33}^T is the relative permittivity at constant stress, ε_0 is the permittivity of free space. It can be concluded from the above equations that piezoelectric materials with low relative permittivity and high piezoelectric charge coefficient possess high figures of merit. Therefore, porous piezoelectric materials exhibit great potential for piezoelectric sensors and SONAR applications.

Piezoelectric materials can convert mechanical vibrations such as ultrasonic waves and human motions into electricity via the *direct piezoelectric effect*, and supply power for low-power electronic devices. Thus, piezoelectric energy harvesting has attracted much interest. About piezoelectric energy harvesting applications, the energy generated is important and the piezoelectric energy harvesting performance can be assessed by the following equation [5].

$$FoM_{33} = \frac{d_{33}^2}{\varepsilon_0 \varepsilon_{33}^T} \tag{17.3}$$

where FoM_{33} is a piezoelectric energy harvesting figure of merit and is a measure of the energy generated from stress in the poling direction. It can be seen from the above equation that the harvested piezoelectric energy density strongly depends on the piezoelectric charge coefficient, d_{33}, and the relative permittivity, ε_{33}^T. Therefore, a high d_{33} and low ε_{33}^T are desirable for piezoelectric energy harvesting applications. The introduction of porosities in piezoelectric materials can reduce the relative permittivity, and the piezoelectric charge coefficient can be well maintained with proper design. Thus, porous piezoelectric materials are great candidates for piezoelectric energy harvesting applications.

The existence of pores in piezoelectric materials is believed to have a negative effect on the mechanical strength, since stress is likely to be concentrated near the pores when mechanical load is applied. While pores can be beneficial to mechanical properties to some extent, which is of interest in the fields of piezoelectric transducers [6]. The acoustic impedance of a material is defined as the product of the density and the sound velocity (V_s).

$$Z = \rho V_s \tag{17.4}$$

where Z is the acoustic impedance and is measured in 'rayl', ρ is the density and V_s is the sound velocity. And the sound velocity can be defined by the following equation, where E is the Young's Modulus.

$$V_s = \left(\frac{E}{\rho}\right)^{1/2} \tag{17.5}$$

Hence $Z = (E\rho)^{0.5}$, and the introduction of pores into piezoelectric materials can decrease both the density and stiffness. Therefore, the acoustic impedance can be reduced, which can improve the impedance matching with water, air, biological tissue, or low-stiffness structures [7].

The above fascinating advantages of porous piezoelectric materials make them of great interest in the fields of piezoelectric sensing, energy harvesting, transduces, and SONAR. Research on porous piezoelectric materials has lasted for more than ten years, and the development of porous piezoelectric materials is continuing to grow, there continue to be several challenges for further development and exploitation. As a result, a comprehensive and systematic summary of the most recent developments in porous piezoelectric materials is urgently needed. In this chapter, the types of porous piezoelectric materials are classified, including porous piezoelectric ceramics, polymers, ceramic-based composites, and thin films. The fabrication techniques of porous piezoelectric materials are summarized, and the simulation methods are presented. Finally, the applications of porous piezoelectric materials in the fields of energy harvesting, sensing, catalysis, biomedical, and energy storage are summarized.

17.2 Types of Porous Piezoelectric Materials

There are many classification rules for porous piezoelectric materials, for example, the connection mode between phases can be distinguished according to their functions. When compared to dense piezoelectric materials, the relative permittivity of porous piezoelectric materials is significantly diminished when pores are present. The charge coefficient d_{ij} of porous piezoelectric materials will be lower than that of nonporous materials, but the quality factors *FoM* may be higher. The potential benefits of porous piezoelectric materials in applications such as energy harvesting, sensing, sonar, and catalysis are consequently of interest to researchers.

Porous piezoelectric materials can also be referred to as composites since they consist of at least two phases: a piezoelectric ceramic matrix and an air/polymer pore phase. Newnham et al. developed a method for defining the structure of piezoelectric composites using two numbers based on the connectivity and interconnectivity of the respective phases [8]. Porous piezoelectric materials are classified as 3-0, 3-1, 3-3, and 2-2 depending on the connection between the pore and the ceramic, shown in Figure 17.1.

The first number represents an active piezoelectric phase connection, whereas the second number represents a passive pore phase connection [9]. Porous piezoelectric ceramics with a 3-0 structure consist of isolated pores in a continuous piezoelectric matrix and are advantageous in that they are simple to make and can be fabricated with a broad range of porosity [10]. In 3-3 type porous piezoelectric ceramics, both the ceramic matrix and the pores are fully interconnected, and these porous materials typically have better acoustic properties than dense materials because the introduction of pores can reduce its acoustic impedance, resulting in enhanced wave coupling between a piezoelectric generator, sensor, and the propagating media, such as air, water, or biological tissue [11]. Nevertheless, structural heterogeneities usually limit the mechanical properties of the aforementioned two kinds of porous

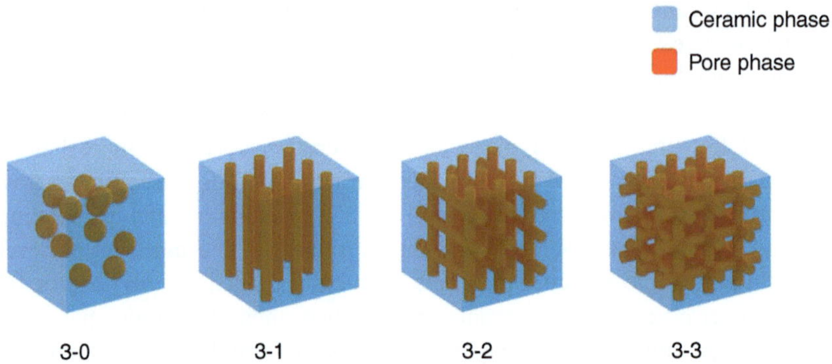

Figure 17.1 Four types of two-phase piezoelectric composites based on connectivity.

piezoelectric ceramics. Due to the orientation of the piezoelectric phase in the polarization direction, 3-1 type materials (aligned pores in a continuous ceramic matrix) and 2-2 type materials (alternating two-dimensional layers of ceramics and pores) demonstrate superior piezoelectric performance [10, 12]. Porous piezoelectric materials can be classified into four varieties based on the recent popular research fields: (i) porous ceramics, (ii) porous polymers, (iii) porous composites, and (iv) porous thin films.

17.2.1 Porous Ceramics

Since the majority of research undertaken so far has been on ceramic-based porous materials, porous piezoelectric ceramics occupy a prominent position in the field of porous piezoelectric materials. In comparison to other ceramic systems, such as bismuth-layer and tungsten bronze structures, the perovskite structure of many piezoelectric ceramics results in enhanced piezoelectric properties [13]. Due to their exceptional piezoelectric properties, the lead zirconate titanate ($Pb[Zr_xTi_{1-x}]O3$, PZT) family provides the basis for the most extensively studied porous perovskite-structure piezoelectric [14]. Lead zirconate titanate (PZT), lead magnesium niobate-lead zirconate titanate (PMN–PZT) [15], lead zirconate titanate-lead cobalt niobate (PZT–PZN), and Nb-doped PZT [16]have all been studied as porous piezoelectric ceramics. Unfortunately, the toxicity of Pb limits its usage, as it is damaging to human health and the environment, and there are increasing restrictions on the use of lead in electronic devices and electronic trash (e-waste) in many countries across the world. Consequently, numerous experiments on porous lead-free piezoelectric ceramics have been done recently [17]. The high piezoelectric d_{33} charge coefficient of porous calcium barium zirconate titanate (BCZT) has attracted a significant deal of interest among these materials [18]. Due to the greater fraction of polarized regions in the ceramic phase, porous BCZT ceramics with a 2-2 structure display more remanent polarization than ceramics with a 3-0 structure [19].

17.2.2 Porous Polymer

Porous piezoelectric polymers have sparked interest in recent years due to their low weight, mechanical flexibility, and biocompatibility for medical sensing and harvesting. Polyvinylidene difluoride (PVDF) and its copolymers are the most common piezoelectric polymers46. PVDF's piezoelectricity is generated from the piezoelectric origin of dipolar crystals, and PVDF and its copolymers exhibit five crystalline phases, namely α, β, γ, δ, and ε [20]. The most typical phase is the α-phase, in which the dipoles are placed in reverse parallel, resulting in no appreciable piezoelectricity. However, the β-phase and γ-phase are polar crystalline phases, since the dipoles in the β-phase and γ-phase are arranged in parallel; therefore, the piezoelectric properties of PVDF are primarily attributed to these two phases [21]. As a result, efforts to enhance the content of β-phase and γ-phase in porous PVDF have been a major focus of study.

In general, porous piezoelectric polymers made from PVDF have two types of structures: a porous surface structure in which pores are situated solely on the polymer surface and a porous bulk structure in which pores are dispersed throughout the entire material. Template-assisted methods and phase-separation methods are used to generate these two types of structures, respectively [22]. Compared to dense piezoelectric polymers, porous structures are more sensitive to mechanical vibrations due to a local increase in stress intensity and the enlargement of the electrode area for charge collection, which makes porous piezoelectric polymers promising candidates for PVDF-based piezoelectric nanodevices.

17.2.3 Porous Ceramic-Based Composite

Typically, piezoelectric composites consist of two phases: an electroactive piezoelectric phase and a non-reactive polymer phase. Apart from the combination of the ceramic phase and air phase mentioned above, a porous ceramic-based piezoelectric composite refers to the combination of a porous piezoelectric ceramic and a polymer, in which the pores are infiltrated with a polymer phase to provide additional functionality, such as mechanical flexibility or toughness [23]. Epoxy resin and polydimethylsiloxane (PDMS) are frequently utilized as a second-phase polymers to infill the pores of porous piezoelectric ceramics due to their low viscosity and desirable flow qualities to impregnate pores prior to curing. Since porous 3-0-based piezoelectric ceramics have isolated pores, impregnation of pores might be challenging. Due to the interconnected pore structure, a polymer can fill the open pores of 3-1, 3-3, and 2-2 type porous piezoelectric ceramics more effectively. Compared to single porous piezoelectric ceramics, porous ceramic-based composites are less brittle, highly ductile, and more flexible, which opens up new application fields for wearable applications or sophisticated applications, such as flexural vibration energy harvesting and conformal sensors for wearable devices and complex geometries.

17.2.4 Porous Thin Film

In addition to the porous piezoelectric materials discussed above, porous piezoelectric thin films are being created. The piezoelectric phase of the piezoelectric film is usually composed of $Pb(Zr,Ti)O_3$, $(Ba,Sr)TiO_3$, $BiFeO_3$, and polyvinylidene difluoride (PVDF) [24]. Radiofrequency (RF) sputtering, metal–organic chemical deposition, electron beam evaporation, pulsed laser deposition, and the sol–gel method are fabrication techniques for piezoelectric thin films. The introduction of porosity into piezoelectric thin films may increase the piezoelectric response because the moderate porosity can minimize the substrate's clamping effects and render the piezoelectric domain walls movable [24a]. Compared to dense piezoelectric films, porous films may demonstrate a threefold increase in piezoelectric coefficient and an improvement in electromechanical response [25]. To Date, PZT-based porous piezoelectric thin films are the most widely investigated porous piezoelectric thin films.

17.3 Fabrication Techniques for Porous Piezoelectric Materials

It is through decades of hard work that various methods for the fabrication of porous ferroelectric materials have been developed. In this section, the current mainstream preparation technologies are summarized, the burnt-out polymer spheres method (BURPS), gel casting, additive manufacturing the replica template method, and freeze casting.

17.3.1 Burnt-out Polymer Spheres (BURPS)

The BURPS method, also known as sacrificial template method, involves mixing ceramic powder, a pore former, and a binder uniformly to produce a green body by uni-axially pressing, isostatic pressing or tape casting methods. Then, the polymer spheres or additives are removed by a heat treatment and high-temperature sintering to obtain the pore structure. It is important to choose a suitable pore-forming agent. The shape and size of the pore former particles determine the shape and size of the pores of the porous ceramic material, as shown in Figure 17.2 [27]. It is essential that the pore-forming agents do not react with the ceramic matrix during the burn-out and sintering process. Several pore-forming agents have been used to fabricate porous ferroelectric ceramics, which are often spherical in shape; these include polymers and organic materials such as polymethyl methacrylate (PMMA) spheres [28] polyethylene oxide (PEO) [10] stearic acid (SA) [29] and polystyrene [30]. In addition, non-spherical pore-forming agents have been used, such as graphene [31]. The pore size of porous ceramics prepared by this method generally ranges from 0.1 to 300 mm, depending on the geometry of pore-forming agents. Notably, the final pore volume fraction is not the same as the initial volume fraction of the additive but a clear correlation between them, since there is some shrinkage of the ceramic

Figure 17.2 Porous PZT fabricated by burned-out polymer spheres method by using (a) irregular and (b) spherical PMMA as pore-forming agents [26]. Source: Zeng et al. [26]/Reproduced with permission of Elsevier.

during the high-temperature sintering process. It is reported that the PMMA-derived pores were approximately spherical, while pores formed by the stearic acid were mostly irregular, and many pore channels and voids can be observed, indicating a strong correlation between the shape and size of the pore former particles and the shape and size of the pores of the porous ceramic material [27].

BURPS is simple and low-cost, and easy to operate. It is currently the most widely used piezoelectric ceramic preparation process. However, polymer spheres cannot be easily mixed uniformly with ferroelectric ceramic powders, resulting in pores being randomly distributed or agglomerated throughout ceramic structure, which leads to cracks in a porous ceramic and shows a reduction in mechanical strength [32]. As a result, this method is suitable for the fabrication of porous ferroelectric ceramics with low porosity (<60%) and randomly distributed pores.

17.3.2 Gel Casting

Gel casting was initially developed to fabricate dense ferroelectric ceramics by Omatete in 1991. In gel casting, a concentrated slurry of ceramic powder in a solution of organic monomers is poured into a mold and then polymerized in situ to form a green body in the shape of the mold cavity. Thus, it is a combination of polymer chemistry with slip processing [33]. Recently, this method has been

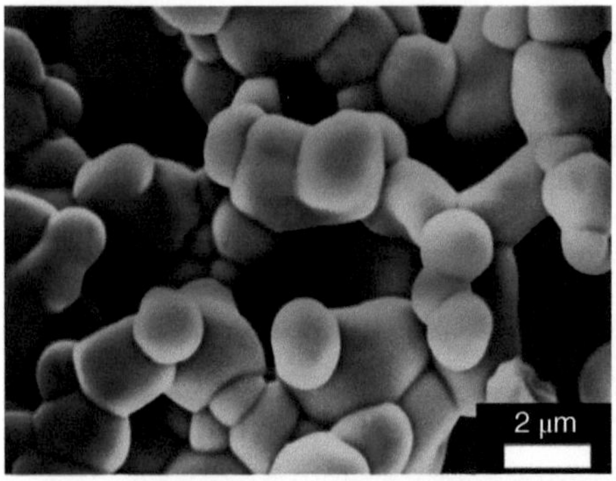

Figure 17.3 Typical scanning electron micrographs of porous PZT ceramics by tert-butyl alcohol-based gel casting [34]. Source: Yang et al. [34]/Reproduced with permission of Elsevier.

employed to manufacture porous ferroelectric ceramics with high strength, uniform composition, defect-free, and complex shapes by the addition of pore formers and decreasing the solid content of the ceramic slurry, as shown in Figure 17.2 [34], tert-butyl alcohol with high saturation vapor pressure and low surface tension force is often selected to replace water as a solvent and pore former, which can be readily evaporated during the drying process [35]. After polymerization, the gel spheres are burned out by the ball during the sintering process to obtain porous ferroelectric ceramics with spherical pores [36].

Gel casting has attracted attention due to its low cost, and its porosity and pore size can be adjusted in a wide range. Piezoelectric constant d_{33} was generally larger for samples by gel casting than for those by the BURPS process mainly induced by the uniform and spherical pore structure (Figure 17.3).

However, tert-butyl alcohol is a high-cost and toxic, and carcinogenic substance to the human body, which hinders its long-term development and wide application. And there is potential for research to seek out alternative alcohol sources [6].

17.3.3 Additive Manufacturing

Additive manufacturing (AM), which is also commonly known as 3D printing, provides flexibility in the manufacturing of complex geometric parts at competitive prices and within a low production time. AM is a novel technique for fabricating porous piezoelectric ceramics without a mold and can be employed to control structure precisely, which cannot be achieved using conventional processes [37]. There are seven major methods for additive manufacturing. Direct ink writing, the most commonly used method to fabricate porous ferroelectric materials, utilizes a moving nozzle that dispenses ink onto a substrate layer by layer to produce integrated multifunctional, multi-material electro-ceramic devices in a rapid, simplistic way [38]. Porous ferroelectric scaffolds with 3-1, 3-2, and 3-3 connectivities can be fabricated by direct ink writing. Smay et al. successfully prepared PZT skeletal structures via direct ink writing and subsequently infiltrated the structures with an epoxy to produce a PZT-polymer composite for hydrophone and ultrasonic imaging applications. These 3-X structures consisted of a three-dimensional lattice of PZT rods (3-3) embedded in a polymer matrix, a PZT lattice/polymer matrix capped with PZT

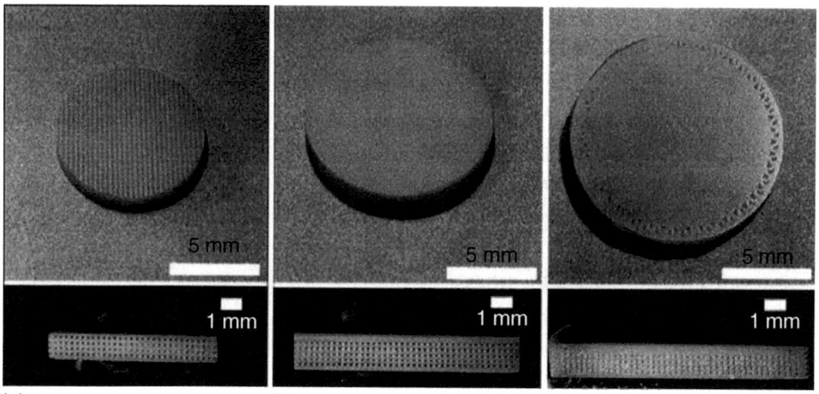

Figure 17.4 Optical images of the top and cross-sectional views of (a) 3-3, (b) 3-2, and (c) 3-1 composites after polymer infiltration and machining [39]. Source: Smay et al. [39]/Reproduced with permission of AIP Publishing LLC.

face plates (3-2), or a PZT lattice/polymer matrix capped with PZT face plates and encircled by a solid PZT ring (3-1), as shown in Figure 17.4 [39]. Nan et al. fabricated macro-porous lead free ferroelectric scaffolds by direct ink writing [40].

It can be seen that different porous structures can be obtained by selecting different manufacturing techniques. Moreover, the direct ink writing method fabricates the complex ceramic-polymer composite grid scaffolds. For example, Ag-coated PNN-PZT ceramic heterojunction particles and dispersed them into the MW-CNTs doped PDMS precursor to form the (0-3) composite slurry with suitable viscosity coefficient, followed by direct ink writing (DIW) method to fabricate the complex ceramic-polymer composite grid scaffolds. It was shown that the porous ferroelectric composite had potential for energy conversion and energy harvesting applications.

Additive manufacturing is an attractive manufacturing method owing to its distinct advantages of fabricating complex geometry and a high degree of freedom in microstructure and pore size design. Porous ferroelectric scaffolds fabricated by additive manufacturing can exhibit multiple geometries. However, in the process of the ink layer being accurately distributed by layer to the substrate, gaps between the printed lines were relatively coarse. As a result, the integrity between layers can be poor [41]. Therefore, the development trends and challenges of porous ferroelectric materials with excellent performance should be paid more attention [42].

17.3.4 Replica Template

The replica template method which impregnates the ceramic suspension into a porous template, and removes the organic matter in the template by drying and sintering [43]. Then, a porous ceramic with the same pore structure as the template is obtained. Sponge is a commonly used template material. On one hand, the sponge possesses suitable strength and elasticity for easy removal of excess slurry after impregnation. On the other hand, a network structure constructed with highly interconnected pores is beneficial for fabrication of pore sizes ranging from 100 to 300 μm [42]. In addition, the shear thinning behavior of the ceramic slurry is also a critical step, the ceramic slurry needs proper rheology to be uniformly impregnated in the sponge. Porous structures with different mechanical properties and structural properties can be obtained through the regulation of templates and slurries. For example, increasing the frequency of impregnations with the ceramic slurry can result in a ceramic scaffold with higher mechanical properties. Although porous ceramics with high porosity (up to 40–95%) can be prepared by the replication template method, the volatilization of the polymer tends to generate pores and defects in the internal structure of the porous framework during the high-temperature sintering process. Therefore, the mechanical strength of the prepared porous ceramics is relatively poor.

17.3.5 Freeze Casting

The freeze-casting method, also known as the ice template method, is a biomimetic technology that can obtain an oriented porous structure with relatively high mechanical strength. It utilizes ice solidified in a directional low-temperature field as a template, freezes and solidifies the ceramic slurry in a unidirectional temperature gradient, and finally removes the ice during the sublimation process at low temperature and low pressure to obtain an oriented porous structure [44]. The use of water-based freeze casting results in a layered-oriented pore structure with a pore morphology of flat oval pores [45]. The pore structure and pore morphology of porous ceramics can be tuned by the ceramic slurry, freezing conditions, and template type. For example, the porosity can be reduced by increasing the solid content of the ceramic slurry, and the pore size can be adjusted by changing the freezing rate, which plays an important role in the nucleation and growth of solidified crystals. Besides water, tert-butanol, and camphene are also commonly used freezing templates. When tert-butanol is used as a freezing template, a unidirectionally arranged through-hole structure can be obtained. When camphene is used as a template, a dendritic structure with a dense pore wall structure can be obtained. At present, freeze casting has become an important method for preparing porous structures. This method has the advantages of low cost, simple process, environmental friendliness, and easy mass production.

17.4 Modeling and Characterization of Porous Piezoelectric Materials

17.4.1 Modeling

A large number of modeling approaches have been carried out to elucidate the mechanisms giving rise to superior performance. Simulation data has significant instructions for designing and manufacturing porous piezoelectric materials. Therefore, the porous piezoelectric materials modelings are summarized in this section, including an analysis of the electric field distribution, polarization-electric field loops, dielectric and piezoelectric properties and piezoelectric potential.

17.4.1.1 Modeling of Electric Field Distribution

The poling process of piezoelectric materials is critical. During the poling process, the remnant polarization achieved can provide a piezoelectric response. When the applied electric field reaches a certain strength, the randomly orientated ferroelectric domains will be aligned, which is called a coercive field (E_c). For dense ferroelectric materials, the electric field distribution is homogeneous due to the excellent connectivity of the material. Hence, dense ferroelectric materials can be completely polarizable under the influence of an electric field. However, due to the introduction of an air phase with a low dielectric constant, the applied electric field will be concentrated in the pores according to Gauss's law, resulting in an inhomogeneous electric field distribution.

Simulation of pores in ferroelectric materials has attracted extensive attention due to the fact that the presence of pores can significantly affect the electric field distribution and poling process. Gheorghiu et al. [46]. calculated the electric field distribution of four porous structures with different porosity by the 3D finite element method (FEM). For porous ferroelectric materials with high porosity, the local electric field became extremely heterogeneous due to the introduction of a low permittivity phase with increasing porosity.

In addition to the porosity, the shape and orientation of the pores also affect the electric field distribution. Khachaturyan et al. [47] constructed finite element models for two types of pores with different aspect ratios to analyze the electric field distribution. Roscow et al. [48] investigated the effect of pore orientation angle on piezoelectric properties through finite element modeling. As the pore orientation angle increased from 0° to 90°, the unpolarized region increased. The highest poled fraction was obtained from the high aspect ratio pores aligned parallel to the applied electric field. Through further expanding the aspect ratio, pore channels would be parallel to the applied electric field when the ferroelectric material was close to the ideal structure of 2-2. Therefore, the porous ferroelectric materials with 2-2 lamellar structures such as those fabricated by freeze-casting can achieve a high degree of poling.

17.4.1.2 Modeling of Polarization–Electric Field Loops

The polarization–field (P–E) loop is an important characteristic parameter of ferroelectric materials and an influential basis for judging whether the materials have ferroelectric properties. It includes the shape of the hysteresis loop, the maximum polarization strength, the remnant polarization strength, and the coercive field.

Zhang et al. [19] developed a finite element model to study the effect of porosity on the P–E loop of ferroelectric ceramics (BCZT), including random pores and aligned pores. As shown in Figure 17.5, when the applied electric field was lower than the coercive field of the dense material, the ferroelectric ceramics remained unpolarized. As the applied field increased, the ferroelectric material was continuously polarized

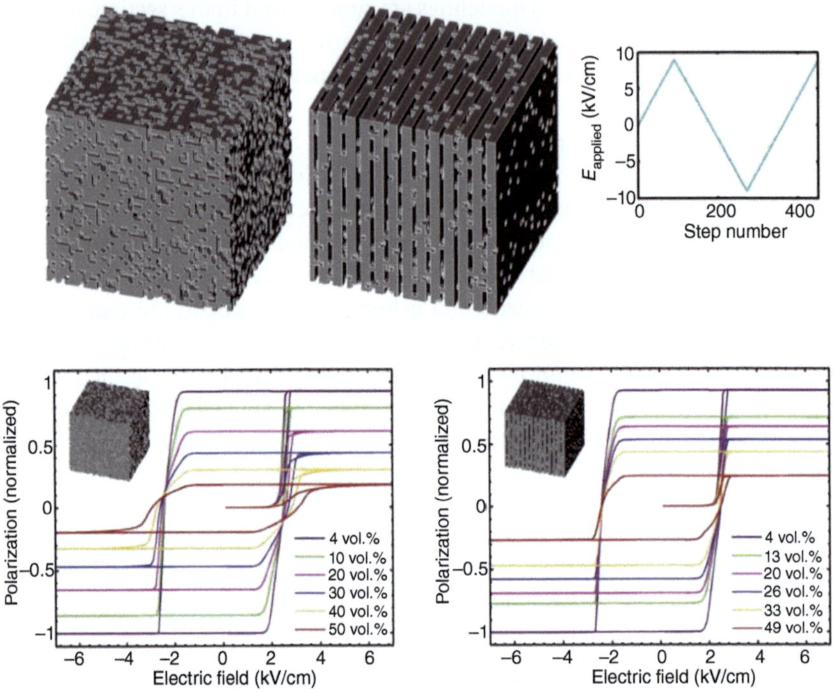

Figure 17.5 Simulation of polarization–field (P–E) loop [19]. Source: Reproduced with permission of Elsevier.

until it was completely polarized. In addition, the tilt of the polarization-field loop and the remnant polarization decreased with an increase in porosity fraction, and the remnant polarization of the random porous ceramics decreased more quickly than that of the aligned porous ceramics, which was consistent with the experimental results.

17.4.1.3 Modelling of Dielectric and Piezoelectric Properties

The dielectric and piezoelectric properties of ferroelectric materials are important, it is significant for ferroelectric materials to understand the associations between the pore and these properties.

Bosse et al. [49] developed a 3D finite element model to characterize the effect of foam shape and porosity aspect ratio on the dielectric and piezoelectric properties of 3-3 type piezoelectric foam structures. The piezoelectric figure of merits such as the piezoelectric charge coefficient (d_h), the piezoelectric voltage coefficient (g_h), and the hydrostatic figure of merit (d_hg_h) can be significantly enhanced by changing the aspect ratio of the porosity. Kar-gupta et al. [50] developed a finite element model to study the effects of porosity on the electromechanical response of 3-1 porous piezoelectric materials. The results showed that the dielectric and piezoelectric properties decreased with increasing porosity volume fraction. Moreover, with the increase of porosity, the acoustic impedance decreased and the hydrostatic charge coefficient increased. Martínez-Ayuso et al. [51] investigated the effect of pore shape on dielectric and piezoelectric properties through computational simulations. Both the dielectric constant and the piezoelectric charge coefficient of the spherical and ellipsoidal models decreased with the increase of porosity volume fraction, while the ellipsoidal model had fewer poled regions, leading to a faster decrease of the piezoelectric charge coefficient.

17.4.1.4 Modelling of Mechanical Properties

According to the positive piezoelectric effect of ferroelectric materials, when an external force is applied, the ferroelectric material will release free charge and generate an open circuit voltage. Therefore, ferroelectric materials are widely used in sensing, transducing, and energy harvesting.

Zhang et al. [52] used the Fourier spectral iterative perturbation method to study 3D interconnected composites and random particle composites using phase field simulations. As shown in Figure 17.6a, under the same applied strain, the stress distribution in the 3D interconnected structure was more homogeneous. Moreover, the average stress and electric field in the 3D interconnected was about four times that in the random particle composite. Hao et al. [53] introduced a simulation using COMSOL to study the distribution of stress and piezoelectric potential in 2-2 and 0-3 composites. As shown in Figure 17.6b, for freeze-cast 2-2 type composites, the interlinked skeleton structure improved the stress transfer, leading to a higher applied net stress and stronger piezoelectric potential.

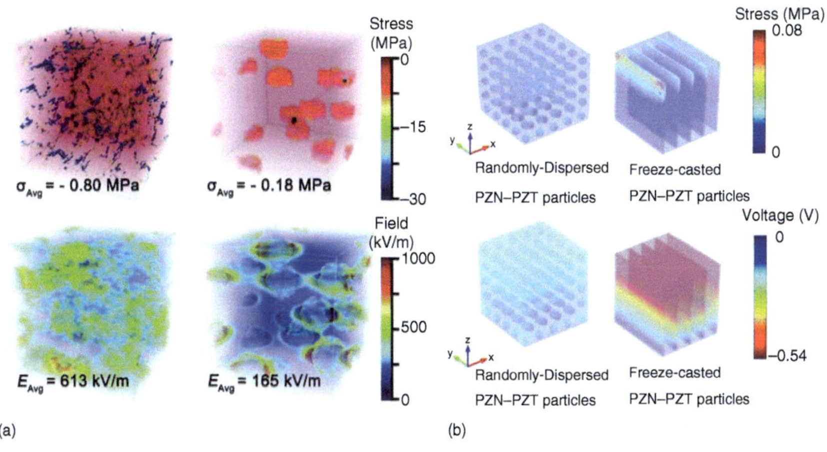

Figure 17.6 Simulation results: (a) the distribution of stress and electric field in the 3D interconnected composite and random particle composite under the same applied strain [52]; (b) the distribution of stress and piezoelectric potential 2-2 and 0-3 composites [53]. Source: (a) Zhang et al. 2018 [52]/Reproduced with permission of Royal Society of Chemistry. (b) Reproduced with permission of Royal Society of Chemistry.

17.5 Applications of Porous Ferroelectric Materials

As a subset of pyroelectric and piezoelectric materials, ferroelectric materials present tremendous potential to convert mechanical or thermal energy into electricity due to the stress-induced piezoelectric effect and thermal-induced pyroelectric effect. Since the development of applications based on ferroelectric materials has been intense, traditional dense ferroelectric materials have some drawbacks, especially the high permittivity ε_{33}^T. Thus, porous ferroelectric materials show lots of benefits compared to dense ferroelectric materials, such as larger surface area, lower permittivity ε_{33}^T and specific heat besides relatively high piezoelectric coefficients d_{33}, pyroelectric coefficient p, and d_h coefficient. Based on the improved piezoelectric and pyroelectric performance, porous ferroelectrics become promising candidates for various applications including energy harvesting, sensing, catalyzing, and biomedical engineering. These applications will be introduced in this section.

17.5.1 Energy Harvesting

Energy harvesting, a powerful technique to remove the dependence on batteries and develop wireless and self-powdered devices, can capture various energy from the ambient environment and store it for useful purposes. According to the diversity of energy sources and different mechanisms of harvesting modes, energy harvesting techniques can be implemented in many ways such as triboelectric, electromagnetic, piezoelectric, and pyroelectric. Among these, energy harvesting based on ferroelectric materials whose polarization switching can be induced by stress or heat is well known for its characteristic of simplicity and practicability. This ability, the functional solid-state coupling between mechanical or thermal and electrical energy, makes developing ferroelectric materials an important exploration for manufacturing unique energy harvesters.

Among two different energy transduction methods of ferroelectric materials, piezoelectric energy harvesting is commonly used because of leveraging vibration energy which is one of the most ubiquitous and accessible energy sources in the ambient environment. It is shown in Eq. (17.6).

$$FOM_{\text{piezo}} = \frac{(d_{33})^2}{\varepsilon_0 \varepsilon_r} \tag{17.6}$$

that the higher piezoelectric charge coefficient d_{33} and lower relative dielectric constant ε_r are two key factors for ferroelectric materials to obtain a high level of piezoelectric energy harvesting figure of merit, which means more vibration energy is captured during the energy harvesting process. Porous ferroelectric materials exhibit the potential to significantly decrease the dielectric constant as well as maintain a relatively high piezoelectric charge coefficient due to the introduction of pores. These materials have received great interest in recent years, covering porous ferroelectric ceramics, polymers, and ceramic-based ferroelectric composites. Roscow et al. produced freeze-cast porous $BaTiO_3$ ceramics to capture vibration energy and obtained a maximum voltage of 234 mV achieved by porous ceramics

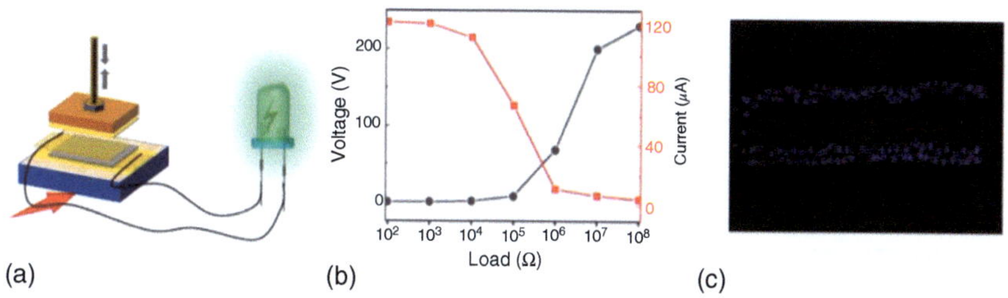

Figure 17.7 PANI/PVDF-TrFE piezoelectric/triboelectric nanogenerator: (a) Connection diagram. (b) The voltage and current density with different resistances. (c) The LED bulbs connected in series are illuminated. Source: Yu et al. [56]. Reprinted with permission of Elsevier.

with 45 vol.% porosity during the process of charging capacitor, which is 138 mV higher than dense $BaTiO_3$ [48]. Although porous ferroelectric ceramic materials are of advantage to weight bearing during variation due to their sufficient mechanical strength and excellent stability, the nature of brittleness and lack of enough toughness are still great challenges to meet the application requirements. Due to the flexibility, processability, and biocompatibility, porous ferroelectric polymers typified by poly (vinylidene fluoride) and its copolymers aroused great interest in recent years. Chen et al. designed a mesoporous PVDF film *via* solvent evaporation, the power density of 8.36 $\mu W/cm^3$ was achieved by porous PVDF film [54]. Mahanty et al. used the etching method to remove ZnO nanoparticles in the composite films and obtained porous poly (vinylidene fluoride-hexafluoropropylene) film. Compared to neat P(VDF-HFP) film, an open-circuit voltage of 9 V, and a short circuit current of 1.3 mA/cm^2 in P(VDF-HFP) films were achieved under repeated stress of 0.36 MPa on the top surface [55]. Yu et al. designed a triboelectric and piezoelectric coupling nanogenerator based on PANI/PVDF-TrFE to power the LED bulbs as shown in Figure 17.7. The maximum voltage and current output reached 246 V and 122 mA respectively, which was able to light up 119 light-emitting diodes [56].

Similarly, capturing energy from thermal fluctuations through the pyroelectric effect is also an important form of energy harvesting of ferroelectric materials. The pyroelectric energy harvesting figure of merit is used to evaluate the ability to convert thermal energy into electric power as shown in Eq. (17.7). From the above equation, three key indicators of high pyroelectric coefficient p, low

$$FOM_{\text{pyro}} = \frac{p^2}{\varepsilon_0 \varepsilon_r C_E^2} \tag{17.7}$$

relative dielectric constant ε_r and low volume-specific heat capacity C_E are beneficial for better pyroelectric energy harvesting performance. Since the introduction of pores plays an important role in decreasing both dielectric constant and volume-specific heat capacity, porous ferrielectric materials are potential to achieve the optimum pyroelectric property of an energy harvester. Hence, freeze-cast porous PZT ceramics was fabricated by Zhang et al. As shown in Figure 17.8, the maximum energy density achieved by porous PZT ceramics with a porosity of 60 vol.% was

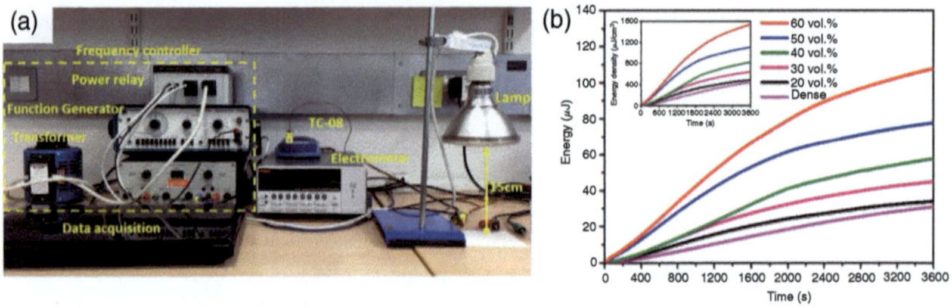

Figure 17.8 Freeze-cast porous PZT ceramics: (a) Test setup platform for pyroelectric energy harvesting. (b) Charging energy of the capacitor and charging energy density. Source: Zhang et al. [57]/Reproduced with permission of Royal Society of Chemistry.

1653 mJ/cm^3 under the thermal fluctuation provided by an IR lamp, which was 374% higher than that of dense PZT ceramics [57].

17.5.2 Sensors

Using ferroelectric materials as sensors to achieve the detection of force, strain, and acceleration, has increasing applications in recent years. Compared to other types of sensors, the unique self-powered properties make sensors based on ferroelectric more portable and practical. On this basis, porous ferroelectric materials are of great concern in the application of sensors because of a relatively high piezoelectric voltage coefficient and good flexibility. For example, Xie et al. developed a porous PZT ceramics with lamellar structure via freeze casting, and impregnated PDMS into the pore channels to obtain a ferroelectric composite. As shown in Figure 17.9, this composite was used to detect finger tapping and shearing at d_{33} and d_{15} mode respectively, whose sensitivity was from ~445 mV/N for a singular shear structure to ~694 mV/N for the longitudinal sensing mode [58].

17.5.3 Catalysis

Ferroelectric materials can not only capture the mechanical vibration to produce piezoelectric charges which can be used to realize piezoelectric catalysis. Compared with dense forms, porous ferroelectric ceramics with a large surface area can provide more active sites, which has great advantages for catalytic reactions. As shown in Figure 17.10, Xu et al. reported a barium strontium titanate-polydimethylsiloxane (BST-PDMS) porous foam, and the degradation efficiency of Rhodamine B reached 97.8% under light irradiation and low-frequency mechanical squeezing [59].

Similarly, it can also harvest thermal energy to change and generate pyroelectric charges which act as an important active substance in pyroelectric catalysis. For example, Min et al. proposed a carbon-polyvinylidene fluoride-barium titanate (C-PVDF-BTO) porous composite to degrade RhB dye solution. It can be seen from Figure 17.11 that a fast average rate of temperature change of 3 °C/s as well as the instantaneous rate of 13 °C/s was achieved, and the color decay of the RhB was observed finally [60].

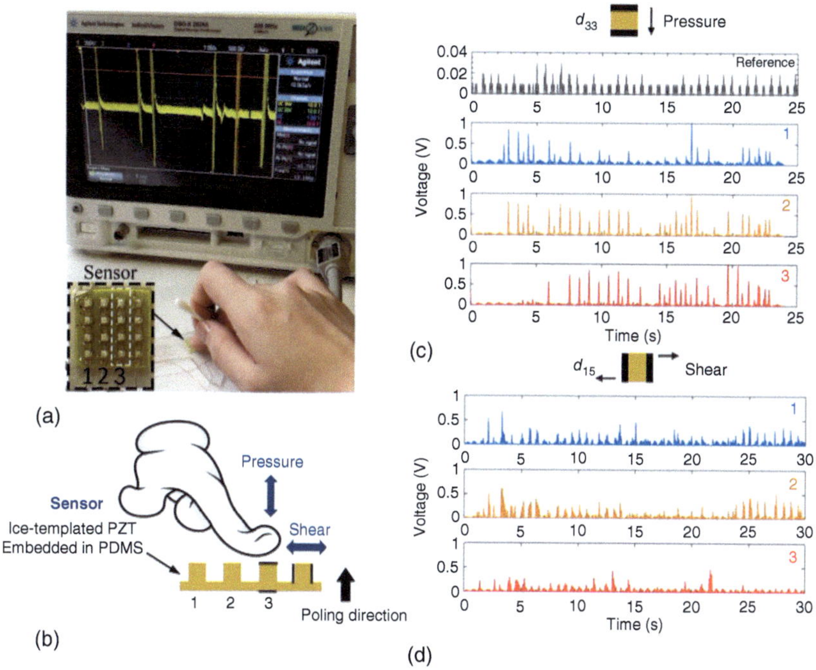

Figure 17.9 Porous PZT ceramics with lamellar structure: (a) Schematic of the finger tapping and shearing measurement. (b) Pressing and shearing mode. (c) The voltage response of the d_{33} pressure sensor. (d) Voltage response of d_{15} shear sensor. Source: Xie et al. [58]/Reproduced with permission of Royal Society of Chemistry.

17.5.4 Biomedical

Fractures and defects of bone tissue are some of the most common medical diseases in clinics. The healing of human bone structure often depends on the regeneration of periosteum and osteoblasts. In small bone defect cases, the formation of periosteum and osteoblasts can make bone regenerate rapidly. In the treatment of large bone tissue injury, due to insufficient regeneration time, even if the operation is carried out in time, the healing is often delayed and damaged. Although autologous and allogeneic bone transplantation is considered an effective method to repair bone tissue and its physiological environment of bone defects, transplantation is seldom taken as an efficient solution because of its complex process and complications. In this case, biomaterials are considered as effective substitutes to help repair bone defects. The most common biomaterials, such as titanium alloy [61] composite bone scaffold, can provide sufficient mechanical strength in the large area of bone defect and maintain the mechanical stability of the damaged area, at the same time, its porous structure can also provide sufficient space for the growth of bone tissue, blood vessels, and other tissues. However, although metal biomaterials such as titanium alloy have the advantages of being non-toxic, good biocompatibility, lightweight, and high strength, they are still inactive materials, which will affect the bonding ability of bone and material surface in the human body, and reduce

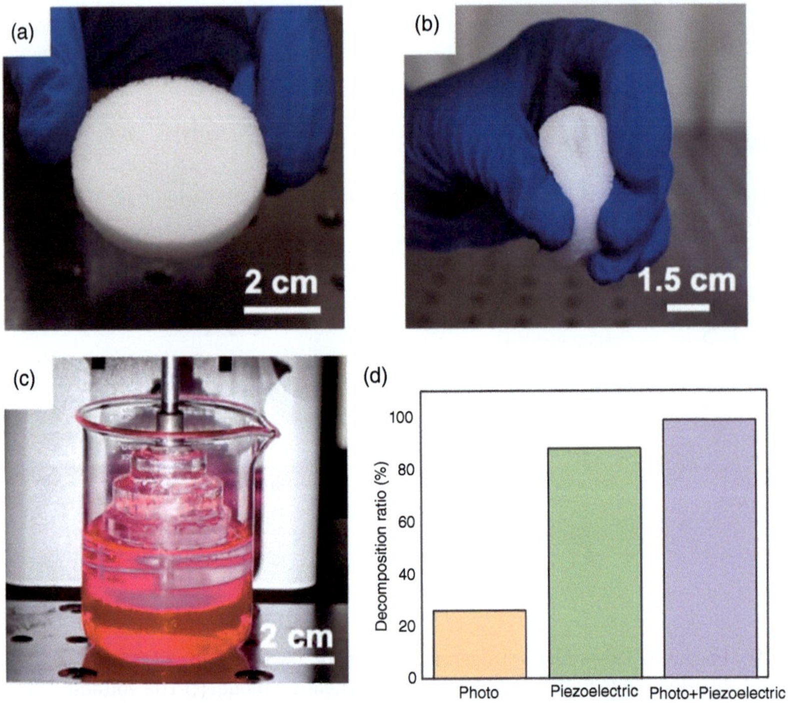

Figure 17.10 BST-PDMS porous foam: (a), (b) Optical image. (c) Degradation process. (d) Decomposition ratio under different catalytic modes. Source: Xu et al. 2020 [59]/Reproduced with permission of Elsevier.

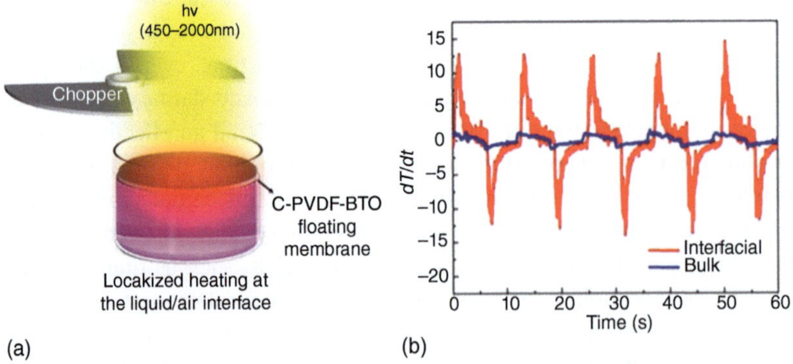

Figure 17.11 C-PVDF-BTO porous composite: (a) Schematic illustration of pyroelectric catalysis. (b) Temperature change rate. Source: Min et al. [60]. Reproduced with permission of American Chemical Society.

the growth of new bone to porous materials, ultimately leading to the material cannot be stable in the human body for a long time [62]. In this case, people started to use porous piezoelectric ceramics as biomedical applications. In the 1960s, Japanese scientists found that bone tissue was a piezoelectric material due to the displacement of hydrogen bonds in a collagen polypeptide chain. It responds to mechanical stresses, causing polarization changes and inducing electric dipoles. Moreover, bioelectricity also plays a very important role in human bone homeostasis [63]. Many research groups use lead-free piezoelectric ceramic materials with good piezoelectric properties and biocompatibility to cooperate with bioelectricity to obtain biomedical materials with better performance. In addition, porous structural materials have been proven to have a very beneficial role in promoting the repair and regeneration of human tissues. Porous ferroelectric ceramic materials have become a good candidate for a new generation of bone tissue repair and biomedical implant materials, which can be widely used in bone tissue repair and human biological function sensing monitoring [64].

The current scientific research has proved that barium titanate-based lead-free piezoelectric ceramics can promote the regeneration of human bone tissue. The material itself is non-toxic and harmless to organisms and has good biocompatibility. At the same time, it can also use additive manufacturing, freezing pouring, and other processes to prepare barium titanate piezoelectric ceramics with oriented porous structures. When used as implant materials in the medical field, it can enhance the biological activity of implants and better induce bone tissue regeneration and repair [65]. This shows that porous ferroelectric materials have great development prospects and application potential in the biomedical field.

17.5.5 Energy Storage

With the progress of science and technology, economic development, and population explosion, energy shortage has gradually become a huge challenge faced by human society. Developing more efficient energy storage media has also become one of the important research directions in the field of materials science. Compared with various types of batteries that are easy to leak electricity and have large volumes, the capacitor devices which have abilities to resist high voltage, high cycle fatigue times, and high power acquire great advantages. The preparation of dielectric materials with high energy storage density and excellent performance is the basis for improving the performance of capacitors. Ferroelectric materials have become the most common solid dielectric materials, including ferroelectric ceramics, ferroelectric polymers, and various kinds of composites based on ferroelectric ceramics, due to their controllable dielectric properties, high power density, ability to charge and discharge quickly and long service life.

Although ferroelectric ceramics have a high dielectric constant, their breakdown field strength is low, which limits their application in energy storage devices to some extent. To combine the high dielectric constant of ceramics with the high breakdown field strength of ferroelectric polymers, the composite of ferroelectric materials with porous structure has become a new research idea [23c]. Among the research achievements in recent years, it is typical that after preparing barium titanate ceramic matrix with a highly oriented layered structure by freezing casting, epoxy resin is used to infiltrate the ceramic matrix [66]. Because of the high orientation of porous ceramic structure and the penetration of epoxy resin, the composite formed by it can obtain high potential shift and discharge energy density under a low electric field, it is several times the corresponding performance of the composite made of barium titanate particles randomly distributed in the epoxy resin matrix. In addition, to improve the low breakdown field strength and other problems of porous ferroelectric materials caused by high porosity, PVDF and its copolymer are used as polymer substrate to composite porous ferroelectric ceramics, which can further improve the energy storage density of materials to a certain extent. When comparing single polymer, single porous ceramics, and composite materials of porous ceramics and epoxy resin. The breakdown field strength has been improved to a certain extent [8b].

These results show that the preparation of ferroelectric materials with oriented porous or layered structure by additive manufacturing and frozen pouring column process is an effective way to achieve high dielectric constant and energy density in ferroelectric ceramic/polymer composites. The penetration or coating of porous ferroelectric materials with epoxy resin or PVDF and other polymers can obtain ferroelectric composites with better performance than pure ceramics or pure polymers, thus improving the energy storage density and use efficiency of energy storage devices [67]. However, although the addition of PVDF can slightly increase the breakdown field strength of the composite, compared with ferroelectric polymers, the low breakdown field strength still limits the potential application of porous ferroelectric materials in energy storage devices. In the future, more suitable polymer materials (such as PVDF copolymers) or more reasonable structural design schemes of porous composites are needed to improve this [42].

17.6 Summary

In recent years, porous piezoelectric materials have attracted great attention in the energy field due to their excellent durability and sensitivity. This book aims to elaborate on the research progress of porous piezoelectric materials in energy fields and introduces the classification, fabrication techniques, analysis model, and energy application of porous piezoelectric materials. Although great progress has been made in the current research on porous piezoelectric materials, many challenges remain in the future.

(1) Porous piezoelectric ceramics are important porous piezoelectric materials with relatively high figures of merit for energy harvesting, but the inherent brittleness of ceramics still limits their applications. At present, the main strategy is to composite porous ceramics and highly flexible polymers to prepare ferroelectric ceramic composites, but their flexibility is still limited, and highly flexible porous ceramic materials may be a hot topic in the future.

(2) Porous ferroelectric thin films have shown great potential in sensing and energy harvesting applications, but the number of studies is relatively small. At present, the preparation process of porous ferroelectric thin films is still relatively complicated, and it is still difficult to obtain high-quality porous ferroelectric thin films. Therefore, further optimization of the preparation process of porous ferroelectric thin films is also one of the future research directions.
(3) At present, porous ceramic matrix ferroelectric composites with 3-3 or 2-2 structures have exhibited higher output voltage than the 0-3 type structure, but due to the complexity of their structure, current research is still relatively few. Therefore, simulation calculation also requires optimizing the structure of porous ceramic matrix ferroelectric composites.

Overall, the research of porous piezoelectric materials in energy applications has made important progress in the past few decades, showing great potential in energy harvesting, sensing, catalysis, biomedical, and energy storage. This book summarizes the current important research progress of porous piezoelectric materials in the field of energy, expecting more inspiring developments in the future.

References

1 (a) Wang, Z. and Song, J. (2006). Piezoelectric nanogenerators based on zinc oxide nanowire arrays. *Science* 312: 242. (b) Wen, K., Qiu, J., Ji, H. et al. (2014). Investigation of phase diagram and electrical properties of xPb(Mg1/3Nb2/3)O3–(12x)Pb(Zr0.4Ti0.6)O3 ceramics. *J. Mater. Sci. Mater. Electron.* 25: 3003–3009.

2 Zhang, Y., Bao, Y., Zhang, D. et al. (2015). Porous PZT ceramics with aligned pore channels for energy harvesting applications. *J. Am. Ceram. Soc.* 98 (10): 2980–2983.

3 Yap, E.W., Glaum, J., Oddershede, J., and Daniels, J.E. (2018). Effect of porosity on the ferroelectric and piezoelectric properties of (Ba0.85Ca0.15)(Zr0.1Ti0.9)O3 piezoelectric ceramics. *Scr. Mater.* 145: 122–125.

4 (a) Rödig, T., Schönecker, A., and Gerlach, G. (2010). A survey on piezoelectric ceramics for generator applications. *J. Am. Ceram. Soc.* 93 (4): 901–912. (b) Zhang, Y., Roscow, J., Xie, M., and Bowen, C. (2018). High piezoelectric sensitivity and hydrostatic figures of merit in unidirectional porous ferroelectric ceramics fabricated by freeze casting. *J. Eur. Ceram. Soc.* 38 (12): 4203–4211.

5 Roscow, J.I., Lewis, R.W.C., Taylor, J., and Bowen, C.R. (2017). Modelling and fabrication of porous sandwich layer barium titanate with improved piezoelectric energy harvesting figures of merit. *Acta Mater.* 128: 207–217.

6 Yang, A., Wang, C.A., Guo, R., and Huang, Y. (2010). Microstructure and electrical properties of porous PZT ceramics fabricated by different methods. *J. Am. Ceram. Soc.* 93: 1984–1990.

7 (a) Chen, Y.-C. and Wu, S. (2004). Piezoelectric composites with 3-3 connectivity by injecting polymer for hydrostatic sensors. *Ceram. Int.* 30: 69–74. (b) Levassort,

F., Holc, J., Ringgaard, E. et al. (2007). Fabrication, modelling and use of porous ceramics for ultrasonic transducer applications. *J. Electroceram.* 19 (1): 127–139.

8 (a) Newnham, R.E., Skinner, D.P., and Cross, L.E. (1978). Connectivity and piezoelectric-pyroelectric composites. *Mater. Res. Bull.* 13 (5): 525–536. (b) Zhang, Z., Luo, S., Yu, S. et al. (2018). Significantly enhanced dielectric and energy storage performance of blend polymer-based composites containing inorganic 3D–network. *Mater. Des.* 142: 106–113.

9 Safari, A. and Akdogan, E.K. (2006). Rapid Prototyping of Novel Piezoelectric Composites. *Ferroelectrics* 331 (1): 153–179.

10 Bowen, C.R., Perry, A., Lewis, A.C.F., and Kara, H. (2004). Processing and properties of porous piezoelectric materials with high hydrostatic figures of merit. *J. Eur. Ceram. Soc.* 24 (2): 541–545.

11 Bowen, C.R. and Kara, H. (2002). Pore anisotropy in 3–3 piezoelectric composites. *Mater. Chem. Phys.* 75 (1): 45–49.

12 Guo, R., Wang, C.-A., and Yang, A. (2011). Effects of pore size and orientation on dielectric and piezoelectric properties of 1–3 type porous PZT ceramics. *J. Eur. Ceram. Soc.* 31 (4): 605–609.

13 Wang, W., Xu, M., Xu, X. et al. (2020). Perovskite oxide based electrodes for high-performance photoelectrochemical water splitting. *Angew. Chem. Int. Ed.* 59 (1): 136–152.

14 (a) Zhang, M., Sun, H., Liu, X. et al. (2020). Structural design of PZT porous ceramics obtained via free-casting by ice-templating and performance exploration. *Mater. Res. Bull.* 127: 110862. (b) Liu, W., Du, L., Wang, Y. et al. (2013). Effects of foam composition on the microstructure and piezoelectric properties of macroporous PZT ceramics from ultrastable particle-stabilized foams. *Ceram. Int.* 39 (8): 8781–8787.

15 Zeng, T., Dong, X., Mao, C. et al. (2006). Preparation and properties of porous PMN–PZT ceramics doped with strontium. *Mater. Sci. Eng. B* 135 (1): 50–54.

16 Lee, S.-H., Jun, S.-H., Kim, H.-E., and Koh, Y.-H. (2007). Fabrication of porous PZT–PZN piezoelectric ceramics with high hydrostatic figure of merits using camphene-based freeze casting. *J. Am. Ceram. Soc.* 90 (9): 2807–2813.

17 (a) Wang, Q., Chen, Q., Zhu, J. et al. (2008). Effects of pore shape and porosity on the properties of porous LNKN ceramics as bone substitute. *Mater. Chem. Phys.* 109 (2): 488–491. (b) Liu, W., Yang, D., Wei, X. et al. (2020). Fabrication of piezoelectric porous BaTiO3 scaffold to repair large segmental bone defect in sheep. *J. Biomater. Appl.* 35 (4–5): 544–552.

18 (a) Liu, W. and Ren, X. (2009). Large piezoelectric effect in Pb-free ceramics. *Phys. Rev. Lett.* 103 (25): 257602. (b) Gao, X., Zheng, M., Yan, X. et al. (2019). The alignment of BCZT particles in PDMS boosts the sensitivity and cycling reliability of a flexible piezoelectric touch sensor. *J. Mater. Chem. C* 7 (4): 961–967.

19 Zhang, Y., Roscow, J., Lewis, R. et al. (2018). Understanding the effect of porosity on the polarisation-field response of ferroelectric materials. *Acta Mater.* 154: 100–112.

20 (a) Chorsi, M.T., Curry, E.J., Chorsi, H.T. et al. (2019). Piezoelectric biomaterials for sensors and actuators. *Adv. Mater.* 31 (1): 1802084. (b) William, R.V.,

Marikani, A., and Madhavan, D. (2016). Dielectric behavior and magnetical response for porous BFO thin films with various thicknesses over Pt/Ti/SiO2/Si substrate. *Ceram. Int.* 42 (6): 6807–6816.

21 Wan, C. and Bowen, C.R. (2017). Multiscale-structuring of polyvinylidene fluoride for energy harvesting: the impact of molecular-, micro-and macro-structure. *J. Mater. Chem. A* 5 (7): 3091–3128.

22 (a) Chen, D., Chen, K., Brown, K. et al. (2017). Liquid-phase tuning of porous PVDF-TrFE film on flexible substrate for energy harvesting. *Appl. Phys. Lett.* 110 (15): 153902. (b) Chen, D. and Zhang, J.X.J. (2015). Microporous polyvinylidene fluoride film with dense surface enables efficient piezoelectric conversion. *Appl. Phys. Lett.* 106 (19): 193901.

23 (a) Luo, S., Shen, Y., Yu, S. et al. (2017). Construction of a 3D-BaTiO 3 network leading to significantly enhanced dielectric permittivity and energy storage density of polymer composites. *Energy Environ. Sci.* 10 (1): 137–144. (b) Kim, D.S., Baek, C., Ma, H.J., and Kim, D.K. (2016). Enhanced dielectric permittivity of BaTiO3/epoxy resin composites by particle alignment. *Ceram. Int.* 42 (6): 7141–7147. (c) Guo, R., Roscow, J.I., Bowen, C.R. et al. (2020). Significantly enhanced permittivity and energy density in dielectric composites with aligned BaTiO 3 lamellar structures. *J. Mater. Chem. A* 8 (6): 3135–3144.

24 (a) Matavž, A., Bradeško, A., Rojac, T. et al. (2019). Self-assembled porous ferroelectric thin films with a greatly enhanced piezoelectric response. *Appl. Mater. Today* 16: 83–89. (b) Stancu, V., Lisca, M., Boerasu, I. et al. (2007). Effects of porosity on ferroelectric properties of Pb(Zr0.2Ti0.8)O3 films. *Thin Solid Films* 515 (16): 6557–6561. (c) Annapu Reddy, V., Pathak, N.P., and Nath, R. (2012). Effect of pore size on ferroelectric properties of multiferroic BiFeO3 films prepared on porous silicon. *Curr. Appl. Phys.* 12 (2): 451–455. (d) Cao, Z., Ding, A., He, X. et al. (2005). Preparation and characteristics of nanoporous PLZT ferroelectric thin films prepared via a one-step CSD process. *J. Cryst. Growth* 273 (3): 489–493.

25 Aleksandrova, M. (2020). Polymeric seed layer as a simple approach for nanostructuring of Ga-doped ZnO films for flexible piezoelectric energy harvesting. *Microelectron. Eng.* 233: 111434.

26 Zeng, T., Dong, X., Mao, C. et al. (2007). Effects of pore shape and porosity on the properties of porous PZT 95/5 ceramics. *J. Eur. Ceram. Soc.* 27 (4): 2025–2029.

27 Zhang, H., Li, J.-F., and Zhang, B.-P. (2007). Microstructure and electrical properties of porous PZT ceramics derived from different pore-forming agents. *Acta Mater.* 55 (1): 171–181.

28 Abdullah, F.F., Nemati, A., and Bagheri, R. (2015). Dielectric and piezoelectric properties of porous PZT–PCN ceramics sintered at different temperatures. *Mater. Lett.* 151: 85–88.

29 Sun, L., Li, Z., Su, R. et al. (2018). Phase-transition induced conversion into a photothermal material: quasi-metallic $WO_{2.9}$ nanorods for solar water evaporation and anticancer photothermal therapy. *Angew. Chem. Int. Ed. Eng.* 57 (33): 10666–10671.

30 Tan, J. and Li, Z. (2015). Fabrication and electrical properties of porous BS–0.64 PT high temperature piezoceramics using polystyrene microsphere. *Ceram. Int.* 41: S414–S420.

31 Stanculescu, R., Ciomaga, C.E., Padurariu, L. et al. (2015). Study of the role of porosity on the functional properties of (Ba, Sr) TiO3 ceramics. *J. Alloys Compd.* 643: 79–87.

32 (a) Piazza, D., Capiani, C., and Galassi, C. (2005). Piezoceramic material with anisotropic graded porosity. *J. Eur. Ceram. Soc.* 25 (12): 3075–3078. (b) Praveenkumar, B., Kumar, H., and Kharat, D. (2006). Study on microstructure, piezoelectric and dielectric properties of 3-3 porous PZT composites. *J. Mater. Sci. Mater. Electron.* 17 (7): 515–518.

33 Omatete, O.O., Janney, M.A., and Nunn, S.D. (1997). Gelcasting: from laboratory development toward industrial production. *J. Eur. Ceram. Soc.* 17 (2–3): 407–413.

34 Yang, A.-K., Wang, C.-A., Guo, R., and Huang, Y. (2011). Effects of porosity on dielectric and piezoelectric properties of porous lead zirconate titanate ceramics. *Appl. Phys. Lett.* 98 (15): 152904.

35 Yang, A., Wang, C.A., Guo, R. et al. (2010). Porous PZT ceramics with high hydrostatic figure of merit and low acoustic impedance by TBA-based gel-casting process. *J. Am. Ceram. Soc.* 93 (5): 1427–1431.

36 Liu, W., Xu, J., Wang, Y. et al. (2013). Processing and properties of porous PZT ceramics from particle-stabilized foams via gel casting. *J. Am. Ceram. Soc.* 96 (6): 1827–1831.

37 Dommati, H., Ray, S.S., Wang, J.-C., and Chen, S.-S. (2019). A comprehensive review of recent developments in 3D printing technique for ceramic membrane fabrication for water purification. *RSC Adv.* 9 (29): 16869–16883.

38 Morissette, S.L., Lewis, J.A., Clem, P.G. et al. (2001). Direct-write fabrication of Pb (Nb, Zr, Ti) O3 devices: influence of paste rheology on print morphology and component properties. *J. Am. Ceram. Soc.* 84 (11): 2462–2468.

39 Smay, J.E., Cesarano, J. III,, Tuttle, B.A., and Lewis, J.A. (2002). Piezoelectric properties of 3-X periodic Pb (Zr_x Ti_{1-x}) O_3–polymer composites. *J. Appl. Phys.* 92 (10): 6119–6127.

40 Nan, B., Olhero, S., Pinho, R. et al. (2019). Direct ink writing of macroporous lead-free piezoelectric Ba0.85Ca0.15Zr0.1Ti0.9O3. *J. Am. Ceram. Soc.* 102 (6): 3191–3203.

41 Chen, Y., Wang, N., Ola, O. et al. (2021). Porous ceramics: light in weight but heavy in energy and environment technologies. *Mater. Sci. Eng. R. Rep.* 143: 100589.

42 Yan, M., Xiao, Z., Ye, J. et al. (2021). Porous ferroelectric materials for energy technologies: current status and future perspectives. *Energy Environ. Sci.* 14: 6158–6190.

43 Zhou, X., Zhou, K., Zhang, D. et al. (2022). Perspective on porous piezoelectric ceramics to control internal stress. *Nanoenergy Adv.* 2 (4): 269–290.

44 Shao, G., Hanaor, D.A.H., Shen, X., and Gurlo, A. (2020). Freeze casting: from low-dimensional building blocks to aligned porous structures—a review of novel materials, methods, and applications. *Adv. Mater.* 32 (17): e1907176.

45 Hao, Y., Hou, Y., Xu, H. et al. (2021). Enhancing the energy harvesting performance of PDMS based piezocomposites via a synergistic effect. *J. Mater. Chem. C* 9 (40): 14303–14308.

46 Gheorghiu, F., Padurariu, L., Airimioaei, M. et al. (2016). Porosity-dependent properties of Nb-Doped Pb(Zr,Ti)O3 ceramics. *J. Am. Ceram. Soc.* 100 (2): 647–658.

47 Khachaturyan, R., Zhukov, S., Schultheiß, J. et al. (2017). Polarization-switching dynamics in bulk ferroelectrics with isometric and oriented anisometric pores. *J. Phys. D. Appl. Phys.* 50 (4): 045303.

48 Roscow, J.I., Zhang, Y., Kraśny, M.J. et al. (2018). Freeze cast porous barium titanate for enhanced piezoelectric energy harvesting. *J. Phys. D. Appl. Phys.* 51 (22): 225301.

49 Bosse, P.W., Challagulla, K.S., and Venkatesh, T.A. (2012). Effects of foam shape and porosity aspect ratio on the electromechanical properties of 3-3 piezoelectric foams. *Acta Mater.* 60 (19): 6464–6475.

50 Kargupta, R. and Venkatesh, T. (2006). Electromechanical response of porous piezoelectric materials. *Acta Mater.* 54 (15): 4063–4078.

51 Martínez-Ayuso, G., Friswell, M.I., Haddad Khodaparast, H. et al. (2019). Electric field distribution in porous piezoelectric materials during polarization. *Acta Mater.* 2019 (173): 332–341.

52 Zhang, Y., Jeong, C.K., Yang, T. et al. (2018). Bioinspired elastic piezoelectric composites for high-performance mechanical energy harvesting. *J. Mater. Chem. A* 6 (30): 14546–14552.

53 Hao, Y., Hou, Y., Fu, J. et al. (2020). Flexible piezoelectric energy harvester with an ultrahigh transduction coefficient by the interconnected skeleton design strategy. *Nanoscale* 12 (24): 13001–13009.

54 Chen, D., Sharma, T., and Zhang, J.X.J. (2014). Mesoporous surface control of PVDF thin films for enhanced piezoelectric energy generation. *Sensors Actuators A Phys.* 216: 196–201.

55 Mahanty, B., Ghosh, S.K., Garain, S., and Mandal, D. (2017). An effective flexible wireless energy harvester/sensor based on porous electret piezoelectric polymer. *Mater. Chem. Phys.* 186: 327–332.

56 Yu, S., Zhang, Y., Yu, Z. et al. (2021). PANI/PVDF-TrFE porous aerogel bulk piezoelectric and triboelectric hybrid nanogenerator based on in-situ doping and liquid nitrogen quenching. *Nano Energy* 80: 105519.

57 Zhang, Y., Xie, M., Roscow, J. et al. (2017). Enhanced pyroelectric and piezoelectric properties of PZT with aligned porosity for energy harvesting applications. *J. Mater. Chem. A Mater.* 5 (14): 6569–6580.

58 Xie, M., Zhang, Y., Krasny, M.J. et al. (2018). Flexible and active self-powered pressure, shear sensors based on freeze casting ceramic-polymer composites. *Energy Environ. Sci.* 11 (10): 2919–2927.

59 Xu, S., Qian, W., Zhang, D. et al. (2020). A coupled photo-piezo-catalytic effect in a BST-PDMS porous foam for enhanced dye wastewater degradation. *Nano Energy* 77: 105305.

60 Min, M., Liu, Y., Song, C. et al. (2018). Photothermally enabled pyro-catalysis of a BaTiO3 nanoparticle composite membrane at the liquid/air interface. *ACS Appl. Mater. Interfaces* 10 (25): 21246–21253.

61 Wang, Y.B., Li, H.F., Cheng, Y. et al. (2013). In vitro and in vivo studies on Ti-based bulk metallic glass as potential dental implant material. *Mater. Sci. Eng. C* 33 (6): 3489–3497.

62 Polley, C., Distler, T., Detsch, R. et al. (2020). 3D printing of piezoelectric barium titanate-hydroxyapatite scaffolds with interconnected porosity for bone tissue engineering. *Materials* 13 (7): 1773.

63 (a) Fukada, E. and Yasuda, I. (1957). On the piezoelectric effect of bone. *J. Phys. Soc. Jpn.* 12 (10): 1158–1162. (b) Park, J.B. and Lakes, R.S. (1992). *Biomaterials: An Introduction*, 2e, 394. New York: Plenum Press *Biomaterials* **1994,** *15* (1), 76.

64 Zhang, Y., Chen, L., Zeng, J. et al. (2014). Aligned porous barium titanate/hydroxyapatite composites with high piezoelectric coefficients for bone tissue engineering. *Mater. Sci. Eng. C* 39: 143–149.

65 Zhujun, J., Liying, C., Yong, Z. et al. (2022). 3D printing of porous scaffolds BaTiO3 piezoelectric ceramics and regulation of their mechanical and electrical properties. *Ceram. Int.* 48 (5): 6477–6487.

66 (a) Hao, Y., Wang, X., Bi, K. et al. (2017). Significantly enhanced energy storage performance promoted by ultimate sized ferroelectric BaTiO3 fillers in nanocomposite films. *Nano Energy* 31: 49–56. (b) Zhao, P., Wang, H., Wu, L. et al. (2019). High-performance relaxor ferroelectric materials for energy storage applications. *Adv. Energy Mater.* 9 (17): 1803048.

67 Yang, Y., Jinliang, H., Qi, L. et al. (2019). Self-healing of electrical damage in polymers using superparamagnetic nanoparticles. *Nat. Nanotechnol.* 14 (2): 151–155.

18

Piezoelectric Flexible Devices

Yejing Dai and Zhihao Zhao

School of Materials, Sun Yat-sen University, No. 66 Gongchang Road, Guangming District, Shenzhen 518107, China

18.1 Introduction

More aspects of our daily lives will involve communication and interaction between humans and smart devices with the development of modern society. Smart devices based on artificial intelligence and the Internet of Things technologies can not only communicate with humans but also exchange information with other smart devices to better serve our daily lives [1, 2]. Wearable smart devices, as a kind of smart device, are developing rapidly for realizing various functions, such as human health monitoring and exercise monitoring [3–5]. These wearable electronic devices should possess various characteristics, including high flexibility, high reliability, lightweight, small size, high sensitivity, low power consumption, and even self-powered capability [6, 7]. Among them, the power supply is the biggest challenge. Although this problem can be alleviated by reducing power consumption and increasing the energy density of the energy storage element, with the increase in complexity and versatility of devices, the number of power-consuming components will also inevitably increase, making it difficult to exchange/charge the energy storage components. Therefore, sustainable/self-powered systems that provide power by harvesting the energy generated by human movement are a very attractive solution for the long-term operation of wearable electronic devices.

Because of their unique electromechanical conversion capabilities, low cost, lightweight, and ease of fabrication, piezoelectric flexible devices have attracted much attention, which can exploit the piezoelectric effect to harvest various types of mechanical energy. Since Wang's group reported the first piezoelectric flexible device in 2006 [8], many piezoelectric flexible devices with different piezoelectric materials have been investigated with enhanced output performance, such as ZnO [9, 10], MoS_2 [11], lead-based/lead-free perovskite materials [12–14], organic–inorganic lead halide perovskite materials [15, 16], and poly(vinylidene fluoride)-based (PVDF-based) organic piezoelectric polymer [2, 13, 14, 17–21]. Among these materials, the inorganic piezoelectric ones are inherently brittle and

Piezoelectric Materials: From Fundamentals to Emerging Applications, First Edition. Edited by Jiagang Wu.

difficult to produce large deformations under stress. However, when combined with flexible organic matrix such as polydimethylsiloxane (PDMS), PVDF, and silicone rubber, piezoelectric flexible devices based on these materials can easily obtain good stretchable, twistable, and foldable features, which are ideal as energy supply units or biomechanical monitoring units in wearable electronics. The piezoelectric flexible devices can power wearable electronics by harvesting different types of biomechanical energy, including walking, running, knocking, and joint movement [7]. However, as a power generator, achieving high power output is still a major challenge for applications in wearable electronics.

In this chapter, we summarize the development of flexible piezoelectric devices, especially focusing on materials and structure design, and offer guidance for future research. Firstly, we will introduce the working mechanism of a piezoelectric device. Subsequently, the representative works of piezoelectric flexible devices about the various inorganic/organic piezoelectric materials and the structure design are introduced in detail. Then, a variety of applications of piezoelectric flexible devices are displayed. Finally, we outline the advantages, prospects, and challenges of piezoelectric flexible devices in electronics.

18.2 General Working Mechanism of Piezoelectric Flexible Devices

The working mechanism of the piezoelectric flexible device, taking the conventional ferroelectric PVDF polymer device as an example, is shown in Figure 18.1. The ferroelectric PVDF film with appropriate poling or macroscopic polarization by fabrication procedure is coated with electrodes on both sides, forming a conventional plate capacitor (Figure 18.1a). The polarization direction is perpendicular to the film plane, and the compensation charges accumulate at the two electrodes to screen the polarization. Thus, when there is no external force applying on the device, there is no output signal in the external circuit (Figure 18.1a). As shown in Figure 18.1b, when a compressive force is applied to the film, the deformation of the film will decrease the density of dipoles as well as the density of compensating charges. Therefore, the transferred compensation charges flow from one electrode to the other electrode through the external circuit due to the changed piezoelectric potential [22], as well as the piezoelectric response signal. Under tensile stress, the output signal is

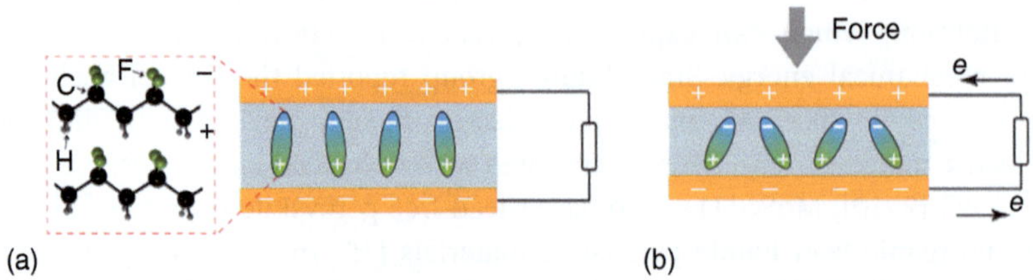

Figure 18.1 Working mechanism of piezoelectric flexible device. (a) Force-free condition, and (b) under compression force. Source: Zhihao Zhao.

opposite. In addition, like the working mechanism under compress mode, when the piezoelectric flexible device is under more complex stress conditions, such as bending or twisting mode, the variation of dipole density caused by deformation can lead to the piezoelectric output.

Generally, conventional inorganic piezoelectric materials, such as $BaTiO_3$, $Pb(Zr,Ti)O_3$ (PZT), and ZnO, have been used in piezoelectric flexible devices due to their relatively high piezoelectric constant (d_{33}). However, the inorganic piezoelectric materials cannot withstand the large deformation required for the piezoelectric flexible devices. Combining inorganic piezoelectric materials with flexible organic substrates or directly utilizing flexible organic piezoelectric materials (e.g. PVDF, poly(vinylidene fluoride-trifluoroethylene) (PVDF-TrFE)) is an effective approach to prepare piezoelectric flexible devices. Therefore, based on the piezoelectric effect, the development of high-performance piezoelectric materials and advanced device structural design are the keys to piezoelectric flexible devices for achieving high power output.

18.3 Materials and Structure Design of Flexible Devices

18.3.1 Inorganic Piezoelectric Material-Based Flexible Devices

Inorganic piezoelectric materials, as one of the most important piezoelectric materials, have been widely used as sensors, actuators, and energy harvesters due to their high piezoelectric constant, large electromechanical coupling coefficient, and relatively high Curie temperature. However, the inherent brittleness of inorganic piezoelectric materials makes it difficult to fabricate piezoelectric flexible devices. Therefore, they are often embedded in flexible organic matrices as piezoelectric fillers. In this section, state-of-the-art processes for piezoelectric flexible devices with different types of inorganic piezoelectric materials will be reviewed.

18.3.1.1 Inorganic Perovskite Structure Piezoelectric Materials

Perovskite-structured piezoelectric materials are widely used in commercial applications due to their excellent piezoelectric properties. For example, lead-based commercial PZT ceramics show a high d_{33} up to ~500 pC/N. Recently, researchers have reported that the relaxor PMN–PT ($Pb(Mg_{1/3}Nb_{2/3})O_3$–$PbTiO_3$) ceramics can provide a high d_{33} value, on the order of 1250–1500 pC/N (the d_{33} value of PMN–PT single crystal is being on the order of 3400–4100 pC/N) [23, 24]. In addition, the lead-free perovskite systems also exhibit favorable piezoelectric constants when compared to wurtzite-structured materials. For example, the $BaTiO_3$-based piezoelectric ceramics have a d_{33} of 150–600 pC/N [25–27], the $Na_{0.5}Bi_{0.5}TiO_3$-based ceramics have a d_{33} of 150–300 pC/N [28, 29], and the $K_{0.5}Na_{0.5}NbO_3$-based ceramics have a d_{33} of 150–570 pC/N [30, 31]. Therefore, these well-crystallized perovskite-structured materials with nano/micro size can be used as key components of piezoelectric flexible devices.

Perovskite-structured piezoelectric materials in the form of 0-dimensional particles are the most commonly used piezoelectric components for piezoelectric flexible devices [32–34]. For example, evenly dispersed PZT particles in a solid silicone rubber without aggregation were used to fabricate the stretchable piezoelectric device, which exhibited excellent electrical and mechanical properties with a high power density of ~81.25 μW/cm^3 and a high stretchability of 30% [35]. In another study, binary $0.72Pb(Zr_{0.47}Ti_{0.53})O_3$–$0.28Pb[(Zn_{0.45}Ni_{0.55})_{1/3}Nb_{2/3}]O_3$ (PZT–PZNN) particles were dispersed in PDMS to form a composite and used as a flexible energy harvester [36]. With a high transduction coefficient ($d_{33} \times g_{33}$, where g_{33} is the piezoelectric voltage constant), the piezoelectric flexible device based on the PZT–PZNN material showed a high open-circuit voltage (V_{oc}) of 55.24 V and a large current density of 2.76 μA/cm^2, accompanied by a high power density of 554 μW/cm^3. In order to address the nonuniform dispersion problem commonly faced by piezoelectric nanoparticles in flexible matrixes, Lee et al. [18] designed a unique composite by mixing amine-modified PZT nanoparticles with triblock copolymers grafted with maleic anhydride to achieve well-distributed PZT nanoparticles in polymer and enhance the stress applied on the PZT particles (Figure 18.2a). Benefiting from the well-distributed PZT nanoparticles in polymer, the prepared piezoelectric flexible device (Figure 18.2b) showed a high V_{oc} of 65 V and an output short current (I_{sc}) of 1.6 μA.

Although PZT-based piezoelectric flexible devices have excellent performance, the high content of lead (Pb) in these PZT materials makes them unsuitable for use in bio-related electronic devices. Therefore, lead-free piezoelectric materials, such as $BaTiO_3$, $K_{0.5}Na_{0.5}NbO_3$, and $Na_{0.5}Bi_{0.5}TiO_3$, with perovskite structures, have been

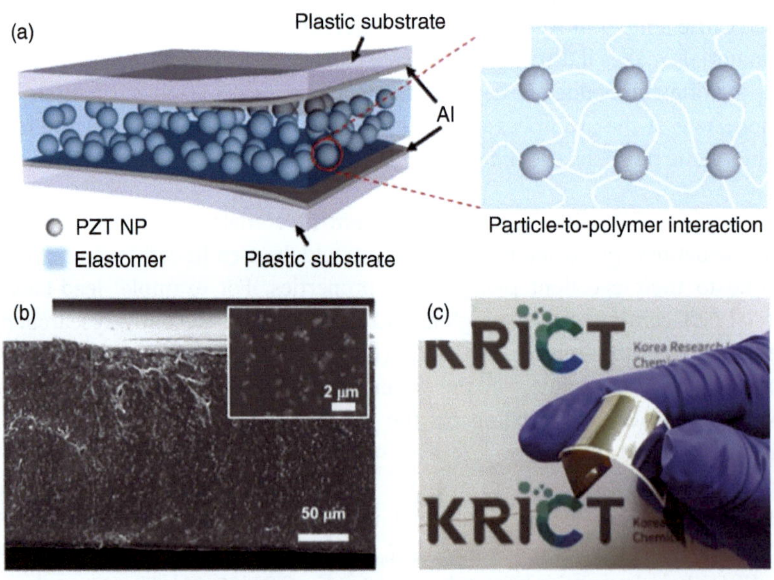

Figure 18.2 The structure and performance of piezoelectric flexible device based on chemically reinforced PZT composite. (a) Schematic illustration of piezoelectric flexible device. (b) The scanning electron microscope (SEM) and (c) optical images of piezoelectric flexible device. Source: Lee et al. [18]/Reproduced with permission of Royal Society of Chemistry.

developed and utilized for piezoelectric flexible devices [37–46]. For example, the nanocomposite of polyurethane and PVDF-TrFE nanofibers embedded with $BaTiO_3$ nanoparticles was used to fabricate stretchable piezoelectric devices for harvesting human kinematics energy [47]. The piezoelectric flexible device presented a high stretchability (~40%) and mechanical durability (9000 stretching cycles with the 30% strain) with V_{oc} of ~9.3 V and I_{sc} of ~189 nA because of its unique structure with a 3D micro-pattern substrate and the free-standing stacked fibers. In addition, Zhou et al. [37] proposed a piezoelectric flexible device using a composite of $BaTiO_3$ nanocubes self-assembled on glass fibers. The self-assembled technology avoided the high-temperature treatment of piezoelectric particles and thus did not decrease the mechanical properties of glass fibers. Based on the high performance of $BaTiO_3$ nanocubes and the well-flexible mechanism of glass fibers, the piezoelectric flexible sensor could possess a high sensitivity of 101.09 nA/kPa in the force range of 0–10 N, as well as a short response time of 19 ms.

The $(K,Na)NbO_3$-based (KNN-based) piezoelectric particles were also used for fabricating piezoelectric flexible devices, but the output energy was relatively low because of their inferior piezoelectric properties [48–51]. Thus, the device structure and composition optimization are very necessary to realize high-performance piezoelectric flexible device. Through an in situ photoreduction reaction, an $Ag/(K,Na)NbO_3$ heterostructure was constructed, which produced an output performance that is two orders of magnitude greater than pure KNN-based devices (V_{oc}: 240 V versus 3.5 V; I_{sc}: 23 μA versus 0.3 μA) due to the improvement of partial poling electric field on the KNN particles [52]. Recently, KNN-based piezoelectric particles with an ultrahigh d_{33}, the polynary $(Na, K)(Nb, Sb)O_3$–$BaZnO_3$–$(Bi, K)ZrO_3$, were embedded in the P(VDF-TrFE) fiber matrix to prepare the piezoelectric flexible device. To reduce the agglomeration effect between KNN particles, surface treatment with tetradecyl phosphonic acid was utilized. Taking advantage of the well-distributed KNN particles in matrix, the composite fibers embedded with KNN particles by electrospinning showed a high mechanical property with an effective modulus of 170 MPa. Taking the composite fibers as piezoelectric active layer, the piezoelectric flexible device could achieve a V_{oc} of 12.2 V and a peak power of 33.2 nW, accompanied by a long-term stability for bending over 10 000 cycles.

The two/three-dimensional perovskite-structured piezoelectric materials, such as nanowires, nanofibers, flakes, films, and 3D frameworks, can further enhance the deformation ratio under the applied stress compared with the 0-dimensional materials, and thus improve the piezoelectric potential due to the asymmetric shape change. For example, the 1D $BaTiO_3$ nanowires were dispersed into the P(VDF-TrFE) substrate [21]. The crosslinked 1D $BaTiO_3$ nanowires were embedded in the P(VDF-TrFE) matrix to form a 3D piezoelectric nanowire network, which improved the output performance effectively (V_{oc} of ~14 V and I_{sc} of ~4 μA) under the applied stress.

Besides the pure 1D $BaTiO_3$ materials, compositionally modified 1D $BaTiO_3$ materials could also be used as the piezoelectric filler in a flexible matrix for the piezoelectric flexible device, such as $0.93(Na_{0.5}Bi_{0.5})TiO_3$–$0.07BaTiO_3$ fiber [53], $(Ba, Ca)(Zr, Ti)O_3$ (BCZT) fiber [54, 55], and $0.96(K_{0.48}Na_{0.52})(Nb_{0.95}Sb_{0.05})O_3$–$0.04Bi_{0.5}$

$(Na_{0.82}K_{0.18})_{0.5}ZrO_3$ fiber [12]. Electrospinning is a simple method to prepare one-dimensional polycrystalline nanofibers with complex compositions, which consist of nano-sized crystallites, but the output performance has two limitations. On the one hand, in order to improve the output performance, the polycrystalline piezoelectric nanofiber should preferably be poled. On the other hand, the piezoelectric properties are closely related to the grain/domain size [26, 56, 57]. The nano-sized grains will inhibit the mobility of domains due to a decrease in domain density and an increase in grain boundary and domain wall volume, thereby greatly compromising the piezoelectric properties [58]. Therefore, how to balance the relationship between grain size and piezoelectric property of polycrystalline nanofibers deserves further investigation. In addition, the fabrication of the one-dimensional single-crystal materials is very difficult due to their complex compositions. In order to balance the performance of piezoelectricity, flexibility, and dielectric property, a hierarchical structure that is built by various PZT yarns knitted together was proposed [59], and the corresponding structure is shown in Figure 18.3a. By the template-assisted method, the PZT sol–powder suspension was dropped into a nylon textile, and then, after drying and sintering, a PZT hierarchical structure constituted with PZT fiber-like units was prepared (Figure 18.3b,c). The P(VDF-TrFE) solution was added to enhance the flexibility and avoid the brittleness of pure PZT textile (Figure 18.3d). The composite shows a well-balanced performance with piezoelectricity, air permeability, flexibility, and toughness (Figure 18.3e). The fabricated samples could generate an ultra-high V_{oc} of 128 V, an I_{sc} of 120 μA, and an instantaneous power density of 0.75 mW/cm^2.

For the 2D perovskite piezoelectric materials, thin film can be grown coherently on a flexible substrate, thus achieving high performance and enhancing the flexibility of the flexible device [60]. By a simple dipping method, He et al.

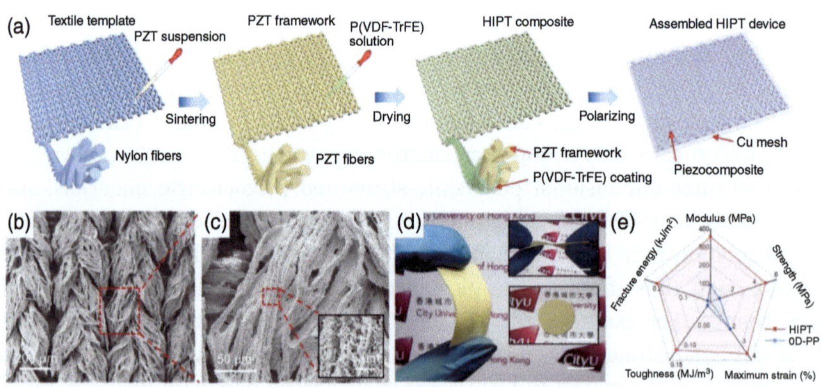

Figure 18.3 The structure and electrical performance of hierarchically interconnected piezoceramic textile (HIPT). (a) Schematic illustration of the preparation of HIPT device with the template-assisted sol–gel method. (b) SEM image of the piezoelectric framework and its (c) enlarge SEM image with hierarchical structure. (d) Optical picture of piezoelectric flexible device (scale bar: 1 cm). (e) The comprehensive mechanical properties of HIPT device with 20% P(VDF-TrFE). Source: Hong et al. [59]/Reproduced with permission of John Wiley & Sons.

[61] fabricated a PZT thin film (~100 nm) on a glass fiber fabric substrate and assembled a large-scale flexible piezoelectric nanogenerator (PENG), which exhibited efficient energy harvesting behaviors (V_{oc} of ~60 V and I_{sc} of ~500 nA) and the pressing/bending multimode energy harvest capability. Wang et al. reported a low-cost flexible PENG with good output characteristics (V_{oc} of ~120 V and current density of ~150 μA/cm^2) by growing the PZT thin film on a 2D mica substrate. After being bent 40 000 times, it displayed a long service life without any obvious performance degradation [62]. In order to alleviate the mismatch between the perovskite piezoelectric thin films and the flexible substrates, a $LaNiO_3$ buffer layer was coated on an ultrathin metal foil substrate, which is conducive to the uniform nucleation and crystallization of PZT. Thus, a high-density and uniform PZT film can be formed on the buffer layer, thereby improving the output performance [63].

Recently, 3D piezoceramic-polymer composites, consisting of 3D interconnected piezoceramic frameworks in a flexible polymer matrix, were prepared as piezoelectric flexible devices by exploiting their enhanced load/stress transfer capability from the matrix to the active piezoceramic fillers [13]. As shown in Figure 18.4, embedding a 3D cellular-structured PZT ceramic foam into a PDMS matrix to fabricate a piezoelectric flexible device significantly improves the piezoelectric performance in various deformation modes (e.g. the V_{oc} of ~85 V under 8% compressive stress), meanwhile exhibiting excellent pyroelectric performance [13]. The lead-free BCZT ceramic could also be fabricated into a 3D framework with an interconnected porous structure through appropriate porous templates (the scanning electron microscope [SEM] images insert in Figure 18.4b), which was used as an energy harvester [64, 65]. From Zhang's study, the compressible/tensile piezoelectric composite flexible generators based on such porous 3D BCZT exhibited a high V_{oc} of 25 V and a power density of 2.6 μW/cm^2, which was 16 times higher than that of conventional composite generators with piezoelectric particle materials [65].

18.3.1.2 Wurtzite Structure Piezoelectric Materials

The wurtzite-type materials, which are as non-ferroelectric materials, such as ZnO, AlN, and GaN, present polarization along the *c*-axis under applied stress. In 2006, Wang's group first reported a PENG based on ZnO nanowires, which was the beginning of piezoelectric energy harvester research [8]. Although the wurtzite-type materials do not show a high piezoelectric constant like $BaTiO_3$ or PZT, they have low dielectric constants (ε_r) and moderate piezoelectric strain constants (e), resulting in higher piezoelectric voltage constants (g). Therefore, they are suitable to be fabricated into different nanostructures for piezoelectric flexible devices [66–69].

As one of the classical semiconductors, ZnO possesses highly diverse nanostructures and designability, making it useful for piezoelectric flexible devices with various structures [70, 71]. The ZnO nanowires usually grow along the <001> (*c*-axis) direction, which is the same as the polarization direction, and thus can improve the output performance of piezoelectric flexible devices. Lee et al. fabricated the ZnO nanowire arrays on an ultrathin metal foil to fabricate a super-flexible piezoelectric device, which was used to harvest the wavy motion energy, for example, on a waving flag [72]. A piezoelectric flexible device was developed based on ZnO

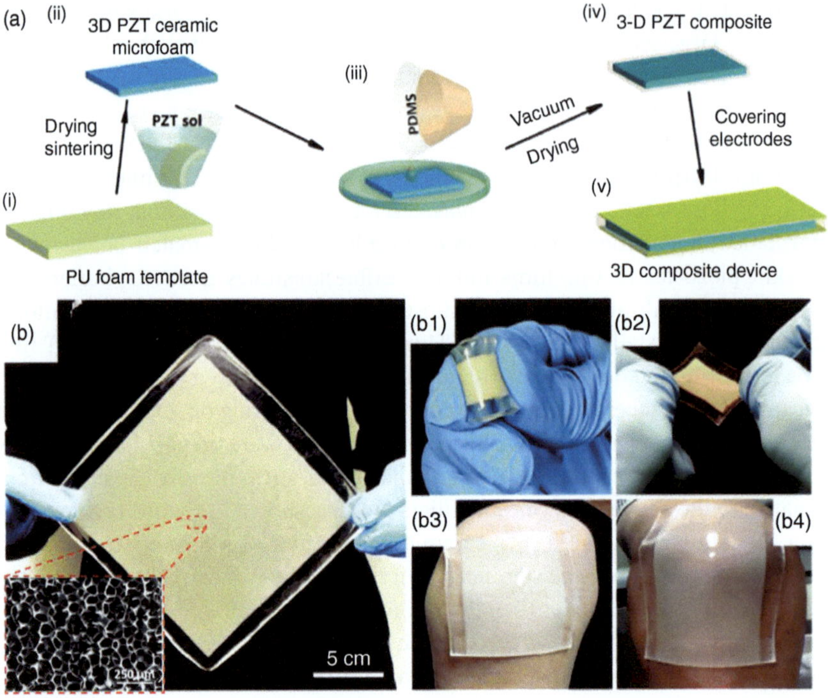

Figure 18.4 The structure and properties of 3D PZT and polymer composites. (a) The preparation procedure of the 3D composites. (b) The optical images of the 3D composites under different stress models (the inset is SEM images of 3D PZT framework). Source: Zhang et al. [13]/Reproduced with permission of Royal Society of Chemistry.

nanorods grown on carbon paper, with V_{oc} and I_{sc} about 3.6–6.8 V and 0.79–1.45 μA, respectively [73].

To optimize the piezoelectric performance of the ZnO, chemical doping of pure ZnO is a common method. The ZnO materials doped with different elements (e.g. halogen dopants: F, Cl, Br, I [74]; transition metal dopants: Fe, V, Ag, Nb [75–78]; rare earth metal dopants: La, Y [79, 80]; and others: Li, Na [81], and S [82]) have been investigated to enhance the output performance of flexible ZnO-based piezoelectric flexible devices.

The wurtzite-structured Group-III-N (III-N) materials (e.g. AlN, GaN, and (Al, Ga)N) have received increasing attention for their application in piezoelectric flexible devices due to their biocompatibility. The piezoelectric flexible device based on AlN thin film displayed a V_{oc} of 1.4 V and an I_{sc} of 1.6 μA, accompanied by a power density of 1.57 μW [83]. A highly stable piezoelectric flexible device was prepared by using a single-crystalline GaN thin film, which showed a high V_{oc} of ~28 V, a current density of ~1 μA/cm^2, and a power density of ~6 μW/cm^2 [84]. The *c*-axis GaN nanowires were also used to assemble piezoelectric flexible device [85], as shown in Figure 18.5a. The Mg-doped semi-insulating GaN nanowires (Figure 18.5b) with enhanced piezoelectric properties were grown on a thin tungsten foil. Benefiting from the synergy effect of reduced electron density by doping

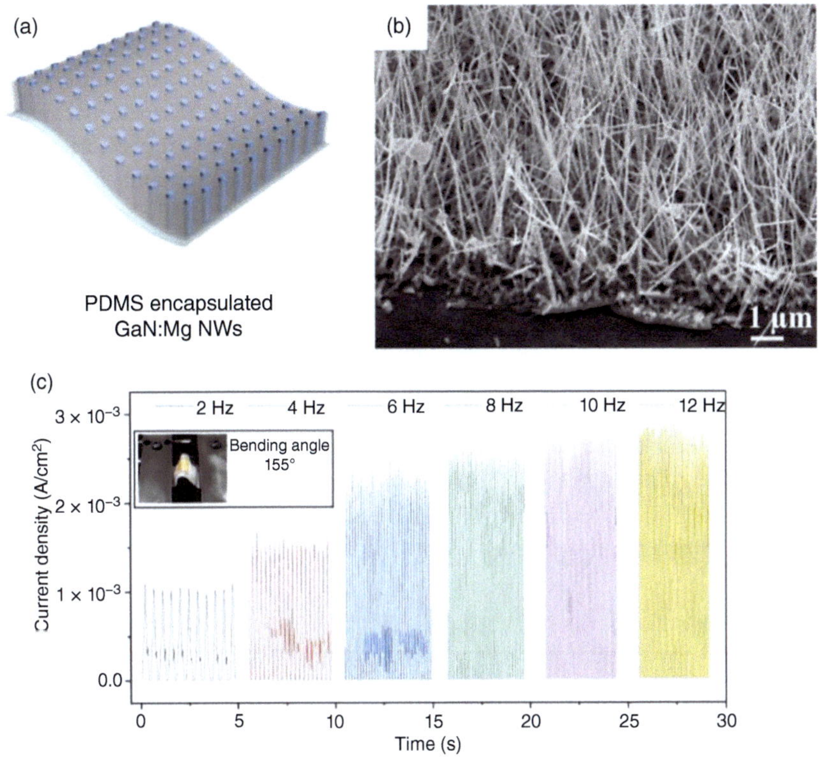

Figure 18.5 The structure and electrical performance of GaN-based piezoelectric flexible device. (a) Schematic diagram of GaN-based piezoelectric flexible device with Mg-doped GaN nanowires on tungsten foil. (b) SEM images of Mg-doped GaN nanowires. (c) Output current density of GaN-based piezoelectric flexible device under periodic bending at frequencies from 2 to 12 Hz (bending angle: 155°). Source: Waseem et al. [85]/Reproduced with permission of John Wiley & Sons.

and excellent conductivity of tungsten foil, the output current of piezoelectric flexible device could reach to 2.5 mA/cm^2 (Figure 18.5c), accompanied by an output power density of 13 mW/cm^2. The flexibility and sensitivity of this GaN-based device can be utilized as a sensor to monitor the movement on joints, e.g. fingers, hands, elbows, and knees. In 2019, Chen et al. reported a multilayer III-N film consisting of an AlN buffer layer, an $Al_xGa_{1-x}N$ interlayer and a top GaN layer and fabricated a flexible PENG by a layer transfer method [86]. This device could produce a V_{oc} of ~50 V, an I_{sc} of ~15 μA, and a maximum power density of ~167 μW under a load resistance of 5 MΩ, with good stability for over 30 000 cycles.

18.3.1.3 2D Piezoelectric Materials

Since the discovery of graphene in 2004 [87], plenty of 2D materials, e.g. graphene oxide, transition metal dichalcogenides, transition metal carbides/carbonitrides (MXenes), and black phosphorus, have been extensively studied due to their unique chemical/physical properties. The weak interlayer Van der Waals forces enable them to be fabricated into atomically thick few-layer 2D materials, which are flexible enough for flexible electronic devices [88]. Some of these 2D materials (e.g. transition metal dichalcogenides such as MoS_2 and WS_2) show layer-dependent piezoelectricity behavior. Although their bulk materials do not exhibit piezoelectricity, their monolayered or few-layered counterparts possess piezoelectricity due to the broken inversion symmetry [11]. Along with flexibility and piezoelectricity, these monolayered or few-layered 2D materials have been researched for their application in piezoelectric flexible devices. For example, the thinly bent MoS_2 2D nanosheets with odd layers could produce electric output due to the piezoelectric effect (Figure 18.6a,b), while the MoS_2 sheets with even layers have no piezoelectric effect. Under a 0.53% bending strain, a single-layer MoS_2 flexible device displayed a V_{oc} of ~ 15 mV, an I_{sc} of ~65 pA, and a power density of ~2 mW/m^2, as shown in Figure 18.6c,d [11].

Besides the 2D transition metal dichalcogenide materials, other 2D materials also exhibit piezoelectric properties and can be chosen to fabricate piezoelectric flexible devices. The monolayer $MoSe_2$ sheet could also be used to prepare the

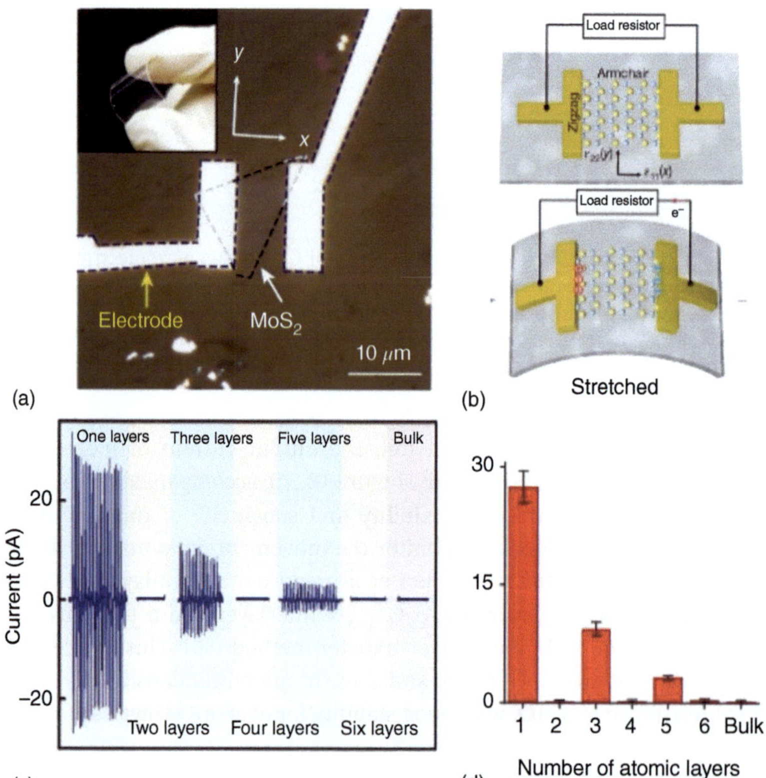

Figure 18.6 The (a) piezoelectric flexible device with single-layer MoS_2 sheet as functional unit and (b) corresponding operation scheme under periodic stretching and releasing. (c,d) The output properties of piezoelectric flexible device with the increase of layer numbers in MoS_2 sheet. Source: Wu et al. [11]/Reproduced with permission of Springer Nature.

piezoelectric flexible device. Wang et al. [89] prepared the monolayer $MoSe_2$ with an atmospheric pressure chemical vapor method, transferred it to the flexible polyethylene glycol terephthalate substrate, and packaged it with PDMS film to fabricate a piezoelectric flexible device. Under the periodic stretching and release, the monolayer $MoSe_2$-based flexible device outputs a V_{oc} of 35 mV and a power density of 42 mW/m^2. Through pasting the device on the different positions of human body, for example, the finger joints, wrists, throat, and neck, the device would provide different outputs with the movement of joints or even sounds (when attaching the device on throat). The mono- and multilayered hexagonal α-In_2Se_3 materials also exhibited the out-of-plane and in-plane piezoelectricity reported by Xue et al. [90]. The as-fabricated flexible device presented a V_{oc} of ~35.7 mV and an I_{sc} of ~47.3 pA under a strain of 0.76%. For another study, the piezoelectric flexible device based on 2D PbI_2 nanosheets, which always show piezoelectricity regardless of the number of layers, displayed a V_{oc} of ~29.4 mV, an I_{sc} of ~20 pA, and a loading power of 0.12 pW, respectively [91]. Moreover, large-area and monolayered SnS was synthesized using a liquid metal-based synthesis technique with a high V_{oc} of 150 mV and a power density of 24 mW/m^2 [92].

18.3.1.4 Other Inorganic Piezoelectric Materials

In addition to the perovskite-structured, wurtzite-structured, and 2D piezoelectric materials, other types of inorganic piezoelectric materials have also been utilized for fabricating piezoelectric flexible devices. For example, the ferroelectric $BaTi_2O_5$ material, a metastable phase of BaO–TiO_2 system, exhibits lattice asymmetry along the *b*-axis. Hence, $BaTi_2O_5$ presents the piezoelectricity feature only along the *b*-axis, accompanied by a high dielectric permittivity [93]. Based on this feature, Lu et al. grew $BaTi_2O_5$ nanorods with the molten-salt method, and then embedded them into a PVDF matrix by hot-pressing to fabricate the $BaTi_2O_5$-based textured flexible PENG device [94]. The PENG had a high output performance with a V_{oc} of ~53.2 V, a high power density of ~27.4 μW/cm^3, and a long working life that could keep stable for 330 000 cycles.

Furthermore, bismuth layer-structured piezoelectric materials are often adopted due to their ultrahigh Curie temperature [95–97]. They are a huge family, including $Bi_4Ti_3O_{12}$, $CaBi_4Ti_4O_{15}$, $BaBi_4Ti_4O_{15}$, $SrBi_4Ti_4O_{15}$, $K_{0.5}Bi_{4.5}Ti_4O_{15}$, and $Na_{0.5}Bi_{4.5}Ti_4O_{15}$, and exhibit unique piezoelectric properties. Among them, the $Bi_4Ti_3O_{12}$ nanosheets/nanoparticles were used as piezoelectric components to composite with different organic matrixes to fabricate piezoelectric flexible devices. For example, a piezoelectric flexible device was investigated by compounding $Bi_4Ti_3O_{12}$ with PDMS, presenting a V_{oc} of ~12.5 V, an I_{sc} of ~100 nA, and a power density of ~562 μW/cm^2 under a force of 10 N [98]. By contrast, the $Bi_4Ti_3O_{12}$-based flexible yarn-based piezoelectric nanogenerator (FY-PNG) showed a much higher output performance when composited with PVDF, with *a* V_{oc} of ~60 V, an I_{sc} of ~400 nA, and a power density of ~18.5 mW/cm^2 under a mechanical load of only 1 N [99].

18.3.2 Organic Piezoelectric Material-Based Flexible Devices

Compared with the above inorganic piezoelectric materials, the d_{33} of the organic piezoelectric materials, such as PVDF-based materials, is not very high, but their excellent flexibility and lightweight properties make them very suitable for wearable PENG devices.

18.3.2.1 PVDF-Based Piezoelectric Materials

The PVDF-based piezoelectric material is the most commonly used one for flexible piezoelectric devices because it can achieve both flexibility and piezoelectricity. There are five polymorphs of PVDF, namely α, β, γ, δ, and ε (Figure 18.7) [100]. Among them, the β-phase PVDF exhibits the strongest piezoelectricity, but non-piezoelectric α-phase PVDF is the most stable phase due to its lowest energy. Therefore, usually, the α-phase PVDF in the matrix is produced by the solution casting of PVDF or cooling the molten PVDF. To obtain higher output performance for the PVDF-based piezoelectric flexible devices, the content of β phase should be increased as much as possible by different fabrication methods, such as the electrospinning method [101–103], mechanical stretching/pressing or annealing [104, 105]. Meanwhile, the addition of inorganic/organic filler materials could also provide more nucleation sites and promote the formation of the β-phase PVDF, thereby effectively improving the output performance of PVDF-based piezoelectric flexible devices. On the other hand, PVDF, as a polymer, can be easily fabricated into various structures, e.g. nanofibers, hollow nanofibers, ultrathin membranes, and porous membranes. The unique characteristics of PVDF make it suitable to be used as the flexible organic matrix or the core piezoelectric component, especially in piezoelectric flexible devices.

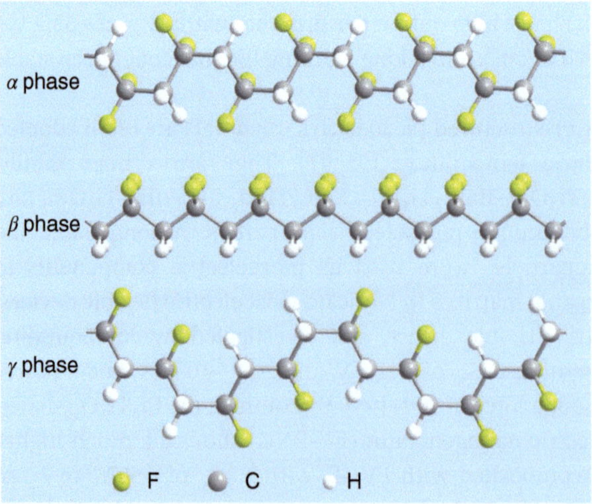

Figure 18.7 Schematic diagram of crystalline α, β, and γ phase of PVDF.

18.3.2.2 Organic–Inorganic Lead Halide Perovskite Piezoelectric Materials

In recent years, organic–inorganic hybrid lead halide perovskite materials (ABX_3) have attracted much attention in optoelectronic applications due to their high photoelectric conversion efficiency [106, 107]. Generally, for the ABX_3 materials, A is CH_3NH_3 (MA) or $CH(NH_2)_2$, B is the divalent metal cation (e.g. Pb or Sn), and X is the I, Br, or Cl. This organic–inorganic hybrid lead halide perovskite material also possesses spontaneous polarization and phase transition behavior between tetragonal and cubic phases, like the inorganic perovskite ABO_3 materials.

The Fe^{2+}-doped $MAPbI_3$ perovskite thin film has a tetragonal–cubic phase transition around room temperature and thus presents a high relative dielectric permittivity of ~107, which contributes to the improved output performance of $MA(Pb, Fe)I_3$ film-based PENG with a V_{oc} of ~7.29 V and a current density of ~0.88 μA/cm^2 [108]. After adding $MAPbI_3$ into the PVDF matrix, the composite film exhibited higher dielectric permittivity and larger remanent polarization than the pure PVDF film [15]. The enhanced output performance of the piezoelectric flexible device could reach a V_{oc} of ~45.6 V and a current density of ~4.7 μA/cm^2. As with other lead-based piezoelectric ceramics, the toxicity of Pb raises environmental concerns. Based on this consideration, Ippili et al. [109] used Sn to completely replace the Pb species in $MAPbI_3$ to prepare the lead-free halide perovskite materials ($MASnI_3$), which were composited with PVDF and used for piezoelectric flexible devices. Such PVDF-$MASnI_3$ PENGs possessed a V_{oc} of ~12.0 V and a current density of 4.0 μA/cm^2.

18.3.2.3 Bio-Based Piezoelectric Materials

Compared to the artificially synthesized inorganic/organic piezoelectric materials, the bio-based piezoelectric materials, e.g. collagen fibrils, cellulose, chitosan, and chitin [110–112], possess unique inherent advantages, such as non-toxicity, biodegradability, flexibility and biocompatibility. These characteristics make them suitable for the preparation of biocompatible piezoelectric flexible devices for the applications such as *in vivo* energy harvesting, health monitoring, and self-powered biomedical sensing.

Collagen fibrils are abundant in animal tissues (e.g. skin, cartilage, and tendon), and are biodegradable and biocompatible. The piezoelectric properties of collagen fibers result from the polar uniaxial orientation of their molecular dipoles [113]. By using the fish swim bladder as the piezoelectric materials, Ghosh and Mandal prepared a piezoelectric flexible device, and the V_{oc} and I_{sc} could reach ~10 V and ~51 nA under the slight stress of a human finger, respectively [112]. In addition, due to the abundant source of collagen fibril, a variety of raw materials can be chosen for bio-based piezoelectric flexible devices, such as the collagen-deposited cotton fabric and bio-wasted eggshell membrane [110, 114]. Chitin is another kind of polysaccharide with chemical structure similar to cellulose, and its piezoelectricity arises from the hydrogen bonds between polysaccharide chains. Thus, the chitin nanofibers were used as the piezoelectric materials in piezoelectric flexible devices, which provided a V_{oc} of ~22 V and an I_{sc} of ~0.12 μA. When embedded into PVDF, its output performance could be further improved, with a V_{oc} up to ~49 V and an

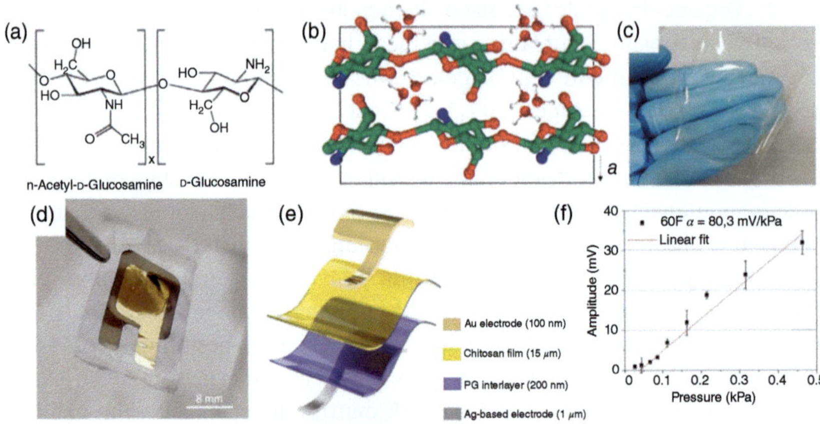

Figure 18.8 The structure of chitosan and the performance of chitosan-based piezoelectric flexible device. (a) Chitosan derives from the *N*-deacetylation of Chitin. (b) The noncentrosymmetric cell structure of the simplified scheme of supermolecular arrangement of chitosan chains. (c) Thin and transparent film prepared with chitosan. (d) Optical photo and (e) scheme of chitosan-based piezoelectric flexible device. (f) Output voltage of chitosan-based piezoelectric flexible device under various pressure forces. Source: de Marzo et al. [115]/Reproduced with permission of Elsevier.

I_{sc} up to ~1.9 μA [111]. As a low-cost natural biopolymer from chitin, chitosan, a nonimmunogenic carbohydrate (Figure 18.8a), has the noncentrosymmetric crystal structure (Figure 18.8b), which makes it able to exhibit piezoelectric effect. Meanwhile, contributing to its easy processability, it can be processed in thin and flexible film with high transparency (Figure 18.8c). To improve its piezoelectricity, de Marzo et al. [115] used a controlled simple solvent casting technique and post-casting conditions to obtain a tuned crystallographic alignment, which improved the piezoelectric constant of chitosan film to 15 pC/N. Thanks to the enhanced piezoelectricity, the flexible device (Figure 18.8d,e) prepared by chitosan film shows a good sensitivity of 80 mV/kPa under applied force (Figure 18.8f).

18.3.2.4 Molecular Ferroelectric Materials

Molecular ferroelectrics are conventionally hybrids consisting of organic molecular groups and inorganic atomic frameworks. Their ferroelectricity/piezoelectricity arises from the asymmetry of the lattice structures and relative displacement of ions/polar groups (Figure 18.9a). Compared to inorganic ferroelectrics, molecular ferroelectrics have their own inherent advantages, e.g. lightweight, structural tunability, and easy processability. In addition, the molecular ferroelectrics can directly crystallize in the solution, which enables them to be embedded in different substrates. Li et al. [116] presented a method to prepare the molecular ferroelectric-based flexible device by depositing the guanidinium perchlorate ($[C(NH_2)_3]ClO_4$) onto the polyurethane (PU) sponge (Figure 18.9b,c). Combined with the piezoelectric property of $[C(NH_2)_3]ClO_4$ and flexibility of sponge, the $[C(NH_2)_3]ClO_4$-PU hybrid device possesses the characteristics of high sensitivity, self-powering, and long-term stability. This device could sense a tiny stress of 3 Pa

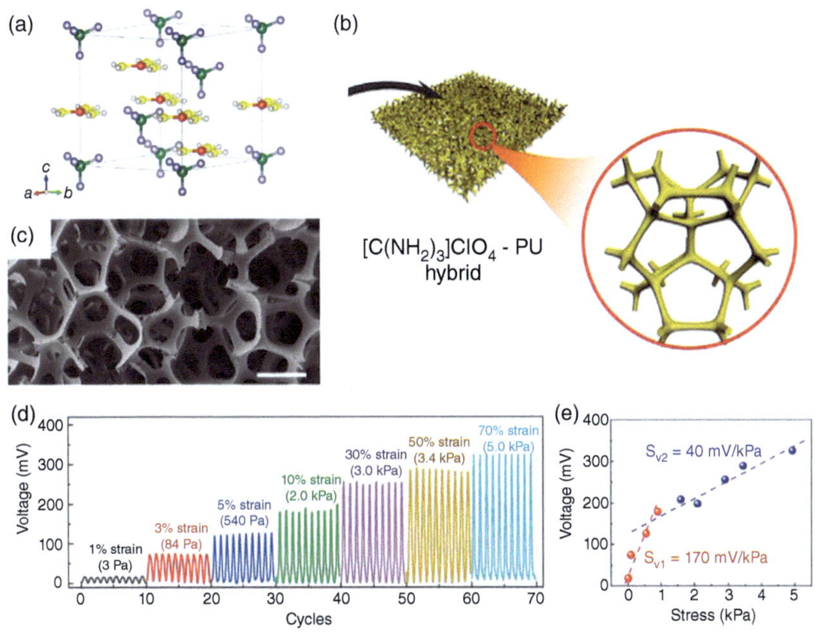

Figure 18.9 The structure and output performance of molecular ferroelectric-based flexible device. (a) Schematic diagram of molecular structure of $[C(NH_2)_3]ClO_4$. (b) Schematic illustration of $[C(NH_2)_3]ClO_4$ and sponge hybrid structure and (c) corresponding SEM image. (d) Piezoelectric voltage of the device under various stresses and (e) corresponding sensitivity. Source: Li et al. [116]. Reproduced with permission of WILEY-VCH.

(corresponding to 0.0013 N) and monitor weak deformation of 1% (Figure 18.9d,e). Lu et al. [117] reported a molecular ferroelectric-based flexible device with biodegradable and recyclable features. Taking the bacterial cellulose hydrogel as matrix and imidazolium perchlorate ($ImClO_4$) as functional unit, the hybrid sensor showed a sensitivity of 4 mV/kPa and wide sensing range (from 0.2 to 31.25 kPa). In addition, the bio-based bacterial cellulose matrix can be degraded into environmentally friendly products (e.g. glucose and oligosaccharide), and the rest $ImClO_4$ could be utilized again to fabricate the other flexible device.

18.4 Application of Piezoelectric Flexible Devices

18.4.1 Piezoelectric Flexible Devices Route Toward Smart Sensors

Based on the piezoelectric effect, which can convert mechanical energy into electrical signals, piezoelectric flexible devices have been developed as detection sensors for biomechanical excitations, such as the heart rate, heart rate variability, breath, body motion, and acoustic vibration. The piezoelectric flexible devices can easily monitor the motion of limbs, joints, hands, and feet. When a piezoelectric flexible device is integrated into the insole of shoes, its output signal changes with the different forms of human body movements [118]. The piezoelectric flexible devices could generate electric signals due to their ability to periodically compress and release with the tester up and down. Similar signals can also be obtained under walking and running conditions [118]. According to different strengths and frequencies, human body movement situations can be real-time monitored. Furthermore, piezoelectric flexible devices can also be integrated into different positions of the insole to monitor different forces and pressures in different foot positions during daily walking/running of patients with movement disorders (Figure 18.10a) [119]. These output signals can reflect the contact force distribution of the patient's foot during daily activities and provide posture misalignments for guiding rehabilitation training. Piezoelectric flexible devices can also be applied to human joints, such as the front elbow and finger joints. The movement of the joints could provide a bending force on a piezoelectric flexible device, which generates electric signals to detect the motion of the human arm/hand. Different bending positions of the joint produce different bending forces and thus different electrical signal strengths (Figure 18.10b) [92]. Meanwhile, the flexible sensor could also be used to identify the multimode mechanical force of a robot wrist, such as stretching, bending, or twisting [116]. The multi-sensor array was attached on the different positions of robot wrist (Figure 18.10c). Taking advantage of the sensitivity and uniformity features of piezoelectric flexible device, the sensor array could convert the mechanical load of stretching, pressing, shearing, bending, and twisting into a 2D voltage distribution map to clarify the stress distribution when the robot wrist moved. Through the analysis of sensor signals in different arrays, the stress state would be confirmed, accompanied by the movement of the wrist. For example,

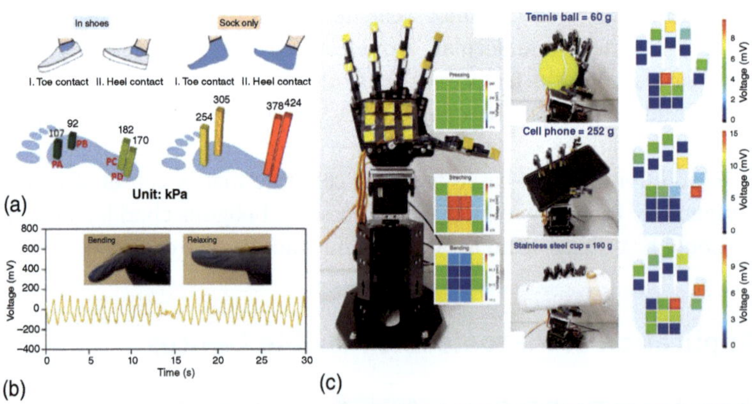

Figure 18.10 Application of the flexible PENG device for human/robot movement monitoring. (a) Pressure mapping of shoes and socks based on the measured output from different foot positions. (b) The output performance at bending and releasing finger. (c) Piezoelectric flexible sensor array on robot hand, and the optical image of robot hand grasping different items and their corresponding signal distinguish maps. Source: (a) Zhu et al. [119]. Reproduced with permission of American Chemical Society. (b) Khan et al. [92]/Reproduced with permission of Springer Nature. (c) Li et al. [116]/Reproduced with permission of Springer Nature.

under compression, uniform voltage distribution would be generated on the array pixels, but under stretching and bending, the middle/border pixels in the array would show high voltage signal. Moreover, the piezoelectric sensor array would monitor the loading condition in the robot hand as well as the recognition of the shape of objects, such as the tennis ball, cell phone, cup, as shown in Figure 18.10c. Thus, the sensor shows potential applications in human-robot interaction systems to realize real-time control of robotic hands or arms.

Due to their high sensitivity, piezoelectric flexible devices can be used as physiological monitoring sensors and can be attached to faces to monitor blink speed and force [107]. In addition, the vibration of the throat skin also caused the piezoelectric flexible devices to generate electrical signals when coughing or yawning, which were closely related to the intensity of the cough [107]. It can also detect weak physiological vibrations, such as pulse in real time, and measure the corresponding changes in heart rate. Meanwhile, due to the good sensitivity of piezoelectric flexible device (Figure 18.11a), electrical signals in the form of multimodal could be obtained (Figure 18.11b), which displayed high-resolution physiological information, including percussion (P-wave), tidal (T-wave), and diastolic (D-wave), as shown in Figure 18.11c [120].

Inspired by the auditory system of human ears, Professor Yoel Fink's team [121] designed an acoustic fabric that can be used as a sensitive audible microphone while retaining the traditional qualities of fabric, such as machine washable and durability. The fabric medium consisted of textile yarns with high Young's modulus,

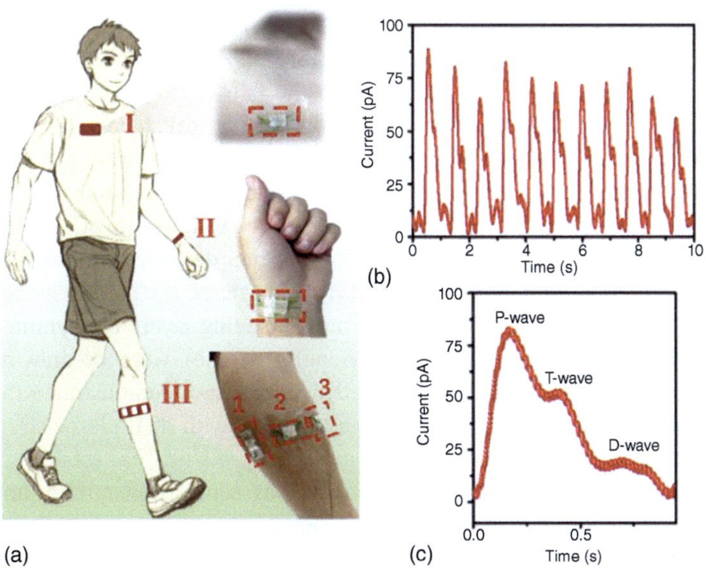

Figure 18.11 Application of the flexible PENG device for human health monitoring. (a) Schematic illustration of the flexible piezoelectric sensors attached on different positions of the human body. (b) The output signals and (c) corresponding enlarged pulse waves of wrist pulses. Source: Yang et al. [120]/Reproduced with permission of Elsevier.

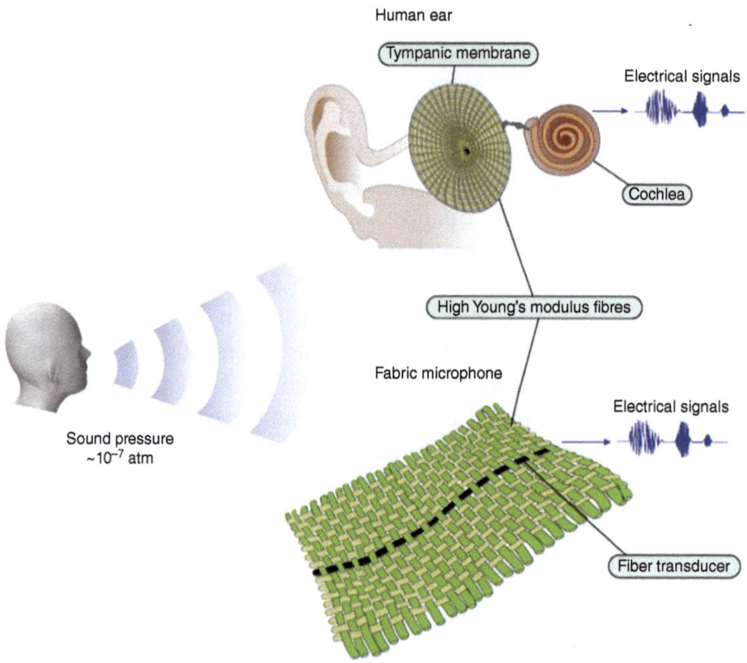

Figure 18.12 Design and principles of acoustic fabrics: the working mechanism of human ear and acoustic fabrics to convert sound into electrical signals. Source: Yan et al. [121]/ Reproduced with permission of Springer Nature.

which could convert weak pressure waves of 10^{-7} atm of audible frequency into low-order mechanical vibration modes (Figure 18.12). The heat-stretched composite flexible piezoelectric fibers are woven into the fabric and combined with the fabric patch to convert mechanical vibrations into electrical signals. In order to enhance the sensitivity of the piezoelectric fibers, the structure, materials, and fabrication process had been well designed to concentrate the mechanical stress on piezoelectric fiber, which had a high piezoelectric constant of 46 pC/N. The flexible piezoelectric fiber was prepared with the thermal drawing method. The piezoelectric composite of P(VDF-TrFE) and $BaTiO_3$ particles, the carbon-loaded polyethylene, the copper wires, and the innermost cladding layers of asymmetrically elastic poly(styrene-*b*-(ethylene-*co*-butylene)-*b*-styrene) were co-drawing into a single fiber with high flexibility. Some applications have demonstrated the usefulness of acoustic fabric, e.g. a shirt of acoustic fabric with dual piezoelectric fibers that could measure the direction of sound source, a shirt that establishes two-way communication between two fabrics used as sound transmitters and receivers, and a shirt that monitors cardiac sound signals.

18.4.2 Piezoelectric Flexible Devices Route Toward Energy Harvesters

A self-power strategy is a very attractive feature for smart devices to avoid the dependence on additional power supplies. Piezoelectric flexible devices can be

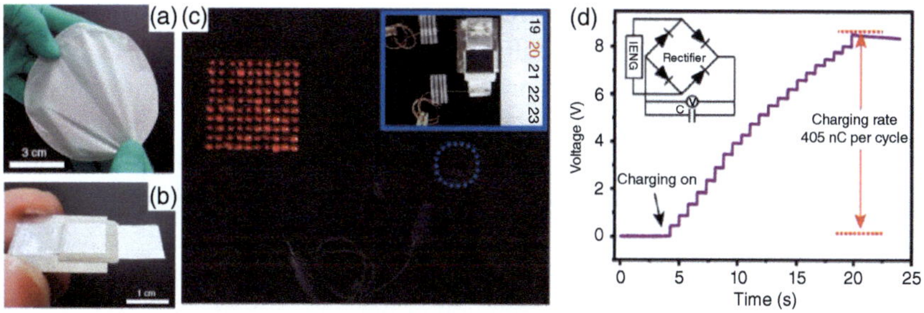

Figure 18.13 Application of the piezoelectric flexible device for biomechanical harvester. (a) Optical image of Sm-PMN–PT/PVDF composite film and (b) corresponding multilayer piezoelectric device, which can (c) drive over 100 LED bulbs and (d) charge 1 μF capacitor. Source: Gu et al. [14]/Reproduced with permission of Springer Nature/CC BY 4.0.

used as both the sensing part and the energy supply of wearable electronic devices, converting naturally dissipated energy into electric energy. Under the action of human motion or vibration, a piezoelectric flexible device can convert mechanical energy into electric energy through the piezoelectric effect [106]. In addition, since some core materials of piezoelectric flexible devices are pyroelectric materials, thermal energy from sources such as human body/ambient temperature can also be harvested and converted into electricity [13, 122]. Compared with traditional piezoelectric ceramics, piezoelectric flexible energy harvesters are more advantageous in wearable electronics due to their advantages of light weight, flexibility, and elasticity. At the same time, high reliability can also be achieved on such devices. For example, the output performance of a piezoelectric flexible device with a 3D electroceramic framework in PDMS did not degrade after more than 3000 pressing and releasing cycles in the durability test [64]. To improve the output performance, multiple piezoelectric devices can be assembled in parallel or multilayer for practical applications in charging various electronic devices [14, 110]. In order to improve the output performance of piezoelectric flexible device in unit area, the multiple flexible Sm-PMN–PT/PVDF composite films (Figure 18.13a) were utilized to fabricate integrated multilayer piezoelectric flexible device (Figure 18.13b). It could output a maximum peak I_{sc} of 320 μA with a high current density of 290 μA/cm^2. The high output could drive 100 LEDs (Figure 18.13c) and charge a 1 μF capacitor from 0 to 8 V within only 21 operation cycles (Figure 18.13d). In addition, the piezoelectric flexible device can also be combined with other energy harvesters, e.g. triboelectric nanogenerators, to fabricate hybrid nanogenerators to improve energy harvesting efficiency and output performance [4, 123].

18.4.3 Piezoelectric Flexible Devices Route Toward Electrical Stimulation Therapy

In addition to be utilized in sensing and energy harvesting, the piezoelectric flexible device is also a potential approach to achieve electric therapeutic applications. Cellular activities, including differentiation, proliferation, and migration, have been confirmed to be affected by electric field. The piezoelectric flexible device can generate electric signals from human body movements, vibrations, or external ultrasounds, and thus it can be utilized as electric source to regulate cellular activities. Meanwhile, it is also appropriate for implanted or non-implanted medical devices owing to its flexibility. Chapter 20 will give a detailed and comprehensive description of the biological aspects of piezoelectric applications, so in this section, we will briefly introduce the applications of piezoelectric flexible devices in electrical stimulation therapy.

Electrical signals regulate various processes in human physiology, such as the contraction of cardiomyocytes, the secretion of hormones by endocrine cells, and the release of neurotransmitters by neurons [124]. These effects are mediated by a number of proteins at the plasma membrane, including ion channels that allow ions to cross the membrane. Inspired by action potentials in electroactive cells, Zhao et al. provided a rapid release of insulin, which was activated by a piezoelectric flexible device that is constituted of PVDF film as core piezoelectric unit. The electric charges generated by piezoelectric device would stimulate the opening of voltage-gated CaV1.2 channels, resulting in the release of insulin stored in secretory vesicles. When the piezoelectric flexible device outputs a voltage of

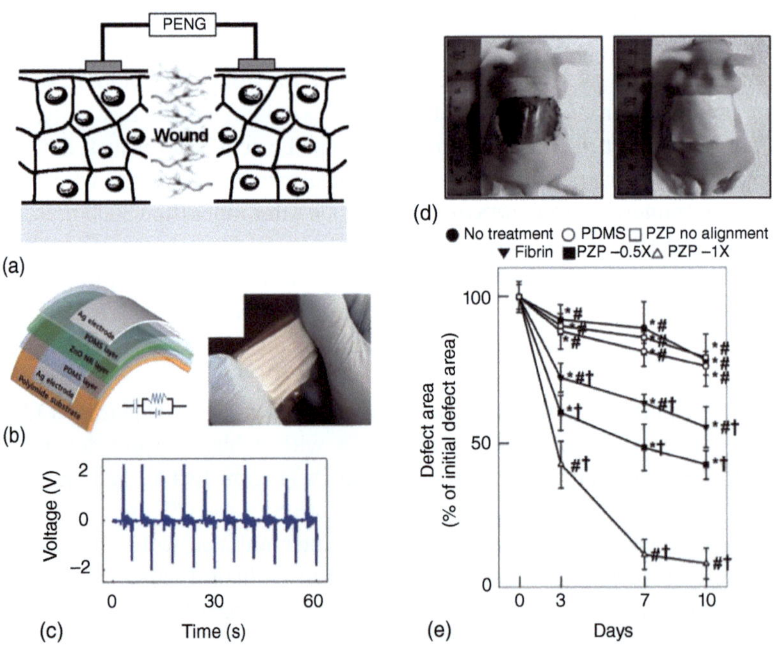

Figure 18.14 Application of the piezoelectric flexible device for wound healing. (a) The schematic diagram of wound healing process based on piezoelectric device. (b) The schematic and optical image of ZnO nanorod-based piezoelectric flexible device, and (c) its corresponding output voltage signals. (d) The piezoelectric flexible device attached to the skin wound on the back of a mouse. (e) Wound closure rates versus the different treatment methods (the PZP-1X sample is the piezoelectric flexible device with 95.2% ZnO nanorod filling density). Source: Bhang et al. [125]/Reproduced with permission of John Wiley & Sons.

1–2 V and a press frequency of 1 Hz, after 3 minutes, the amount of insulin secreted by stimulated cells is over 20-fold greater than that of nonstimulated cells, which is comparable to the value of the cells depolarized by 40 mM KCl. With the further extension of stimulation time, the amount of insulin increased, and when the time reached 30 minutes, the secreted insulin was at a stable level, indicating that all of the vesicle-stored insulin was released.

Electrical stimulation can also be used to imitate endogenous electric field and promote the wound healing process [125]. As shown in Figure 18.14a, the working mechanism of wound healing is that the piezoelectric device attached on the wound generates the electric field between two electrodes, which imitates the endogenous electric field and thereby promotes wound healing. Bhang et al. provided a one-directionally aligned ZnO nanorods embedded in a piezoelectric flexible device that could be applied on skin wound (Figure 18.14b). With the applied mechanical deformations during animal motion, the electric field generated (Figure 18.14c) at the wound bed promoted the wound healing process because of the enhancement of cellular metabolism, migration, and protein synthesis. As the piezoelectric flexible device (PZP-1X) was attached to the wound on the back of the mice (Figure 18.4d),

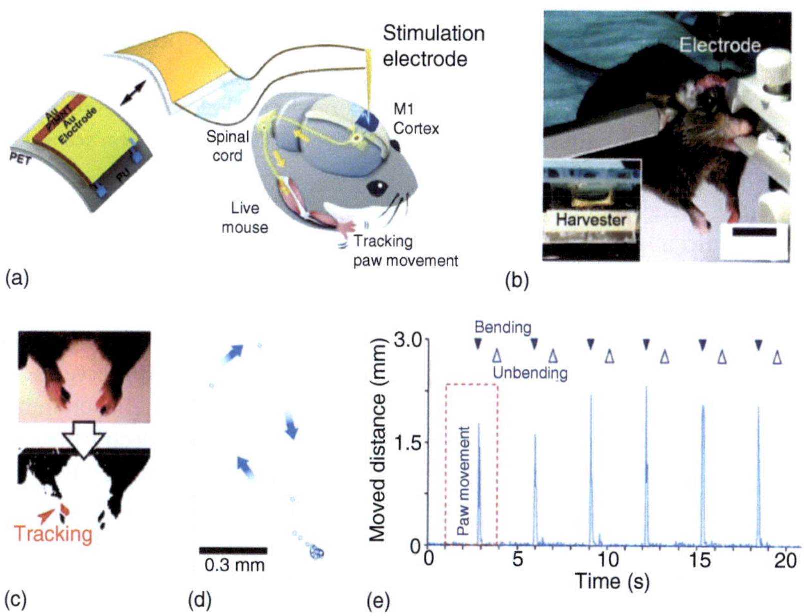

Figure 18.15 Application of the piezoelectric flexible device for neural stimulation. (a) Schematic diagram of animal experiment for brain stimulation of mouse by $Pb(In_{1/2}Nb_{1/2})O_3$–$Pb(Mg_{1/3}Nb_{2/3})O_3$–$PbTiO_3$ (PIMNT)-based flexible device. (b) The optical image of stimulation test on a mouse. (c) The captured video image of the movement of the legs and forelimbs of the mouse during the electrical stimulation. (d) Scattered plot of trajectory data of the mouse's right legs. (e) The distance of mouse's leg movement during electric brain stimulation. Source: Hwang et al. [20]/Reproduced with permission of Royal Society of Chemistry.

after 10 days of recovery, the wound was almost completely closed. As a contrast, the wound of no-treatment mice only recovered just 20% of its initial wound area.

In addition, the electric simulation can also be used to treat neurological diseases, or to recover lost functions during rehabilitation training/treatment, e.g. deep brain stimulation to alleviate the symptoms of Parkinson's disease and cochlear nerve stimulation to regain hearing. Huang et al. [20] demonstrated the self-powered $Pb(In_{1/2}Nb_{1/2})O_3$–$Pb(Mg_{1/3}Nb_{2/3})O_3$–$PbTiO_3$ thin film brain stimulator to convert small mechanical motions into electric signals (Figure 18.15a). Under slight bending, the piezoelectric flexible device outputs a V_{ov} and I_{sc} of 11 V and 285 μA, respectively, which was sufficient to meet the requirement of brain stimulation. Thus, when electrical signals were connected to the M1 cortex of the mouse's brain (Figure 18.15b) to generate muscle contraction of the forelimbs, the movement of the legs and forelimbs of the treated mouse could be clearly detected under electrical stimulation (Figure 18.15c–e).

18.5 Summary and Prospects

In this chapter, the fundamental mechanism, material and structural optimization, and applications of piezoelectric flexible devices are summarized. The piezoelectric flexible devices exhibit versatility, flexibility, stability, ease of processing, and low cost and could be used as energy harvesters or self-powered wearable sensors for smart and wearable electronics in the future. Although continuous researches have provided many optimization methods for piezoelectric flexible devices, some key issues should still be paid attention.

Firstly, the energy output performance and energy conversion efficiency of the piezoelectric flexible devices should be further improved. The increase in complexity and functionality of smart/wearable electronic devices has resulted in an increase in power consumption. Biomechanical energy is often a low-frequency excitation source. Although the instantaneous power output of piezoelectric flexible device is high, the average power output is still small, making it difficult to serve as the sole energy supply for wearable electronics. The combination of material optimization and device structure optimization, e.g. multidimensional piezoelectric material filler or multilayer electrode structure, is the fundamental way to enhance the output performance of piezoelectric flexible devices.

Second, the power management and integration of piezoelectric flexible devices should also be further investigated in wearable electronic devices. The power output of piezoelectric flexible devices has the characteristics of alternating current and transient pulses and needs to be rectified into step-down direct current to supply power for the power consumption unit or to charge the energy storage part. Thus, power management technology is necessary, but how to achieve the flexibility of the power management circuit is another issue for realizing the practical applications of piezoelectric flexible devices.

Thirdly, in recent years, despite the encouraging progress in the field of flexible electronics, there are still huge challenges. The design and fabrication of highly stretchable, highly conductive, and fatigue-free electrodes are critical for the development of highly piezoelectric flexible devices and wearable electronic devices as well.

Finally, as a part of wearable electronics, the output stability, sensibility, and lifetime of piezoelectric flexible devices under harsh environmental conditions, e.g. high temperature, high humidity, and even washing conditions, are also of interest.

Despite these challenges, piezoelectric flexible devices have great potential for applications as energy harvesters and sensors in wearable electronics. Especially through the customization of materials and structures, piezoelectric flexible devices with high output performance will make their way toward the energy supply parts to charge the batteries or capacitors and simultaneously serve as movement/health monitoring sensors for detecting biomechanical energy when integrated with skin/cloth.

References

1 Li, Z., Liu, P., Ji, X. et al. (2020). Bioinspired simultaneous changes in fluorescence color, brightness, and shape of hydrogels enabled by AIEgens. *Adv. Mater.* 32 (11): 1906493.

2 Stadlober, B., Zirkl, M., and Irimia-Vladu, M. (2019). Route towards sustainable smart sensors: ferroelectric polyvinylidene fluoride-based materials and their integration in flexible electronics. *Chem. Soc. Rev.* 48 (6): 1787–1825.

3 Yi, F., Ren, H., Shan, J. et al. (2018). Wearable energy sources based on 2D materials. *Chem. Soc. Rev.* 47 (9): 3152–3188.

4 Maiti, S., Karan, S.K., Kim, J.K., and Khatua, B.B. (2019). Nature driven bio-piezoelectric/triboelectric nanogenerator as next-generation green energy harvester for smart and pollution free society. *Adv. Energy Mater.* 9: 1803027.

5 Du, X., Zhou, Z., Zhang, Z. et al. (2022). Porous, multi-layered piezoelectric composites based on highly oriented PZT/PVDF electrospinning fibers for high-performance piezoelectric nanogenerators. *J. Adv. Ceram.* 11 (2): 331–344.

6 Gupta, K., Brahma, S., Dutta, J. et al. (2019). Recent progress in microstructure development of inorganic one-dimensional nanostructures for enhancing performance of piezotronics and piezoelectric nanogenerators. *Nano Energy* 55: 1–21.

7 Fan, F.R., Tang, W., and Wang, Z.L. (2016). Flexible nanogenerators for energy harvesting and self-powered electronics. *Adv. Mater.* 28 (22): 4283–4305.

8 Wang, Z.L. and Song, J. (2006). Piezoelectric nanogenerators based on zinc oxide nanowire arrays. *Science* 312 (5771): 242.

9 Sultana, A., Alam, M.M., Ghosh, S.K. et al. (2019). Energy harvesting and self-powered microphone application on multifunctional inorganic-organic hybrid nanogenerator. *Energy* 166: 963–971.

10 Xu, Z., Jin, C., Cabe, A. et al. (2020). Flexible energy harvester on a pacemaker lead using multibeam piezoelectric composite thin films. *ACS Appl. Mater. Interfaces* 12 (30): 34170–34179.

11 Wu, W., Wang, L., Li, Y. et al. (2014). Piezoelectricity of single-atomic-layer MoS_2 for energy conversion and piezotronics. *Nature* 514 (7523): 470–474.

12 Zhu, R., Wang, Z., Ma, H. et al. (2018). Poling-free energy harvesters based on robust self-poled ferroelectric fibers. *Nano Energy* 50: 97–105.

13 Zhang, G., Zhao, P., Zhang, X. et al. (2018). Flexible three-dimensional interconnected piezoelectric ceramic foam based composites for highly efficient concurrent mechanical and thermal energy harvesting. *Energy Environ. Sci.* 11 (8): 2046–2056.

14 Gu, L., Liu, J., Cui, N. et al. (2020). Enhancing the current density of a piezoelectric nanogenerator using a three-dimensional intercalation electrode. *Nat. Commun.* 11 (1): 1030.

15 Jella, V., Ippili, S., Eom, J.-H. et al. (2018). Enhanced output performance of a flexible piezoelectric energy harvester based on stable $MAPbI_3$-PVDF composite films. *Nano Energy* 53: 46–56.

16 Khan, A.A., Rana, M.M., Huang, G. et al. (2020). Maximizing piezoelectricity by self-assembled highly porous perovskite-polymer composite films to enable the internet of things. *J. Mater. Chem. A* 8 (27): 13619–13629.

17 Zapsas, G., Patil, Y., Gnanou, Y. et al. (2020). Poly(vinylidene fluoride)-based complex macromolecular architectures: from synthesis to properties and applications. *Prog. Polym. Sci.* 104: 101231.

18 Lee, E.J., Kim, T.Y., Kim, S.-W. et al. (2018). High-performance piezoelectric nanogenerators based on chemically-reinforced composites. *Energy Environ. Sci.* 11 (6): 1425–1430.

19 Yuan, X., Gao, X., Yang, J. et al. (2020). The large piezoelectricity and high power density of a 3D-printed multilayer copolymer in a rugby ball-structured mechanical energy harvester. *Energy Environ. Sci.* 13 (1): 152–161.

20 Hwang, G.-T., Kim, Y., Lee, J.-H. et al. (2015). Self-powered deep brain stimulation *via* a flexible PIMNT energy harvester. *Energy Environ. Sci.* 8 (9): 2677–2684.

21 Jeong, C.K., Baek, C., Kingon, A.I. et al. (2018). Lead-free perovskite nanowire-employed piezopolymer for highly efficient flexible nanocomposite energy harvester. *Small* 14: 1704022.

22 Wang, Z.L., Yang, R., Zhou, J. et al. (2010). Lateral nanowire/nanobelt based nanogenerators, piezotronics and piezo-phototronics. *Mater. Sci. Eng. R* 70 (3): 320–329.

23 Li, F., Cabral, M.J., Xu, B. et al. (2019). Giant piezoelectricity of Sm-doped $Pb(Mg_{1/3}Nb_{2/3})O_3$-$PbTiO_3$ single crystals. *Science* 364 (6437): 264.

24 Li, F., Lin, D., Chen, Z. et al. (2018). Ultrahigh piezoelectricity in ferroelectric ceramics by design. *Nat. Mater.* 17 (4): 349–354.

25 Gao, J., Xue, D., Wang, Y. et al. (2011). Microstructure basis for strong piezoelectricity in Pb-free $Ba(Zr_{0.2}Ti_{0.8})O_3$-$(Ba_{0.7}Ca_{0.3})TiO_3$ ceramics. *Appl. Phys. Lett.* 99 (9): 092901.

26 Zhao, Z., Li, X., Ji, H. et al. (2015). Microstructure and electrical properties in Zn-doped $Ba_{0.85}Ca_{0.15}Ti_{0.90}Zr_{0.10}O_3$ piezoelectric ceramics. *J. Alloys Compd.* 637: 291–296.

27 Liu, W. and Ren, X. (2009). Large piezoelectric effect in Pb-free ceramics. *Phys. Rev. Lett.* 103 (25): 257602.

28 Zhao, Z.-H., Ye, M.-Y., Ji, H.-M. et al. (2018). Enhanced piezoelectric properties and strain response in <001> textured BNT-BKT-BT ceramics. *Mater. Des.* 137: 184–191.

29 Maurya, D., Zhou, Y., Wang, Y. et al. (2015). Giant strain with ultra-low hysteresis and high temperature stability in grain oriented lead-free $K_{0.5}Bi_{0.5}TiO_3$-$BaTiO_3$-$Na_{0.5}Bi_{0.5}TiO_3$ piezoelectric materials. *Sci. Rep.* 5: 8595.

30 Wang, X., Wu, J., Xiao, D. et al. (2014). Giant piezoelectricity in potassium-sodium niobate lead-free ceramics. *J. Am. Chem. Soc.* 136 (7): 2905–2910.

31 Saito, Y., Takao, H., Tani, T. et al. (2004). Lead-free piezoceramics. *Nature* 432: 84.

32 Zhou, Z., Zhang, Z., Zhang, Q. et al. (2020). Controllable core-shell $BaTiO_3$@carbon nanoparticle-enabled P(VDF-TrFE) composites: a cost-effective approach to high-performance piezoelectric nanogenerators. *ACS Appl. Mater. Interfaces* 12 (1): 1567–1576.

33 Zhang, Y., Jeong, C.K., Wang, J. et al. (2018). Flexible energy harvesting polymer composites based on biofibril-templated 3-dimensional interconnected piezoceramics. *Nano Energy* 50: 35–42.

34 Liu, K., Choi, H.J., Kim, B.K. et al. (2018). Piezoelectric energy harvesting and charging performance of $Pb(Zn_{1/3}Nb_{2/3})O_3$-$Pb(Zr_{0.5}Ti_{0.5})O_3$ nanoparticle-embedded P(VDF-TrFE) nanofiber composite sheets. *Compos. Sci. Technol.* 168: 296–302.

35 Niu, X., Jia, W., Qian, S. et al. (2019). High-performance PZT-based stretchable piezoelectric nanogenerator. *ACS Sustain. Chem. Eng.* 7 (1): 979–985.

36 Kim, K.-B., Cho, J.Y., Jeon, D.H. et al. (2018). Enhanced flexible piezoelectric generating performance *via* high energy composite for wireless sensor network. *Energy* 159: 196–202.

37 Zhou, P., Zheng, Z., Wang, B., and Guo, Y. (2022). Self-powered flexible piezoelectric sensors based on self-assembled 10 nm $BaTiO_3$ nanocubes on glass fiber fabric. *Nano Energy* 99: 107400.

38 Wang, Z., Liu, Z., Zhao, G. et al. (2022). Stretchable unsymmetrical piezoelectric $BaTiO_3$ composite hydrogel for triboelectric nanogenerators and multimodal sensors. *ACS Nano* 16 (1): 1661–1670.

39 Wang, C., Gao, X., Zheng, M. et al. (2021). Two-step regulation strategy improving stress transfer and poling efficiency boosts piezoelectric performance of 0-3 piezocomposites. *ACS Appl. Mater. Interfaces* 13 (35): 41735–41743.

40 Hyeon, D.Y., Lee, G.-J., Lee, S.-H. et al. (2022). High-temperature workable flexible piezoelectric energy harvester comprising thermally stable $(K,Na)NbO_3$-based ceramic and polyimide composites. *Composites, Part B* 234: 109671.

41 Dudem, B., Kim, D.H., Bharat, L.K., and Yu, J.S. (2018). Highly-flexible piezoelectric nanogenerators with silver nanowires and barium titanate embedded composite films for mechanical energy harvesting. *Appl. Energy* 230: 865–874.

42 Gao, X., Zheng, M., Yan, X. et al. (2019). The alignment of BCZT particles in PDMS boosts the sensitivity and cycling reliability of a flexible piezoelectric touch sensor. *J. Mater. Chem. C* 7 (4): 961–967.

43 Hu, X., Yan, X., Gong, L. et al. (2019). Improved piezoelectric sensing performance of P(VDF-TrFE) nanofibers by utilizing BTO nanoparticles and penetrated electrodes. *ACS Appl. Mater. Interfaces* 11 (7): 7379–7386.

44 Fu, J., Hou, Y., Zheng, M., and Zhu, M. (2020). Flexible piezoelectric energy harvester with extremely high power generation capability by sandwich structure design strategy. *ACS Appl. Mater. Interfaces* 12 (8): 9766–9774.

45 Batra, K., Sinha, N., and Kumar, B. (2019). Flexible lead-free piezo-/ferroelectric $Bi_{0.5}(Na_{0.6}K_{0.4})_{0.5}TiO_3$ ceramic incorporated PDMS polymer composites for energy harvesting application. *J. Mater. Sci.-Mater. Electron.* 30 (6): 6157–6165.

46 Wang, Z., Cheng, J., Hu, R. et al. (2021). An approach combining additive manufacturing and dielectrophoresis for 3D-structured flexible lead-free piezoelectric composites for electromechanical energy conversion. *J. Mater. Chem. A* 9 (47): 26767–26776.

47 Siddiqui, S., Lee, H.B., Kim, D.-I. et al. (2018). An omnidirectionally stretchable piezoelectric nanogenerator based on hybrid nanofibers and carbon electrodes for multimodal straining and human kinematics energy harvesting. *Adv. Energy Mater.* 8: 1701520.

48 Bairagi, S. and Ali, S.W. (2019). A unique piezoelectric nanogenerator composed of melt-spun PVDF/KNN nanorod-based nanocomposite fibre. *Eur. Polym. J.* 116: 554–561.

49 He, Y., Wang, Z., Hu, X. et al. (2017). Orientation-dependent piezoresponse and high-performance energy harvesting of lead-free (K,Na)NbO_3 nanorod arrays. *RSC Adv.* 7 (28): 16908–16915.

50 Zhuang, Y., Xu, Z., Li, F. et al. (2015). Fabrication of flexible energy harvesting device based on $K_{0.5}Na_{0.5}NbO_3$ nanopowders. *J. Alloys Compd.* 629: 113–117.

51 Kim, S.-R., Yoo, J.-H., Kim, J.H. et al. (2021). Mechanical and piezoelectric properties of surface modified (Na,K)NbO_3-based nanoparticle-embedded piezoelectric polymer composite nanofibers for flexible piezoelectric nanogenerators. *Nano Energy* 79: 105445.

52 Huan, Y., Zhang, X., Song, J. et al. (2018). High-performance piezoelectric composite nanogenerator based on Ag/(K, Na)NbO_3 heterostructure. *Nano Energy* 50: 62–69.

53 Liu, B., Lu, B., Chen, X. et al. (2017). A high-performance flexible piezoelectric energy harvester based on lead-free $(Na_{0.5}Bi_{0.5})TiO_3$-$BaTiO_3$ piezoelectric nanofibers. *J. Mater. Chem. A* 5 (45): 23634–23640.

54 Jin, C.C., Liu, X.C., Liu, C.H. et al. (2018). High-performance BCTZ nanowires-based energy harvesting device and self-powered bio-compatible flexion sensor. *Mater. Des.* 144: 55–63.

55 Jin, C.C., Liu, C.H., Liu, X.C. et al. (2018). Experimental and simulation study on BCTZ-based flexible energy harvesting device filled with Ag-coated Cu particles. *Ceram. Int.* 44 (14): 17391–17398.

56 Tan, Y., Zhang, J., Wu, Y. et al. (2015). Unfolding grain size effects in barium titanate ferroelectric ceramics. *Sci. Rep.* 5: 9953.

57 Zheng, P., Zhang, J.L., Tan, Y.Q., and Wang, C.L. (2012). Grain-size effects on dielectric and piezoelectric properties of poled $BaTiO_3$ ceramics. *Acta Mater.* 60 (13): 5022–5030.

58 Huan, Y., Wang, X., Fang, J., and Li, L. (2013). Grain size effects on piezoelectric properties and domain structure of $BaTiO_3$ ceramics prepared by two-step sintering. *J. Am. Ceram. Soc.* 96 (11): 3369–3371.

59 Hong, Y., Wang, B., Long, Z. et al. (2021). Hierarchically interconnected piezoceramic textile with a balanced performance in piezoelectricity, flexibility, toughness, and air permeability. *Adv. Funct. Mater.* 31 (42): 2104737.

60 Park, K.-I., Son, J.H., Hwang, G.-T. et al. (2014). Highly-efficient, flexible piezoelectric PZT thin film nanogenerator on plastic substrates. *Adv. Mater.* 26 (16): 2514–2520.

61 He, S., Dong, W., Guo, Y. et al. (2019). Piezoelectric thin film on glass fiber fabric with structural hierarchy: an approach to high-performance, superflexible, cost-effective, and large-scale nanogenerators. *Nano Energy* 59: 745–753.

62 Wang, D., Yuan, G., Hao, G., and Wang, Y. (2018). All-inorganic flexible piezoelectric energy harvester enabled by two-dimensional mica. *Nano Energy* 43: 351–358.

63 Won, S.S., Seo, H., Kawahara, M. et al. (2019). Flexible vibrational energy harvesting devices using strain-engineered perovskite piezoelectric thin films. *Nano Energy* 55: 182–192.

64 Zhang, Y., Sun, H., and Jeong, C.K. (2018). Biomimetic porifera skeletal structure of lead-free piezocomposite energy harvesters. *ACS Appl. Mater. Interfaces* 10 (41): 35539–35546.

65 Zhang, Y., Jeong, C.K., Yang, T. et al. (2018). Bioinspired elastic piezoelectric composites for high-performance mechanical energy harvesting. *J. Mater. Chem. A* 6 (30): 14546–14552.

66 Chowdhury, A.R., Abdullah, A.M., Hussain, I. et al. (2019). Lithium doped zinc oxide based flexible piezoelectric-triboelectric hybrid nanogenerator. *Nano Energy* 61: 327–336.

67 Liu, B., Wang, M., Chen, M. et al. (2019). Effect of TC(002) on the output current of a ZnO thin-film nanogenerator and a new piezoelectricity mechanism at the atomic level. *ACS Appl. Mater. Interfaces* 11 (13): 12656–12665.

68 Pandey, R., Maria Joseph Raj, N.P., Singh, V. et al. (2019). Novel interfacial bulk heterojunction technique for enhanced response in ZnO nanogenerator. *ACS Appl. Mater. Interfaces* 11 (6): 6078–6088.

69 Deng, W., Yang, T., Jin, L. et al. (2019). Cowpea-structured PVDF/ZnO nanofibers based flexible self-powered piezoelectric bending motion sensor towards remote control of gestures. *Nano Energy* 55: 516–525.

70 Wang, Q., Qiu, Y., Yang, D. et al. (2018). Improvement in piezoelectric performance of a ZnO nanogenerator by modulating interface engineering of CuO-ZnO heterojunction. *Appl. Phys. Lett.* 113 (5): 053901.

71 Wang, Q., Yang, D., Qiu, Y. et al. (2018). Two-dimensional ZnO nanosheets grown on flexible ITO-PET substrate for self-powered energy-harvesting nanodevices. *Appl. Phys. Lett.* 112 (6): 063906.

72 Lee, S., Bae, S.-H., Lin, L. et al. (2013). Super-flexible nanogenerator for energy harvesting from gentle wind and as an active deformation sensor. *Adv. Funct. Mater.* 23 (19): 2445–2449.

73 Hasan Ul Banna, G.M. and Park, I.-K. (2017). Flexible ZnO nanorod-based piezoelectric nanogenerators on carbon papers. *Nanotechnology* 28 (44): 445402.

74 Zhang, Y., Liu, C., Liu, J. et al. (2016). Lattice strain induced remarkable enhancement in piezoelectric performance of ZnO-based flexible nanogenerators. *ACS Appl. Mater. Interfaces* 8 (2): 1381–1387.

75 Zhu, D., Hu, T., Zhao, Y. et al. (2015). High-performance self-powered/active humidity sensing of Fe-doped ZnO nanoarray nanogenerator. *Sensors Actuators B Chem.* 213: 382–389.

76 Shin, S.-H., Kwon, Y.H., Lee, M.H. et al. (2016). A vanadium-doped ZnO nanosheets-polymer composite for flexible piezoelectric nanogenerators. *Nanoscale* 8 (3): 1314–1321.

77 Lee, S., Lee, J., Ko, W. et al. (2013). Solution-processed Ag-doped ZnO nanowires grown on flexible polyester for nanogenerator applications. *Nanoscale* 5 (20): 9609–9614.

78 Batra, K., Sinha, N., Goel, S. et al. (2018). Enhanced dielectric, ferroelectric and piezoelectric performance of Nd-ZnO nanorods and their application in flexible piezoelectric nanogenerator. *J. Alloys Compd.* 767: 1003–1011.

79 Kang, L., An, H., Park, J.Y. et al. (2019). La-doped p-type ZnO nanowire with enhanced piezoelectric performance for flexible nanogenerators. *Appl. Surf. Sci.* 475: 969–973.

80 Sinha, N., Goel, S., Joseph, A.J. et al. (2018). Y-doped ZnO nanosheets: gigantic piezoelectric response for an ultra-sensitive flexible piezoelectric nanogenerator. *Ceram. Int.* 44 (7): 8582–8590.

81 Jin, C., Hao, N., Xu, Z. et al. (2020). Flexible piezoelectric nanogenerators using metal-doped ZnO-PVDF films. *Sens. Actuators, A* 305: 111912.

82 Hsu, C.-L., Su, I.L., and Hsueh, T.-J. (2015). Sulfur-doped-ZnO-nanospire-based transparent flexible nanogenerator self-powered by environmental vibration. *RSC Adv.* 5 (43): 34019–34026.

83 Algieri, L., Todaro, M.T., Guido, F. et al. (2018). Flexible piezoelectric energy-harvesting exploiting biocompatible AlN thin films grown onto spin-coated polyimide layers. *ACS Appl. Energy Mater.* 1 (10): 5203–5210.

84 Chen, J., Oh, S.K., Zou, H. et al. (2018). High-output lead-free flexible piezoelectric generator using single-crystalline GaN thin film. *ACS Appl. Mater. Interfaces* 10 (15): 12839–12846.

85 Waseem, A., Bagal, I.V., Abdullah, A. et al. (2022). High performance, stable, and flexible piezoelectric nanogenerator based on GaN:Mg nanowires directly grown on tungsten foil. *Small* 18 (20): 2200952.

86 Chen, J., Oh, S.K., Nabulsi, N. et al. (2019). Biocompatible and sustainable power supply for self-powered wearable and implantable electronics using

III-nitride thin-film-based flexible piezoelectric generator. *Nano Energy* 57: 670–679.

87 Novoselov, K.S., Geim, A.K., Morozov, S.V. et al. (2004). Electric field effect in atomically thin carbon films. *Science* 306 (5696): 666.

88 Zhang, H. (2015). Ultrathin two-dimensional nanomaterials. *ACS Nano* 9 (10): 9451–9469.

89 Wang, D., Zhang, D., Li, P. et al. (2021). Electrospinning of flexible poly(vinyl alcohol)/mxene nanofiber-based humidity sensor self-powered by monolayer molybdenum diselenide piezoelectric nanogenerator. *Nano-Micro Lett.* 13 (1): 57.

90 Xue, F., Zhang, J., Hu, W. et al. (2018). Multidirection piezoelectricity in mono- and multilayered hexagonal alpha-In_2Se_3. *ACS Nano* 12 (5): 4976–4983.

91 Song, H., Karakurt, I., Wei, M. et al. (2018). Lead iodide nanosheets for piezoelectric energy conversion and strain sensing. *Nano Energy* 49: 7–13.

92 Khan, H., Mahmood, N., Zavabeti, A. et al. (2020). Liquid metal-based synthesis of high performance monolayer SnS piezoelectric nanogenerators. *Nat. Commun.* 11 (1): 3449.

93 Akashi, T., Iwata, H., and Goto, T. (2003). Dielectric property of single crystalline $BaTi_2O_5$ prepared by a floating zone method. *Mater. Trans.* 44 (8): 1644–1646.

94 Fu, J., Hou, Y., Gao, X. et al. (2018). Highly durable piezoelectric energy harvester based on a PVDF flexible nanocomposite filled with oriented $BaTi_2O_5$ nanorods with high power density. *Nano Energy* 52: 391–401.

95 Horn, J.A., Zhang, S.C., Selvaraj, U. et al. (1999). Templated grain growth of textured bismuth titanate. *J. Am. Ceram. Soc.* 82 (4): 921–926.

96 Jardiel, T., Caballero, A.C., and Villegas, M. (2008). Aurivillius ceramics: $Bi_4Ti_3O_{12}$-based piezoelectrics. *J. Ceram. Soc. Jpn.* 116 (1352): 511–518.

97 Li, T., Li, X., Zhao, Z. et al. (2015). Evolution of textured $Ca_{0.85}(LiCe)_{0.075}Bi_4Ti_4O_{15}$ ceramics *via* templated grain growth using a rolling-extended method. *J. Mater. Sci. Mater. Electron.* 26 (4): 2082–2089.

98 Raj, N.P.M.J., Alluri, N.R., Khandelwal, G., and Kim, S.-J. (2019). Lead-free piezoelectric nanogenerator using lightweight composite films for harnessing biomechanical energy. *Composites, Part B* 161: 608–616.

99 Raj, N.P.M.J., Alluri, N.R., Vivekananthan, V. et al. (2018). Sustainable yarn type-piezoelectric energy harvester as an eco-friendly, cost-effective battery-free breath sensor. *Appl. Energy* 228: 1767–1776.

100 Lu, L., Ding, W., Liu, J., and Yang, B. (2020). Flexible PVDF based piezoelectric nanogenerators. *Nano Energy* 78: 105251.

101 Li, T., Feng, Z.-Q., Yan, K. et al. (2018). Pure OPM nanofibers with high piezoelectricity designed for energy harvesting *in vitro* and *in vivo*. *J. Mater. Chem. B* 6 (33): 5343–5352.

102 Guo, Y., Zhang, X.-S., Wang, Y. et al. (2018). All-fiber hybrid piezoelectric-enhanced triboelectric nanogenerator for wearable gesture monitoring. *Nano Energy* 48: 152–160.

103 You, M.-H., Wang, X.-X., Yan, X. et al. (2018). A self-powered flexible hybrid piezoelectric-pyroelectric nanogenerator based on non-woven nanofiber membranes. *J. Mater. Chem. A* 6 (8): 3500–3509.

104 Hussain, N., Zhang, M.-H., Zhang, Q. et al. (2019). Large piezoelectric strain in sub-10 nanometer two-dimensional polyvinylidene fluoride nanoflakes. *ACS Nano* 13: 4496–4506.

105 Vijayakumar, R.P., Khakhar, D.V., and Misra, A. (2010). Studies on α to β phase transformations in mechanically deformed PVDF films. *J. Appl. Polym. Sci.* 117 (6): 3491–3497.

106 Mondal, S., Paul, T., Maiti, S. et al. (2020). Human motion interactive mechanical energy harvester based on all inorganic perovskite-PVDF. *Nano Energy* 74: 104870.

107 Kim, D.B., Park, K.H., and Cho, Y.S. (2020). Origin of high piezoelectricity of inorganic halide perovskite thin films and their electromechanical energy-harvesting and physiological current-sensing characteristics. *Energy Environ. Sci.* 13 (7): 2077–2086.

108 Ippili, S., Jella, V., Kim, J. et al. (2018). Enhanced piezoelectric output performance via control of dielectrics in Fe^{2+}- incorporated $MAPbI_3$ perovskite thin films: flexible piezoelectric generators. *Nano Energy* 49: 247–256.

109 Ippili, S., Jella, V., Eom, J.-H. et al. (2019). An eco-friendly flexible piezoelectric energy harvester that delivers high output performance is based on lead-free $MASnI_3$ films and $MASnI_3$-PVDF composite films. *Nano Energy* 57: 911–923.

110 Karan, S.K., Maiti, S., Paria, S. et al. (2018). A new insight towards eggshell membrane as high energy conversion efficient bio-piezoelectric energy harvester. *Mater. Today Energy* 9: 114–125.

111 Hoque, N.A., Thakur, P., Biswas, P. et al. (2018). Biowaste crab shell-extracted chitin nanofiber-based superior piezoelectric nanogenerator. *J. Mater. Chem. A* 6 (28): 13848–13858.

112 Ghosh, S.K. and Mandal, D. (2016). Efficient natural piezoelectric nanogenerator: electricity generation from fish swim bladder. *Nano Energy* 28: 356–365.

113 Athenstaedt, H. (1970). Permanent longitudinal electric polarization and pyroelectric behaviour of collagenous structures and nervous tissue in man and other vertebrates. *Nature* 228 (5274): 830–834.

114 Vivekananthan, V., Alluri, N.R., Purusothaman, Y. et al. (2018). Biocompatible collagen nanofibrils: an approach for sustainable energy harvesting and battery-free humidity sensor applications. *ACS Appl. Mater. Interfaces* 10 (22): 18650–18656.

115 de Marzo, G., Mastronardi, V.M., Algieri, L. et al. (2022). Sustainable, flexible, and biocompatible enhanced piezoelectric chitosan thin film for compliant piezosensors for human health. *Adv. Electron. Mater.* 9: 2200069.

116 Li, W., Li, C., Zhang, G. et al. (2021). Molecular ferroelectric-based flexible sensors exhibiting supersensitivity and multimodal capability for detection. *Adv. Mater.* 33 (44): 2104107.

117 Lu, J., Hu, S., Li, W. et al. (2022). A biodegradable and recyclable piezoelectric sensor based on a molecular ferroelectric embedded in a bacterial cellulose hydrogel. *ACS Nano* 16 (3): 3744–3755.

118 Guan, X., Xu, B., and Gong, J. (2020). Hierarchically architected polydopamine modified $BaTiO_3$@P(VDF-TrFE) nanocomposite fiber mats for flexible piezoelectric nanogenerators and self-powered sensors. *Nano Energy* 70: 104516.

119 Zhu, M., Shi, Q., He, T. et al. (2019). Self-powered and self-functional cotton sock using piezoelectric and triboelectric hybrid mechanism for healthcare and sports monitoring. *ACS Nano* 13 (2): 1940–1952.

120 Yang, T., Pan, H., Tian, G. et al. (2020). Hierarchically structured PVDF/ZnO core-shell nanofibers for self-powered physiological monitoring electronics. *Nano Energy* 72: 104706.

121 Yan, W., Noel, G., Loke, G. et al. (2022). Single fibre enables acoustic fabrics *via* nanometre-scale vibrations. *Nature* 603 (7902): 616–623.

122 Li, Y., Xu, M.-H., Xia, Y.-S. et al. (2020). Multilayer assembly of electrospun/electrosprayed PVDF-based nanofibers and beads with enhanced piezoelectricity and high sensitivity. *Chem. Eng. J.* 388: 124205.

123 Kim, D.H., Dudem, B., and Yu, J.S. (2018). High-performance flexible piezoelectric-assisted triboelectric hybrid nanogenerator *via* polydimethylsiloxane-encapsulated nanoflower-like ZnO composite films for scavenging energy from daily human activities. *ACS Sustain. Chem. Eng.* 6 (7): 8525–8535.

124 Krawczyk, K., Xue, S., Buchmann, P. et al. (2020). Electrogenetic cellular insulin release for real-time glycemic control in type 1 diabetic mice. *Science* 368 (6494): 993–1001.

125 Bhang, S.H., Jang, W.S., Han, J. et al. (2017). Zinc oxide nanorod-based piezoelectric dermal patch for wound healing. *Adv. Funct. Mater.* 27 (1): 1603497.

19

Ferroelectrics Under Compression and Applications

Zhipeng Gao[1,2], Zhengwei Xiong[2], Hongliang He[1], Jun Li[1], and Ke Jin[1]

[1] *China Academy of Engineering Physics, Institute of Fluid Physics, 64 Mianshan Road, Mianyang 621900, China*
[2] *Southwest University of Science and Technology, School of Mathematics and Physics, 59 Qinglong Avenue, Mianyang 621010, China*

19.1 Introduction

Ferroelectric (FE) materials are characterized by their switchable spontaneous polarizations, which could be oriented by an applied electric field. After the poling process, FE materials retain bound charge associated with remnant polarizations (P_r). The poled FE materials show piezoelectric properties when they are operated at low strain, and no large-scale disorientation of the FE domains occurs. When the applied stress/strain is high enough to re-orientate the polarizations, the depolarization of the FEs will happen, and the release of bound charge during depolarization can generate an electrical current. This process results in a loss of P_r and a decrease or elimination of the piezoelectric effect. Sudden release of this charge by rapid depolarization through compression (high strain rate loading) has been proposed as a source of rapid energy transfer. This depolarization behavior is known as force-electrical effects or ferroelectric generator (FEG). Different from batteries and electrochemical capacitors, the energy stored in these FE materials is stable, while the charge-release process is ultrafast due to the charge-storage/release by this force-electric effect, which is determined by P_r changing and polarization switching. This behavior is important for applications that require portability, compactness, and high-power density.

To maximize the power output density based on force-electric effect, it is necessary to maximize P_r, increase the electrical breakdown field (E_b), and minimize the density of the ferroelectric solid. The P_r determines the charge shored/released on the materials; the E_b is critical for increasing output voltage; while the density of the materials represents the mass. The force-electric effect in FEs was first reported for barium titanate ($BaTiO_3$) in the 1950s. Since the 1960s, lead zirconate titanate (PZT)-based compounds have largely replaced barium titanate and dominated this field for more than half a century. In recent years, researchers have explored the

Piezoelectric Materials: From Fundamentals to Emerging Applications, First Edition. Edited by Jiagang Wu.

force-electric properties of some lead-free FEs and FE single crystals. They made some contributions to design materials and develop high-power applications.

19.2 Lead Zirconate Titivate Compounds

Since PZT has been the ferro-ceramic of choice for use in FEGs, their physical and electrical properties will be introduced in this section. The output of the FEG depends on many factors, including the direction of propagation of the shock wave relative to the direction of polarization, shock pressures, electrical characteristics of the load circuit (short-circuit, open-circuit, capacitive, resistive, or inductive), composition of the PZT, material properties (grain size, crystalline structure), presence of dopants, degree of poling, type of transition the PZT undergoes (polarized → nonpolarized, ferroelectric → antiferroelectric [AFE]), failure modes (fracturing, electrical breakdown), properties of the fill or potting material used, physical constraints (clamped or unclamped), and so on. In other words, the materials used in FEGs require the proper mix of good electrical, shock, and mechanical properties.

19.2.1 PZT

PZT is a member of the family of FE ceramics, and the general chemical formula for perovskite molecules is ABO_3. The piezoelectric, dielectric, and mechanical properties of PZT can be adjusted by replacing A or B cations with other cations. The $PbZrO_3$ and $PbTiO_3$ perovskites form a continuous series of solid solutions ($PbZr_{1-x}Ti_xO_3$), where x can vary from 0 to 1. These solid solutions can undergo a rich variety of FE, paraelectric, and AFE transitions brought on by changes in composition, temperature, electric field, and/or stress. They also exhibit large spontaneous polarization and piezoelectric coefficients.

The rhombohedral and tetragonal phases are separated by an almost temperature-independent morphotropic (composition-dependent) phase boundary (MPB) at which there is an abrupt structural change in the PZT's crystalline structure. The MPB depends on the relative composition of lead zirconate and lead titanate. It is not a sharp boundary but rather a temperature-dependent range of compositions over which there is a mixture of tetragonal and monoclinic phases. At room temperature, the two phases coexist when the concentration of lead titanate lies between 45% and 48% and the piezoelectric activity reaches its peak in the region of MPB.

19.2.2 $Pb_{0.99}(Zr_{0.95}Ti_{0.05})_{0.98}Nb_{0.02}O_3$ (PZT 95/5)

In high-power applications, a particularly important member of the Zr-rich PZT family is a PZT composed of $PbZr_{0.95}Ti_{0.05}O_3$ doped with small amounts (2%) of Nb ($Pb_{0.99}(Zr_{0.95}Ti_{0.05})_{0.98}Nb_{0.02}O_3$). It is commonly known as PZT 95/5, which was developed by Sandia National Laboratory [1]. PZT 95/5 is located on the side of the

FE body, near the AFE orthogonal body and the FE body phase boundary. In this structure, the pole axis is located in one of the eight <111> orientations. The structure elongates in the polar direction and shrinks perpendicular to the polar direction. Under environmental conditions, PZT 95/5 has a FE composition. It can polarize and maintain its residual polarization for several decades. According to the specifications of the manufacturer (TRS Technologies, Inc.), the PZT 95/5 has a high residual polarization of 34–38 $\mu C/cm^2$.

The volume of orthogonal AFE phase is smaller than that of the rhombic AFE phase. The oxygen octahedron also has an alternating tilt, which makes the polarization of adjacent unit cells change 180° in reverse [2]. Compared with the FE phase, the unit cell of the AFE phase is smaller, which provides the possibility for the components located on the FE side near the phase boundary to drive the phase transition from FE to AFE by compressive stress.

Fritz investigated the FE properties of PZT 95/5 under hydrostatic pressures and the response of PZT 95/5 to static uniaxial stress [3]. The results show that in the hydrostatic loading of PZT 95/5, the phase transition from FE rhombohedral to AFE orthorhombic occurs suddenly under a pressure of 0.32 GPa. This pressure-induced phase transition reduces the volume of the material by about 0.1% and depolarizes the material completely. The charge released by the PZT 95/5 sample due to pressure-induced FE to AFE phase transition can be used to generate current, voltage, and power in external loading.

Xiong et al. [4], Fu et al. [5], and Setchell [1] presented the experimental results of depolarization of PZT 95/5 by adiabatic compression of a plane shock wave generated by an accelerating impact projectile with the air gun system. Their experiments proved that PZT 95/5 samples could be completely depolarized in a longitudinal direction (shock wave propagation direction was parallel or anti-parallel to polarization) [6] or transverse shock wave (shock wave propagation direction was perpendicular to polarization) [1, 7]. In these two depolarization modes, the hydrostatic component of the stress tensor induces structural transformation at 0.9 GPa, resulting in complete depolarization of PZT 95/5, and the generated charge density is equal to the residual polarization of the material [1, 6, 7]. Shock-phase transition leads to rapid depolarization of FE materials, which lays the foundation for the development of autonomous high-power FE systems.

PZT 95/5 FE sample is not widely used because of its negligible low piezoelectric coefficient, which is not used in mass production of sensors, actuators, and low-power sensors. In the past 10 years, TRS Technologies has been developing a new technology to produce high-quality PZT 95/5 FE ceramics [8]. The PZT 95/5 ceramic sample was treated by TRS Technology Company at high temperatures, that is, mixed oxide powder, calcination, ball milling, adding binder, pressing, and sintering. The shock wave triggered by the impact projectile accelerated by the air gun system can provide accurate control of the pre-impact geometry and impact pressure [1, 6, 7]. However, these gas guns can only be used for academic research, but not for engineering applications, because of their large size, large mass, and complex control system.

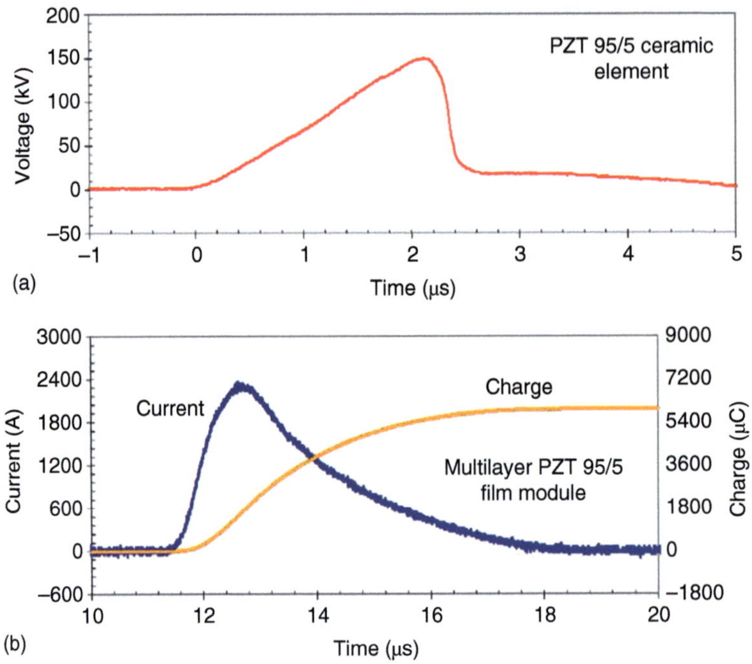

Figure 19.1 (a) Voltage waveform of PZT 95/5. (b) Current waveforms of the 120-layer PZT 95/5 film. Source: (a) Shkuratov et al. [9]. Reproduced with permission of AIP Publishing. (b) Shkuratov et al. [10]. Reproduced with permission of Wiley.

Figure 19.1a displays a typical voltage waveform generated by transverse compression of a cylindrical PZT 95/5 element with a diameter of 22 mm and a thickness of 26 mm in a micro-autonomous device working in high voltage mode [9]. The voltage amplitude is 149 kV. Figure 19.1b is a typical waveform of current and charge generated by 120-layer PZT 95/5 thin film modules connected in parallel [10]. The layer thickness is 32 mm, and the layer area is 1.52 cm^2. The module generates a large current pulse with an amplitude of 2.36 kA and a charge of 5930 mCμC. That is, the PZT 95/5 layer is almost completely depolarized, and the surface charge density of 32.7 mC/cm^2 $\mu C/cm^2$ is released.

19.2.3 $Pb_{0.99}Nb_{0.02}[(Zr_{0.90}Sn_{0.10})_{0.96}Ti_{0.04}]_{0.98}O_3$ (PZST)

Jiang et al. [11] reported the static water and shock wave depolarization of lead zirconate titanate $PbZr_{1-x}Ti_xO_3$ doped with Sn, Nb, and PZST. This FE component is located near the boundary between rhombic FE and orthogonal AFE phases. The hydrostatic test results show that under a pressure of 0.14 GPa [11]. The PZST sample is completely depolarized due to FE–AFE phase transition, and the density of released charge is equal to its residual polarization of 28 $mC/cm^2 \mu C/cm^2$. PZST specimens of 2 mm (thickness) × 30 mm (width) × 10 mm (length) were used for transverse impact depolarization short-circuit experiments (shock wave moves perpendicular to polarization direction) under impact pressures of 1.05, 1.23, and

2.5 GPa [11]. The experimental results of short-circuit depolarization show that the PZST sample produces a quasi-rectangular current pulse, which is completely depolarized at 2.5 GPa due to the impact-induced FE-AFE phase transition.

Under the load of 1000 Ω, the transverse impact depolarization high-voltage experiment was carried out with four PZST specimens stacked in series. Under the transverse impact load of 2.5 GPa, the highest voltage amplitude of 31 kV is obtained. In these experiments, the PZST sample produced 1 MW power. The energy density of the PZST sample under high pressure is 1.5 J/cm^3. These results lead people to conclude that PZST can be successfully used in high-power systems. The author also carried out a longitudinal impact depolarization short circuit experiment with PZST sample (shock wave moves parallel to polarization direction) [11]. They studied a PZST specimen with 5 mm (thickness) × 10 mm (width) × 20 mm (length), and the impact pressure ranged from 0.81 to 4.5 GPa. The experimental results show that the depolarization process of longitudinal shock waves is very complicated. When the current reaches its maximum value, the amplitude of depolarization current, the length of current pulse, and the time are closely related to the impact pressure. Under the longitudinal impact compression of 1.6 GPa, the maximum current amplitude is 48 A and the maximum stress-induced charge density is 26 μC/cm^2. In the high-voltage experiment of longitudinal impact depolarization, the maximum peak voltage of 8.5 kV is obtained at 2.68 GPa.

Jiang et al. [12, 13] reported the mechanism of shock-induced depolarization and repolarization of PZST FE ceramics under a high electric field. The process of shock wave depolarization and repolarization is studied in detail. The theoretical model of electrical system containing polarized FE ceramics was established [12–14]. The calculated results of the model are in good agreement with the experimental results.

19.2.4 $Pb_{0.99}(Zr_{0.52}Ti_{0.48})_{0.99}Nb_{0.01}O_3$ (PZT 52/48)

In PZT phase diagram, the MPB between FE tetragonal phase and FE rhombic phase occurs near $Pb(Zr_{0.525}Ti_{0.475})O_3$ [2]. Another important member of PZT family is the tetragonal composition near MPB. $Pb(Zr_{0.52}Ti_{0.48})O_3$ is doped with 1% Nb ($Pb_{0.99}(Zr_{0.52}Ti_{0.48})_{0.99}Nb_{0.01}O_3$), which is called PZT 52/48. The PZT 52/48 has larger residual polarization and higher Curie temperature. Contrary to the PZT 95/5, the PZT 52/48 has a high piezoelectric coefficient and a high dielectric constant [2, 15]. It is widely used in mass production of piezoelectric transducers, nano-positioner actuators, ultrasonic motors, active scalpels, and many other piezoelectric systems. The PZT 52/48 ceramic samples are cheap and commercially available in various shapes and sizes.

The first study of shock wave depolarization of PZT 52/48 was carried out in Los Alamos National Laboratory. Reynolds and Seay [16] reported the experimental results of depolarization of PZT 52/48 compressed by plane longitudinal shock wave. The results show that the charge density produced by PZT 52/48 ceramic sample is equal to its residual polarization under a longitudinal stress of 1 GPa. Loki Company conducted detailed experimental and theoretical research on the transverse and longitudinal impact depolarization of PZT 52/48 ceramics [3, 9, 12, 14, 17–24], whose components are provided by EDO Company [25]. Different from PZT 95/5, PZT 52/48 did not undergo a phase transition from polar phase to nonpolar phase under transverse impact load but remained in the polar phase and depolarized due to stress-induced domain reorientation [26–28].

19.3 Lead-Free Ferroelectric Compounds

19.3.1 $(Na_{0.5}Bi_{0.5})TiO_3$ (NBT)

In recent years, NBT has garnered significant attention due to the resurgence in the development of lead-free FE materials [4, 29, 30]. Recently developed relaxor AFEs based on NBT exhibit high energy density and efficiency, making them ideal for dielectric energy storage applications.

In 2018, Peng et al. [29] reported on experimental investigations of the shock-induced depolarization of $(Na_{0.5}Bi_{0.5})TiO_3$–$BiAlO_3$ solid solutions. The FE ceramic specimens utilized in this study had a composition of $0.99[0.98(Na_{0.5}Bi_{0.5})(Ti_{0.995}Mn_{0.005})O_3$–$0.02BiAlO_3]$–$0.01NaNbO_3$(NBT–BA–0.01NN), and were in the form of rectangular prisms measuring 2 mm in thickness, 9 mm in width, and 16 mm in length. These specimens were subjected to compression along a direction perpendicular to the polarization axis using planar shock waves generated via a one-stage gas gun facility. The experimental findings demonstrate that the NBT–BA–0.01NN specimens experienced complete depolarization when subjected to transverse shock compression at a pressure of 8.4 GPa. Subsequently, these specimens released an electric charge with a density of 38 $\mu C/cm^2$, which closely approximated their remanent polarization value (P_r = 40 $\mu C/cm^2$). i.e. the utilization of this FE material in high-power systems has shown promising results. In their study, the authors [29] attributed the shock-induced depolarization of NBT–BA–0.01NN to a phase transition from the FE phase to the ergodic relaxor phase. Under high hydrostatic pressure, this transition leads to the formation of polar nanoregions with randomly distributed directions of dipole moments.

In 2019, Gao et al. [30]. conducted a comprehensive experimental investigation on the shock-induced depolarization of NBT FE ceramics, as depicted in Figure 19.2. The experimental setup involved subjecting a NBT FE ceramic specimen, measuring 2 mm in thickness, 8 mm in width, and 16 mm in length, to compression via a planar shock wave directed perpendicular to the polarization axis. The authors [30] employed a gas gun system to initiate a planar transverse shock wave in the NBT specimen by impacting a copper projectile. They conducted two series of shock depolarization experiments. In the first series, the load resistance was set at 1 Ω. These experiments involved subjecting the NBT specimen to a wide range of shock pressures, ranging from 1.8 to 6.7 GPa. The experimental results indicate that NBT FE specimens are not depolarized under 1.8 GPa shock loading. An increase in shock pressure to 2.9 GPa results in the generation of stress-induced current with an amplitude of 20 A. The NBT specimen experiences partial depolarization and releases electric charge with a density approximately half of its remanent polarization. When subjected to a shock pressure of 4.3 GPa, the NBT specimen generates

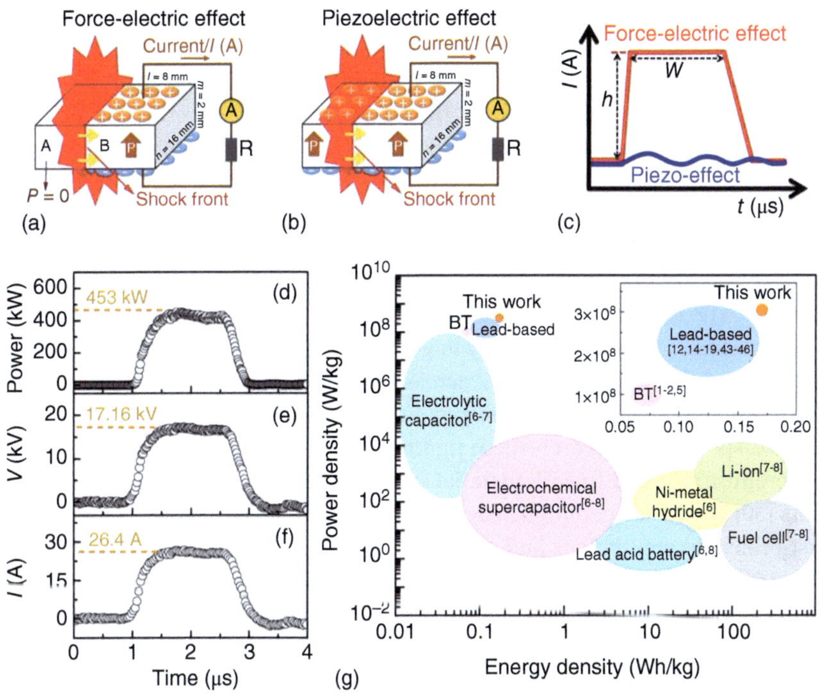

Figure 19.2 Force-electric effect (a) and the piezoelectric effect (b) under shock compression. The corresponding current waveforms in the external circuit are shown in (c). The power (d) and voltage (e) and (f) achieved are 453 kW, 17.16 kV, and 26.4 A, respectively, using a small power source with a volume of 0.256 cm^3 at a shock pressure of 4.93 GPa. The plot (g) comparing specific power and specific energy across various electrical energy storage devices demonstrates the higher power density of the NBT force-electric power source compared to other sources. The NBT power source also exhibits improved power density and energy density when compared to PZT and $BaTiO_3$ FEs. The data have an error margin of ±10%. Source: Gao et al. [30]. Reproduced with permission of APS.

a quasi-rectangular current pulse with an amplitude of 26 A and a duration of 2 μs. The degree of depolarization in this case is 87%. At a pressure of 6.7 GPa, the NBT specimen undergoes complete depolarization and releases electric charge with a density equal to its remanent polarization (P_r = 38 μC/cm^2). The force-electric effect observed in FEs is marked by the liberation of bound charge during pressure or shock-induced depolarization. Unlike other electrical energy storage systems, the charge storage and release in FEs through the force-electric effect rely on processes such as polarization switching or polar–nonpolar phase transitions. This discovery opens up additional possibilities for material design in the field of energy conversion, particularly for high power density applications. In this study, we demonstrate that NBT, a FE ceramic, exhibits a remarkable high-power output (3.04×10^8 W/kg) when subjected to shock compression. This achievement represents one of the highest values achieved through the utilization of the force-electric effect. In-depth investigations using in situ synchrotron X-ray diffraction (XRD) have provided insights into the mechanism behind the high-power output observed in NBT under shock compression. The findings indicate that this power generation predominantly stems from a polar–nonpolar phase transition, specifically the rhombohedral–orthorhombic transition. Through first-principles calculations, it has been determined that this phase transition is of the first-order nature and involves a two-step structural transformation. These results not only expand the scope of applications for the force-electric effect but also represent a significant advancement in our understanding of the phase transition behaviors of NBT under high pressure.

In the second series of shock depolarization experiments, the load resistance was set to 650 Ω. The results indicate that when subjected to a shock pressure of 4.9 GPa, the NBT specimen exhibits the ability to generate a quasi-rectangular high-voltage pulse with an amplitude of 17.16 kV across the load. The power dissipated in the resistive load reached 453 kW, corresponding to a power density of 1.8 MW/cm^3 for the FE specimen. Based on these findings [30], it can be concluded that NBT shows promise as a lead-free FE material suitable for high-power applications. The authors [30] conducted comprehensive investigations on NBT using a combination of first-principles simulations and experimental techniques. First-principles simulations were performed on the NBT unit cell under high pressure using the Vienna Ab initio Simulation Program package. In addition, static compression experiments were conducted using NBT FE powders. To further analyze the behavior of NBT under high pressure, high-pressure angle-dispersive XRD experiments were carried out utilizing a symmetrical diamond anvil cell (DAC) at Beamline 12.2.2 of the Advanced Light Source, Lawrence Berkeley National Laboratory. Based on the results of simulations and static compression experiments, the authors reached the conclusion that the complete shock-induced depolarization of NBT can be attributed to a first-order phase transition from the rhombohedral polar phase to a nonpolar orthorhombic centrosymmetric phase (space group *Pnma*). This phase transition occurs in two steps, involving the shrinkage of the unit cell and rearrangement of the oxygen ion chains. The phase transition initiates at a pressure of 1.5 GPa, with phase coexistence observed until the transition is completed at approximately 4.2 GPa. This explanation satisfactorily accounts for the observed depolarization phenomenon in the shock depolarization experiments.

In 2019, in a study conducted by Gao and Xiong et al. [4], the phase transition behaviors and current outputs of $0.92Bi_{0.5}Na_{0.5}TiO_3$–$0.08BiAlO_3$ (NBT–8BA) materials under high pressures were investigated. Figure 19.3 illustrates the current (black) and released charge (orange) of NBT–8BA ceramics at various shock pressures. To evaluate the proportion of released bound charge, the parameter r was introduced, which is determined by calculating the actual released charge relative to the static bound charge. The actual released charge is determined based on the current, while the bound charge is calculated as the product of P_r and the sample area. As the pressure is gradually increased up to 6.64 GPa, the sample exhibits a maximum current value (I_{max}) of 25 A and a ratio of released bound charge (r) equal to 94%. From this analysis, it can be concluded that the complete depoling of NBT–8BA ceramics occurs within the pressure range of 3.04–3.28 GPa. Under shock compression, the BNT–8BA FE ceramics can undergo

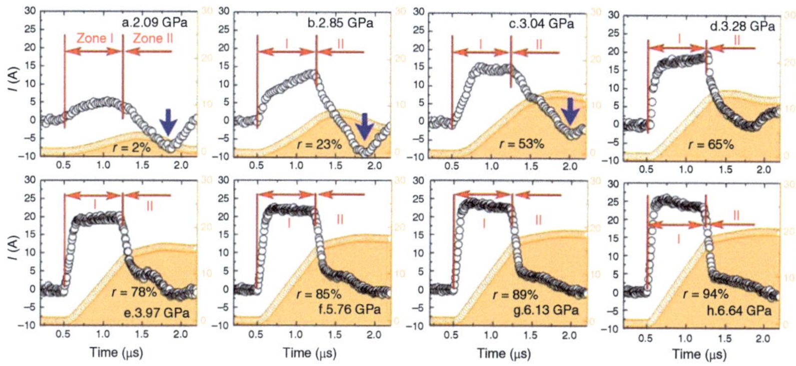

Figure 19.3 Current (black) and released charge (orange) of NBT-8BA ceramics under different shock pressures. The *r* value is the ratio of the released charge in the external circuit to the bound charge calculated from P_r. In general, zone I is the compressed zone when shock wave is passing, and zone II is the pulled zone when shock wave has passed. Blue arrows show the negative current, indicating piezoelectric effect. Source: Xiong et al. [4]. Reproduced with permission of Wiley.

a complete depolarization process, transitioning from a polar state to a nonpolar state. This depolarization process leads to the generation of a current output in the external circuit. In conjunction with in-situ synchrotron XRD experiments using a DAC, the depolarization pressures induced by phase transitions in NBT–8BA ceramics were determined. Notably, these depolarization pressures were found to be lower than those observed in pure NBT ceramics under both dynamic and static high-pressure conditions. This result implies a lower trigger pressure requirement when employing pulse power supply in practical applications.

19.3.2 $(Ag_{0.935}K_{0.065})NbO_3$ (AKN)

$AgNbO_3$ has garnered significant attention in recent studies due to the growing efforts to develop lead-free piezoelectric and FE materials suitable for a wide range of applications [5, 31–37]. Fu et al. [5] successfully verified the AFE-like characteristics of $AgNbO_3$ ceramics by measuring the polarization hysteresis loops, which revealed a double polarization hysteresis. The AFE properties exhibited by $AgNbO_3$ are particularly significant for dielectric capacitor applications. These properties can be further tailored by incorporating dopants [33–35] or by creating AFE solid solutions based on $AgNbO_3$ [30]. Consequently, $AgNbO_3$ shows immense promise as an AFE material for high-energy-density dielectric capacitors.

The remanent polarization of $AgNbO_3$ is deemed negligible [31, 32], but nonetheless, this compound serves as an ideal foundation for developing new lead-free FE solid solutions with the desired properties. Fu et al. [31] conducted a comprehensive experimental study on the FE properties of binary $(Ag_{1-x}K_x)NbO_3$ (AKN) solid solution. The results of their experiments demonstrate that within the range of $0.07 < x < 0.2$, AKN undergoes a transformative phase change from the $AgNbO_3$-type orthorhombic phase with a minimal hysteresis to a new stable FE orthorhombic phase. This new phase exhibits a distinct *P–E* loop characteristic, prominently displaying a high remanent polarization.

Liu et al. [38] published the findings of their experimental investigation into the shock depolarization behavior of AKN FE ceramics. The study involved subjecting the AKN FE ceramics with a composition of $(Ag_{0.935}K_{0.065})NbO_3$ to planar shock waves generated by impacting projectiles accelerated in a one-stage gas gun facility.

The AKN specimens of 5 mm (thickness) × 8 mm (width) × 16 mm (length) were shock-loaded perpendicular to the polarization direction. Experimental evidence shows that when exposed to 6.9 GPa shock compression, the AKN specimens generate depolarization current pulses of quasi-rectangular shape and undergo complete depolarization, resulting in the release of an electric charge with a density of 38 μC/cm^2. These findings highlight the significant potential of AKN FEs for high-power applications. The authors [38] conducted comprehensive static compression experiments and performed in-depth structural analysis of AKN ceramics. Based on the compelling findings, they reached the significant conclusion that AKN FEs undergo complete depolarization as a result of the stress-induced FE to AFE phase transition. Their research, involving electron diffraction and in situ neutron diffraction, revealed that the FE-to-AFE phase transition process is characterized by a pressure-driven irreversible transformation from a $Pmc2_1$ FE phase to a *Pbcm* AFE phase. These discoveries shed new light on the behavior of AKN FEs and contribute to a deeper understanding of their material properties and potential applications.

19.3.3 $BiFeO_3$ (BFO)

A promising FE material for pulse power sources should possess several key characteristics: a large remnant polarization (P_r), a high electrical breakdown field (E_b), and a low density (ρ). These properties correspond to the ability to store or release a significant amount of charge, produce high output voltages, and maintain a lightweight device, respectively [31]. In comparison to lead-based ceramics, lead-free FE ceramics exhibit higher theoretical P_r values and smaller densities (ρ), enabling the generation of higher charge output densities [4, 5, 29, 30, 38–40]. Notable examples of such lead-free FE systems include NBT, AKN, and BFO [4, 5, 29, 30, 38–40]. BNT and AN have demonstrated high-energy output under shock compression [5, 31]. In comparison, BFO exhibits a higher Curie temperature (T_c = 1104 K) and maximum experimental remnant polarization (P_r) of 60 μC/cm^2, surpassing that of BNT and AN [41, 42]. Moreover, the pressure-induced structural phase transition plays a crucial role in the discharge mechanisms of the FE pulse power source [4, 29, 30, 38]. The phase transition of BFO-based ceramics under high pressures exhibits similarities to the pressure-induced polar-to-nonpolar phase transition observed during the complete depolarization of PZT 95/5 ceramics and PIN–PMN–PT crystals [25, 41–44]. Thus, considering the exceptional performance and pressure-induced phase transition mechanisms, it is reasonable to anticipate that BFO-based ceramics can undergo rapid and complete depolarization under shock compression, leading to the generation of high-energy pulses.

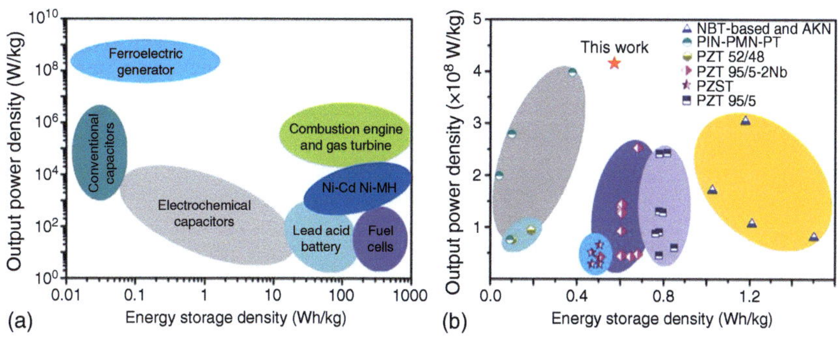

Figure 19.4 Comparison of energy storage density and output power density. (a) Ferroelectric generator compared to other materials. (b) BFO–BT ceramics compared with other ceramics. Source: Zhou et al. [45]. Reproduced with permission of AIP.

Gao et al. revealed on the electrical response of $0.9BiFeO_3$–$0.1BaTiO_3$ (BFO–BT) FE ceramics under shock-wave compression. Their study achieved a groundbreaking result, demonstrating a record-high power density of 4.21×10^8 W/kg. This power density exceeds that of any existing lead-based ceramics and other available energy storage materials, as shown in Figure 19.4. The lead-free BFO–BT material, known for its high performance, exhibits outstanding potential for utilization in pulsed power source applications. The electrical breakdown field (E_b) of the BFO–BT ceramics was reported for the first time, measuring 14.57 kV/mm. This value is nearly twice as high as that of lead-based PZ T ceramics. Notably, it represents the highest E_b reported among FE ceramics under shock compression, highlighting the significant potential of BFO–BT ceramics in high-voltage device applications. In conjunction with in-situ high-pressure neutron diffraction analysis, the BFO–BT ceramics exhibited a structural transition from a polar phase (*R*3*c*) to a nonpolar phase (*Pbam*) at 3.46 GPa. This transition is believed to be responsible for the mechanism of shock-induced depolarization. These significant findings introduce a promising lead-free material system that could potentially replace lead-based materials in pulse power devices, offering new avenues for application development.

19.4 Applications and Properties

19.4.1 Explosive-Driven Ferroelectric Generators

19.4.1.1 Design of Explosive-Driven FEGs

Remarkable advancements have been made in the advancement of autonomous compact FEGs, particularly with the design introduced by Shkuratov et al. [46]. They developed an explosive-driven FEG, as depicted in the schematic diagram presented in Figure 19.5. This FEG design has demonstrated its reliability as a prime power source, exhibiting consistent performance without failure in generating pulsed power across over 230 explosive experiments. The FEG shown in Figure 19.8

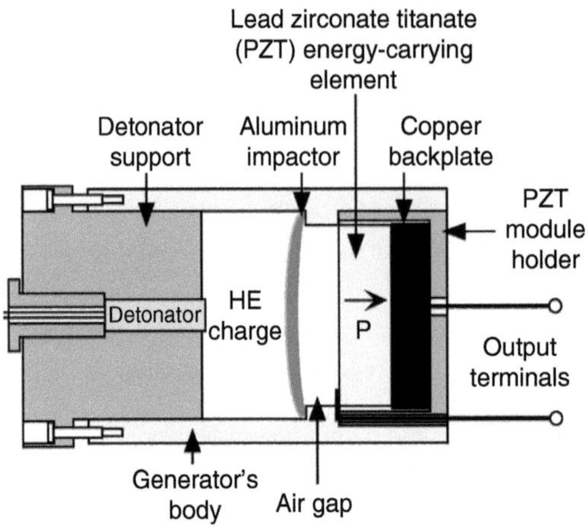

Figure 19.5 Schematic diagram of an explosive-driven longitudinal ferroelectric generator. Source: Shkuratov et al. [41]. Reproduced with permission of AIP.

comprises several components, including a cylindrical body, an explosive chamber equipped with a detonator holder, a FE element holder, and two leads that connect the contact plates on the FE module to its output terminals. In the FEGs described in this chapter, the FE elements used were primarily commercially available PZT ceramic disks supplied by EDO Corporation [47]. The polarization direction (represented by vector *P*) in the FE module is depicted by an arrow in Figure 19.6. Metallic contact plates were deposited on both flat surfaces of the disk. The physical properties of these PZT ceramic disks are as follows: density of $7.5 \times 103\,\text{kg/m}^3$, dielectric constant $\varepsilon = 1300$, Curie temperature of 320 °C, Young's modulus of $7.8 \times 10^{10}\,\text{N/m}^2$, piezoelectric constant $d_{33} = 295 \times 10^{-12}\,\text{C/N}$, and piezoelectric constant $g_{33} = 25 \times 10^{-3}\,\text{m}^2/\text{C}$. While PZT 52/48 was the primary material used for most of the testing, other materials, such as PZT 95/5, were also utilized in some experiments. It was observed that PZT 95/5 exhibited superior performance in FEGs, generating significantly higher voltages.

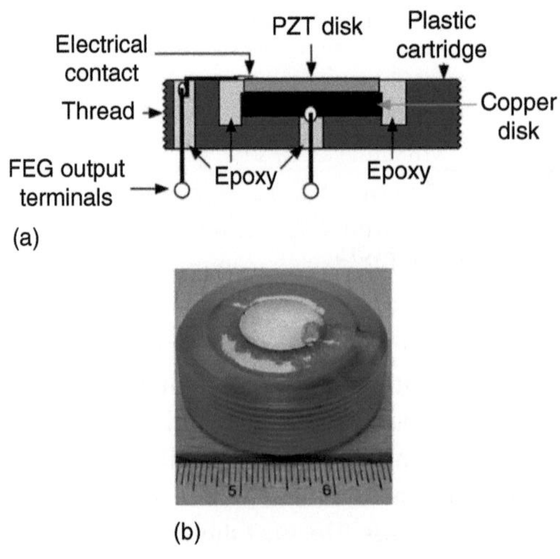

Figure 19.6 (a) Schematic diagram of the ferroelectric module holder. (b) Photo of the arrangement of the $Pb(Zr_{52}Ti_{48})O_3$ energy-carrying elements with a diameter of $D = 26$ mm and $h = 0.65$ mm. Source: Shkuratov et al. [26]. Reproduced with permission of IEEE.

The operation of the FEG can be described as follows: Once the high explosive (HE) charge is detonated, it generates a shock wave and high-pressure gases. These high-pressure gases are utilized to accelerate a flyer plate (referred to as the "aluminum impactor" in Figure 19.8). The collision between the flyer plate and the front surface of the FE element initiates a shock wave within the FE body. This shock wave propagates through the FE element, resulting in its depolarization. During the depolarization process, the induced charge accumulated on the metallic end plates of the poled FE disk is released, resulting in the generation of a voltage pulse across the output terminals of the generator. In the FEG depicted in Figure 19.5, the direction of shock wave propagation is parallel to the polarization vector P. This particular configuration is known as a longitudinal FEG. It is worth noting that there is another variant of the FEG where the shock wave propagates perpendicular or normal to the polarization vector. This type of generator is referred to as a transverse or normal FEG.

The explosive component of the FEG comprised a plastic cylindrical detonator support that housed a detonator, high-explosive charge, and a metallic impactor (flyer plate). The air gap between the flyer plate and the front contact plate of the FE element, which serves as the acceleration path for the flyer plate, had a length of 5 mm. The metallic impactor is designed to initiate the shock wave within the FE element. In order to achieve a planar impact on the FE element, the impactor is constructed as a curved plate that undergoes deformation to assume a quasi-flat structure when subjected to blast loading. To determine the appropriate curvature and thickness of the flyer plate, the arrival times and positions of the explosive shock were estimated using geometric optics and impedance matching principles for detonation waves. Subsequently, the shock transit time through the flyer plate material was calculated to determine the arrival time of the shock wave along concentric circular paths on the target side of the flyer plate. By combining the arrival times and positions of the shock wave on the plate surface, it was possible to calculate the curvature required to achieve a nearly flat flyer plate upon impact with the PZT target.

The cylindrical body of the FEG needs to be constructed using a material that possesses excellent electrical insulating properties to prevent electrical breakdown during generator operation. In a study conducted by Shkuratov et al. [46], a series of experiments were performed to investigate the effects of different body materials and thicknesses on the explosive and electrical performance of the FEG. Three types of plastic materials, namely polyethylene, polyvinylchloride (PVC), and polycarbonate, were employed for the generator bodies. The experimental results demonstrated that all of these materials exhibited satisfactory performance. There was no significant difference in the generation of voltage pulses by FEGs having body-wall thicknesses ranging from 1 to 50 mm. From this, it can be concluded that the generator's body can be a lightweight, thin plastic shell (less than a millimeter larger than the diameter of the FE element). The FEG holds together and does not disintegrate until after the explosive charge has completed detonation.

To assemble the FEG, the FE disk was bonded to a copper back plate with a diameter 1 mm larger than the FE element. The back plate had a thickness of 5 mm. A silver-loaded epoxy was utilized as the adhesive to join the FE disk and the back plate. The copper back plate served to achieve mechanical impedance matching, minimizing the reflection of stress waves when they reached the back face of the FE element. Additionally, the silver-loaded epoxy facilitated electrical contact and reduced the capacitive reactance of the bond to a negligible level. The FE energy, along with the copper back plate, was centralized within the cylindrical plastic holder. The two electrical output terminals of the FEG were connected to two contact leads connected to the metallic contact plates deposited on the surface of the FE disk. One contact lead was bonded to the front plate of the FE disk, which was subjected to flyer plat impact, with silver-loaded epoxy. The other lead was bonded in the same manner to the copper back plate.

19.4.1.2 Electrical Breakdown Problems

Electrical breakdown within the generator is one of the factors that can lead to performance degradation in explosive-driven FEGs [41, 46, 48, 49]. An example of a voltage pulse waveform generated by an earlier FEG [48] is illustrated in Figure 19.6. The waveform exhibits an irregular shape. Notably, the waveforms produced by similar FEGs, utilizing FE disks of the same type and nearly identical dimensions, displayed significant variations from shot to shot. These discrepancies strongly indicated the occurrence of electrical breakdown within the system. Through extensive investigations, including the use of high-speed photography to observe FEG operation, it was discovered that electrical breakdown did not occur as anticipated at the high-voltage terminals. Surprisingly, breakdown occurred along the sidewalls of the FE disk. This occurrence was not caused by surface flashover, as initially suspected, since the sidewalls of the FE disk were meticulously cleaned prior to device assembly. Instead, the breakdown was attributed to the expansion of the high-speed, high-pressure plasma generated by the detonation of the HE.

To mitigate the impact of detonation plasma on the electrical performance of the FEG, the design of the FE element holder was revised [26, 27]. Figure 19.6 illustrates a schematic diagram of the redesigned FE module holder. The gap between the sidewall of the FE disk and the inner wall of the module holder was widened, and a hard insulating material (such as epoxy) was used to fill the gap. A series of experiments were conducted using various epoxy compounds to identify a potting material that provided optimal high-voltage insulation in the FEG [46].

Figure 19.7 highlights the advantages of the redesigned ferroelectric module holder. The voltage pulse amplitudes displayed in Figure 19.7 were obtained from explosive-driven FEGs that utilized the same type of ferroelectric elements: PZT disks with a diameter $D = 25$ mm and a thickness $h = 2.5$ mm. However, there was a significant difference in voltage pulse amplitudes between the two cases, with $U_{max}(t) = 9.1$ kV (Figure 19.6) and $U_{max}(t) = 1.4$ kV (Figure 19.7), representing a factor of 6.5 reduction. The waveform of the voltage pulse depicted in Figure 19.7 exhibits a regular shape and demonstrates no evidence of electrical breakdown.

Studies [41] were conducted to investigate potential statistical variations in the amplitudes and pulse durations (Full Width at Half Maximum or FWHM) of voltage pulses generated by FEGs utilizing identical FE elements. The findings of these studies revealed that the relative standard deviation for the amplitudes and FWHMs of

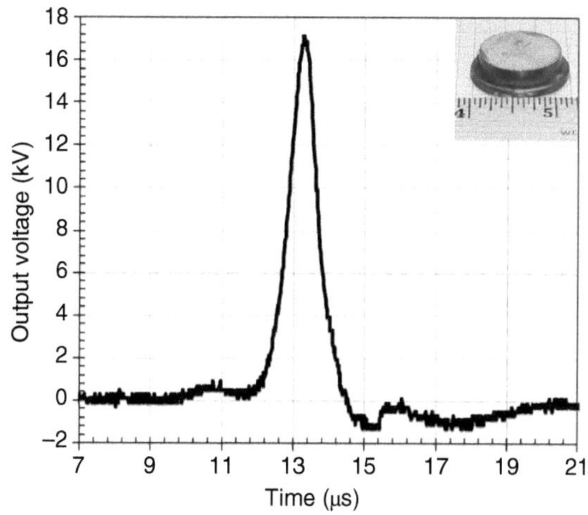

Figure 19.7 A typical waveform of the high-voltage pulse generated by the FEG system, which incorporates a PZT disk with a diameter (*D*) of 25 mm and a thickness (*h*) of 5 mm. The inset photograph presents a visual representation of the PZT disks. Source: Shkuratov et al. [41]. Reproduced with permission of AIP.

the voltage pulses remained below 6–8 % and 6–10 %, respectively. These parameter "spreads" were deemed noncritical for the majority of practical FEG applications. Furthermore, it is highly likely that these statistical variations are associated with industry standard tolerances for the initial polarization, *P*, of FE cylinders (such as PZT 52/48), rather than being attributed to the design of the FEG system. When the output voltages generated by FEGs exceeded 20–23 kV, their reproducibility and stability were compromised. It was hypothesized that this issue arose from a different form of electrical breakdown. In order to identify the source of this electrical breakdown, a series of experimental investigations were conducted, incorporating additional high-speed photography.

19.4.2 FEG-Pulsed Power Generation

High-resistance loads, such as spark gaps, are commonly encountered in various types of pulsed power systems. Delivering pulsed power to high resistance loads poses well-known challenges in explosive-driven pulsed power generators, magnetohydrodynamic generators, and other similar systems. These generators typically face limitations when it comes to directly supplying high voltages to high-resistance loads. However, a series of systematic experimental studies conducted by Shkuratov [26, 27, 41, 46, 47, 50] have demonstrated that longitudinal FEGs stand out as an exception to this list. Explosive-driven FEGs are pulsed-power sources capable of efficiently generating high voltages across high-resistance loads. Furthermore, the output parameters of FEGs are significantly influenced by the geometric dimensions of the FE element.

Figure 19.8 illustrates the experimental setup employed to investigate the operation of the FEG in open-circuit mode. The FEG was positioned within a detonation tank. The high-voltage output terminal, corresponding to the positive plate of the FE element, was directly connected to the input of a Tektronix 6015 A high-voltage probe (100 MΩ and 3 pF), while the negative (front) plate of the PZT disk was grounded. It is important to note that high-voltage diodes and rectifiers were not

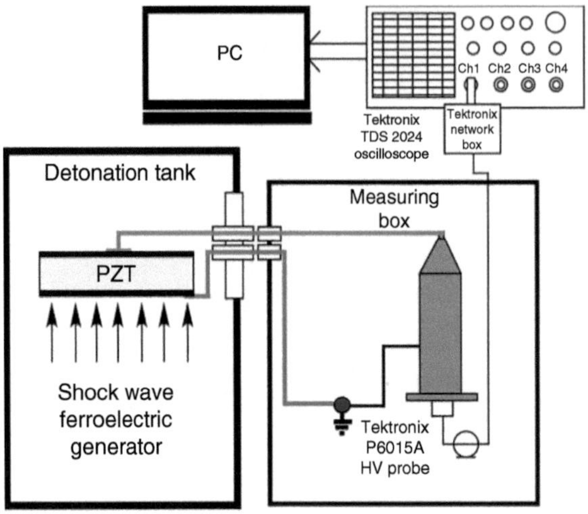

Figure 19.8 Schematic diagrams of the measuring system employed to investigate the operation of the FEG in the open-circuit mode. Source: Shkuratov et al. [26]. Reproduced with permission of IEEE.

utilized in these experimental trials. The FE elements used in the FEG were PZT 52/48 disks polarized along their cylindrical axis. Silver contact plates (electrodes) were deposited on both ends of the PZT disks using electron beam deposition. Each disk was then poled to achieve its full remnant polarization value of $P_0 = 30\,\mu\text{C/cm}^2$. Figure 19.8 displays the waveform of a typical voltage pulse generated by an FEG incorporating a PZT 52/48 FE element with dimensions $D = 25$ mm and $h = 6.5$ mm. The peak amplitude of the voltage pulse was $U_g(t)_{max} = 21.4$ kV, and its Full Width at Half Maximum (FWHM) was measured to be 1.1 μs.

In Figure 19.7, the rise in the voltage pulse from zero to its peak value can be attributed to the depolarization of the FE element. The depolarization, caused by the shock wave, leads to the release of induced electric charge from the contact plates of the FE disk to the load. Due to the high resistance and low capacitance of the load, which is represented by the high-voltage probe, the released charge from the FE element (initially acting as a capacitor) charges the load to a high voltage. The rise time of the voltage pulse corresponds to the time taken for the shock front to propagate through the thickness of the FE disk. In this mode of FEG operation, where no charge transfer occurs from the FE element, one might anticipate that the waveform of the pulse generated by the FEG would resemble a square pulse, featuring a flat top with a duration of several microseconds. This flat top would persist until the mechanical destruction of the FE element was completed. However, experimental observations reveal that, contrary to expectations, the pulse does not maintain its maximum value after reaching it; instead, it rapidly decreases.

In order to comprehend this phenomenon, it is crucial to consider the highly intricate characteristics of the shock wave propagating through the FE element [6, 51]. The shock wave is composed of a combination of elastic and inelastic acoustic and shock waves, which effectively depolarize the FE element and induce substantial changes in its physical properties. Evidently, the swift decline in the voltage pulse subsequent to reaching its peak value can be attributed to either a notable surge in the electrical conductivity of the shock-compressed section of the ceramic material, positioned behind the shock as it traversed through the disk, along with a concurrent escalation in the leakage current within the element, or to an internal electrical breakdown transpiring within the ceramic disk.

References

1 Setchell, R.E. (2005). Shock wave compression of the ferroelectric ceramic $Pb_{0.99}(Zr_{0.95}Ti_{0.05})_{0.98}Nb_{0.02}O_3$: depoling currents. *J. Appl. Phys.* 97 (1): 013507. https://doi.org/10.1063/1.1828215.

2 Jaffe, B., Cook, W.R., and Jaffe, H. (1971). *Piezoelectric Ceramics*. London: Academic Press https://doi.org/10.1016/0022-460X(72)90684-0.

3 Shkuratov, S.I., Baird, J., and Talantsev, E.F. (2013). Extension of thickness-dependent dielectric breakdown law on adiabatically compressed ferroelectric materials. *Appl. Phys. Lett.* 102 (5): 052906. https://doi.org/10.1063/1.4791597.

4 Xiong, Z., Liu, G., Nie, H. et al. (2021). Shock-driven depolarization behaviors and electrical output in $BiAlO_3$-doped $Bi_{0.5}Na_{0.5}TiO_3$ ferroelectric ceramics. *J. Am. Ceram. Soc.* 104 (2): 1169–1177. https://doi.org/10.1111/jace.17531.

5 Fu, D., Endo, M., Taniguchi, H. et al. (2007). $AgNbO_3$: a lead-free material with large polarization and electromechanical response. *Appl. Phys. Lett.* 90 (25): 252907. https://doi.org/10.1063/1.2751136.

6 Halpin, W.J. (1966). Current from a shock-loaded short-circuited ferroelectric ceramic disk. *J. Appl. Phys.* 37 (1): 153–163. https://aip.scitation.org/doi/10.1063/1.1707798.

7 Lysne, P.C. and Percival, C.M. (1975). Electric energy generation by shock compression of ferroelectric ceramics: normal-mode response of PZT 95/5. *J. Appl. Phys.* 46 (4): 1519–1525. https://doi.org/10.1063/1.321803.

8 TRS Technologies Incorporated. (2024). High power hard piezoelectric ceramics, https://www.trstechnologies.com/Materials/High-Power-Hard-Piezoelectric-Ceramics.

9 Shkuratov, S.I., Baird, J., and Antipov, V.G. (2021). Effect of shock vector/polarization vector configuration on the generation of ultrahigh voltage by adiabatically compressed ferroelectric materials. *Appl. Phys. Lett.* 119 (9): 092903. https://doi.org/10.1063/5.0064562.

10 Shkuratov, S.I., Baird, J., Antipov, V.G. et al. (2019). Multilayer PZT 95/5 antiferroelectric film energy storage devices with giant power density. *Adv. Mater.* 31 (48): 1904819. https://doi.org/10.1002/adma.201904819.

11 Jiang, D., Feng, Y., Du, J., and Gu, Y. (2011). Shock-wave- and hydrostatic-pressure-induced depolarization of Pb(Zr, Sn, Ti)O_3 for pulse power applications. *High Pressure Res.* 31 (3): 436–444. https://doi.org/10.1080/08957959.2011.602675.

12 Jiang, D., Du, J., Gu, Y., and Feng, Y. (2012). Shock wave compression and self-generated electric field repolarization in ferroelectric ceramics

$Pb_{0.99}[(Zr_{0.90}Sn_{0.10})_{0.96}Ti_{0.04}]_{0.98}Nb_{0.02}O_3$. *J. Phys. D: Appl. Phys.* 45 (11): 155401. https://doi.org/10.1088/0022-3727/45/11/115401.

13 Jiang, D., Zhang, N., Feng, Y. et al. (2012). Electric response of $Pb_{0.99}[(Zr_{0.90}Sn_{0.10})_{0.968}Ti_{0.032}]_{0.98}Nb_{0.02}O_3$ ceramics to the shock-wave-induced ferroelectric-to-antiferroelectric phase transition. *Mater. Sci. Eng., B* 177 (2): 210–216. https://doi.org/10.1016/j.mseb.2011.12.015.

14 Jiang, D., Du, J., Gu, Y., and Feng, Y. (2012). Phenomenological description of depoling current in $Pb_{0.99}Nb_{0.02}(Zr_{0.95}Ti_{0.05})_{0.98}O_3$ ferroelectric ceramics under shock wave compression: relaxation model. *J. Appl. Phys.* 111 (10): 104102. https://doi.org/10.1063/1.4716461.

15 Lines, M.E. and Glass, A.M. (2001). *Principles and Applications of Ferroelectrics and Related Materials*. Oxford: Classic Text in the Physical Sciences https://doi.org/10.1093/acprof:oso/9780198507789.001.0001.

16 Reynolds, C.E. and Seay, G.E. (1962). Two-wave shock structures in the ferroelectric ceramics barium titanate and lead zirconate titanate. *J. Appl. Phys.* 33 (7): 2234–2241. https://doi.org/10.1063/1.1728934.

17 Shkuratov, S.I., Baird, J., and Talantsev, E.F. (2010). Note: autonomous pulsed power generator based on transverse shock wave depolarization of ferroelectric ceramics. *Rev. Sci. Instrum.* 81 (12): 126102. https://doi.org/10.1063/1.3505489.

18 Shkuratov, S.I., Talantsev, E.F., and Baird, J. (2011). Electric breakdown of longitudinally shocked $Pb(Zr_{0.52}Ti_{0.48})O_3$ ceramics. *J. Appl. Phys.* 110 (2): 024113. https://doi.org/10.1063/1.3609074.

19 Shkuratov, S.I., Baird, J., and Talantsev, E.F. (2011). The depolarization of $Pb(Zr_{0.52}Ti_{0.48})O_3$ ferroelectrics by cylindrical radially expanding shock waves and its utilization for miniature pulsed power. *Rev. Sci. Instrum.* 82 (5): 054701. https://doi.org/10.1063/1.3585145.

20 Shkuratov, S.I., Baird, J., and Talantsev, E.F. (2011). Note: Miniature 120-kV autonomous generator based on transverse shock-wave depolarization of $Pb(Zr_{0.52}Ti_{0.48})O_3$ ferroelectrics. *Rev. Sci. Instrum.* 82 (8): 086107. https://doi.org/10.1063/1.3625276.

21 Shkuratov, S.I., Baird, J., and Talantsev, E.F. (2012). Effect of shock front geometry on shock depolarization of $Pb(Zr_{0.52}Ti_{0.48})O_3$ ferroelectric ceramics. *Rev. Sci. Instrum.* 83 (7): 074702. https://doi.org/10.1063/1.4732809.

22 Shkuratov, S.I., Baird, J., Talantsev, E.F. et al. (2012). Miniature 100-kV explosive-driven prime power sources based on transverse shock-wave depolarization of PZT 95/5 ferroelectric ceramics. *IEEE Trans. Plasma Sci.* 40 (10): 2512–2516. https://doi.org/10.1109/TPS.2012.2202249.

23 Shkuratov, S.I., Baird, J., Antipov, V.G. et al. (2016). High voltage generation with transversely shock-compressed ferroelectrics: breakdown field on thickness dependence. *IEEE Trans. Plasma Sci.* 44 (10): 1919–1927. https://doi.org/10.1109/TPS.2016.2553000.

24 Shkuratov, S.I., Baird, J., and Antipov, V.G. (2021). Generation of giant electric energy density by adiabatically compressed PIN-PMN-PT ferroelectric single crystals. *Appl. Phys. Lett.* 118 (12): 122902. https://doi.org/10.1063/5.0037620.

25 Kozlenko, D.P., Belik, A.A., Belushkin, A.V. et al. (2011). Antipolar phase in multiferroic $BiFeO_3$ at high pressure. *Phys. Rev. B* 84 (9): 094108. https://doi.org/10.1103/PhysRevB.84.094108.

26 Shkuratov, S.I., Talantsev, E.F., Antipov, V.G. et al. (2008). High voltage charging of a capacitor bank. *IEEE Trans. Plasma Sci.* 36 (1): 44–51. https://doi.org/10.1109/TPS.2007.913818.

27 Shkuratov, S.I., Baird, J., Antipov, V.G. et al. (2010). PZT 52/48 depolarization: quasi-static thermal heating versus longitudinal explosive shock. *IEEE Trans. Plasma Sci.* 38 (8): 1856–1863. https://doi.org/10.1109/TPS.2010.2050497.

28 Shkuratov, S.I., Baird, J., Antipov, V.G. et al. (2014). Depolarization mechanisms of $PbZr_{0.52}Ti_{0.48}O_3$ and $PbZr_{0.95}Ti_{0.05}O_3$ poled ferroelectrics under high strain rate loading. *Appl. Phys. Lett.* 104 (21): 212901. https://doi.org/10.1063/1.4879545.

29 Peng, P., Nie, H., Wang, G. et al. (2018). Shock-driven depolarization behavior in BNT-based lead-free ceramics. *Appl. Phys. Lett.* 113 (8): 082901. https://doi.org/10.1063/1.5045392.

30 Gao, Z., Peng, W., Chen, B. et al. (2019). Giant power output in lead-free ferroelectrics by shock-induced phase transition. *Phys. Rev. Mater.* 3 (3): 035401. https://doi.org/10.1103/PhysRevMaterials.3.035401.

31 Fu, D., Itoh, M., and Koshihara, S.Y. (2009). Dielectric, ferroelectric, and piezoelectric behaviors of $AgNbO_3$-$KNbO_3$ solid solution. *J. Appl. Phys.* 106 (10): 104104. https://doi.org/10.1063/1.3259410.

32 Tian, Y., Jin, L., Zhang, H. et al. (2016). High energy density in silver niobate ceramics. *J. Mater. Chem. A* 4 (44): 17279–17287. https://doi.org/10.1039/C6TA06353E.

33 Gao, J., Zhao, L., Liu, Q. et al. (2018). Antiferroelectric-ferroelectric phase transition in lead-free $AgNbO_3$ ceramics for energy storage applications. *J. Am. Ceram. Soc.* 101 (12): 5443–5450. https://doi.org/10.1111/jace.15780.

34 Zhao, L., Gao, J., Li, Q. et al. (2018). Silver niobate lead-free antiferroelectric ceramics: enhancing energy storage density by B-site doping. *ACS Appl. Mater. Interfaces* 10 (1): 819–826. https://doi.org/10.1021/acsami.7b17382.

35 Gao, J., Zhang, Y., Zhao, L. et al. (2019). Enhanced antiferroelectric phase stability in La-doped $AgNbO_3$: perspectives from the microstructure to energy storage properties. *J. Mater. Chem. A* 7 (5): 2225–2232. https://doi.org/10.1039/C8TA09353A.

36 Gao, J., Li, Q., Dong, J. et al. (2020). Local structure heterogeneity in Sm-doped $AgNbO_3$ for improved energy-storage performance. *ACS Appl. Mater. Interfaces* 12 (5): 6097–6104. https://doi.org/10.1021/acsami.9b20803.

37 Gao, J., Li, Q., Zhang, S., and Li, J.F. (2020). Lead-free antiferroelectric $AgNbO_3$: phase transitions and structure engineering for dielectric energy storage applications. *J. Appl. Phys.* 128 (7): 070903. https://doi.org/10.1063/5.0018373.

38 Liu, Z., Lu, T., Xue, F. et al. (2020). Lead-free $(Ag,K)NbO_3$ materials for high-performance explosive energy conversion. *Sci. Adv.* 6 (21): eaba0367. https://www.science.org/doi/full/10.1126/sciadv.aba0367.

39 Yang, L., Kong, X., Li, F. et al. (2019). Perovskite lead-free dielectrics for energy storage applications. *Prog. Mater Sci.* 102: 72–108. https://doi.org/10.1016/j.pmatsci.2018.12.005.

40 Sun, H., Wang, X., Sun, Q. et al. (2020). Large energy storage density in $BiFeO_3$-$BaTiO_3$-$AgNbO_3$ lead-free relaxor ceramics. *J. Eur. Ceram. Soc.* 40 (8): 2929–2935. https://doi.org/10.1016/j.jeurceramsoc.2020.03.012.

41 Shkuratov, S.I., Talantsev, E.F., Menon, L. et al. (2004). Compact high-voltage generator of primary power based on shock wave depolarisation of lead zirconate titanate piezoelectric ceramics. *Rev. Sci. Instrum.* 75 (8): 2766–2769. https://doi.org/10.1063/1.1771490.

42 Wu, Y., Han, X., and Huang, H. (2018). Structural transformation pathways of multiferroic $BiFeO_3$ under high pressures. *J. Phys. Chem. C* 122 (12): 6852–6857. https://doi.org/10.1021/acs.jpcc.8b00977.

43 Shkuratov, S.I., Baird, J., Antipov, V.G. et al. (2018). Complete stress-induced depolarization of relaxor ferroelectric crystals without transition through a non-polar phase. *Appl. Phys. Lett.* 112 (12): 122903. https://doi.org/10.1063/1.5019593.

44 Knee, C.S., Tucker, M.G., Manuel, P. et al. (2014). High pressure crystal and magnetic phase transitions in multiferroic $Bi_{0.9}La_{0.1}FeO_3$. *Chem. Mater.* 26 (2): 1180–1186. https://doi.org/10.1021/cm403558j.

45 Zhou, Z., Gao, Z., Xiong, Z. et al. (2022). Giant power density from $BiFeO_3$-based ferroelectric ceramics by shock compression. *Appl. Phys. Lett.* 121 (11): 113903. https://doi.org/10.1063/5.0102102.

46 Shkuratov, S.I., Baird, J., Talantsev, E.F. et al. (2006). Completely explosive ultra-compact high-voltage pulse generating system. *Rev. Sci. Instrum.* 77 (4): 043904. https://doi.org/10.1063/1.2168674.

47 Setchell, R.E. (2003). Shock wave compression of the ferroelectric ceramic $Pb_{0.99}(Zr_{0.95}Ti_{0.05})_{0.98}Nb_{0.02}O_3$: Hugoniot states and constitutive mechanical properties. *J. Appl. Phys.* 94 (1): 573–588. https://doi.org/10.1063/1.1578526.

48 Novitskii, E.Z. and Sadunov, V.D. (1984). Development of breakdown in a shock-compressed ferroelectric. *Combust. Explos. Shock Waves* 20 (4): 439–441. https://doi.org/10.1007/BF00782397.

49 Lysne, P.C. (1973). Dielectric breakdown of shock-loaded PZT 65/35. *J. Appl. Phys.* 44 (2): 577–582. https://doi.org/10.1063/1.1662227.

50 Shkuratov, S.I., Kristiansen, M., Dickens, J. et al. (2001). Experimental study of compact explosive driven shock wave ferroelectric generators, PPPS-2001 Pulsed Power Plasma Science 2001. *28th IEEE International Conference on Plasma Science and 13th IEEE International Pulsed Power Conference. Digest of Papers* (Cat. No. 01CH37251). IEEE, 2001; 2, 959–962. https://doi.org/10.1109/PPPS.2001.960832.

51 Doran, D.G. (1968). Shock-wave compression of barium titanate and 95/5 lead zirconate titanate. *J. Appl. Phys.* 39 (1): 40–47. https://doi.org/10.1063/1.1655775.

20

Piezoelectricity in Biomedical Applications

Laiming Jiang and Jiagang Wu

Sichuan University, College of Materials Science and Engineering, No. 29 Wangjiang Road, Chengdu 610064, PR China

20.1 Introduction

Advances in biomedicine impacts human well-being. In recent years, with the reform and development of healthcare, the treatment paradigm has gradually shifted from hospital-centric to patient-centric, greatly encouraging the development of portable, miniaturized, and multifunctional bio-electronic products for personalized medicine [1, 2], which has far-reaching socio-economic implications. Emerging material science and processing technology developments have facilitated the development of personalized bioelectronics for monitoring vital information, such as physical activity and vital signs, and providing noninvasive, engaging, and dynamic health assessments and personalized medicine [1].

Due to the inherent properties of piezoelectric materials, their application can allow for the reversible conversion of mechanical and electrical energy, permitting efficient electromechanical coupling, fast response curves, and simple engineering structures. Piezoelectric technology not only acts as an energy harvester but also provides active sensing capabilities in the form of tactile sensing, stress or strain sensing, acoustic sensing, and more [1, 3]. When connected to the human body, piezoelectric devices can monitor weak physiological signals of human activity, including pulse detection, respiration monitoring, tissue elastic modulus calculation, blood flow monitoring, and more. In addition to these applications, the electricity generated is also used extensively for *in situ* electrical stimulation, cellular/neural activity regulation, tissue regeneration, and drug delivery. In recent years, with the advent of new materials and the development of new processing technologies, piezoelectricity has developed rapidly and has made significant advances in biomedically relevant areas of powering, sensing, and therapy [1, 4, 5].

The purpose of this chapter is to present recent research on the use of piezoelectric materials for biomedical applications, with a focus on reviewing and discussing a range of high-performance piezoelectric materials for biomedical applications, including powering (e.g. implantable energy harvesters), sensing (e.g. ultrasound

Piezoelectric Materials: From Fundamentals to Emerging Applications, First Edition. Edited by Jiagang Wu.

imaging transducers), and therapy (e.g. bone/tissue regeneration, direct cell/neural stimulation, and tumor therapy). This chapter also discusses opportunities and challenges and concludes by highlighting the prospects for future research.

20.2 Implantable Energy Harvesters

The latest developments in biomedical systems, especially in microelectronic devices for healthcare and other medical applications, illustrate the potential of bio-implantable devices for monitoring, diagnosis, and therapy [6, 7]. Medical implants with many benefits have become commonplace. Traditionally, the standard powering method relies on batteries, with electronic devices equipped with independent power sources to achieve independent operation. Although battery technologies provide a reliable, independent power option, regular replacement through surgical interventions to enable continuous running is necessary, which results in a non-negligible risk and burden. To alleviate the battery issues, piezoelectric energy harvesting systems are promising for use in the next generation of untethered, safe, and reliable biomedical electronics [8–10]. Advances in materials, nanotechnology, and microelectronics are revolutionizing piezoelectric self-powered and wireless-powered systems [11–13]. Various kinds of piezoelectrics-based energy harvesters (PEHs) have been developed to scavenge energy from various sources of deformation, including body movements [14], heartbeats [15], and even from external ultrasonic induction [16].

20.2.1 Self-Powered Systems

Due to their excellent electromechanical coupling properties, piezoelectric materials can efficiently convert mechanical energy into electrical energy and are therefore primarily developed for self-powered systems, such as cardiac defibrillators, pacemakers, brain stimulators, and cochlear implants. The rich biomechanical activities of the human body, including walking, running, limb movements, and even the beating of the heart, can provide a rich source of passive energy for piezoelectric systems. Over the past few years, researchers have performed extensive research to convert passive biomechanical energy into electrical energy through a range of piezoelectric systems (e.g. inorganic materials-based systems, organic materials-based systems, composites-based systems, and natural piezoelectric materials-based systems) to power low-power electronic devices for human motion sensing, vital signs sensing, and assistive physical therapy.

Self-powered systems based on inorganic piezoelectric materials have been widely investigated in view of their excellent electrical output properties. Most conventional inorganic piezoelectric materials have quite high piezoelectric properties (e.g. piezoelectric coefficient of soft PZT $\sim$ 500–600 pC/N); however, they are bulky, inherently hard, and brittle, and cannot withstand considerable deformation. A common approach is to enhance the flexibility of the device through the design of the structure; for example, the incorporation of piezoelectric films on flexible substrates is an effective strategy. Park et al. [17] have developed a flexible self-powered system based on an eco-friendly $BaTiO_3$ film, which was fabricated by RF magnetron sputtering and soft lithography transfer techniques. The device exhibits a degree of flexibility with an output voltage of ~1.0 V and a current of ~26 nA, providing progressive opportunities for implantable biological devices. To maintain flexibility and enhance output, Li et al. [18] recently investigated a two-layer implantable system in which the same single crystal 0.72PMN–0.28PT, with a thickness of 50 mm, was used as the piezoelectric layer (Figure 20.1). This piezoelectric layer was bonded to an elastic backbone of PET to give it flexibility. A maximum output voltage of ~20 V was generated in series mode, and a maximum output current of ~15 μA in parallel mode. The device is capable of powering a commercial pacemaker, which is an impressive step toward the manufacture of self-powered pacemakers.

In contrast to inorganic piezoelectric materials, piezoelectric polymers are very popular in the field of implantable power systems due to their excellent flexibility.

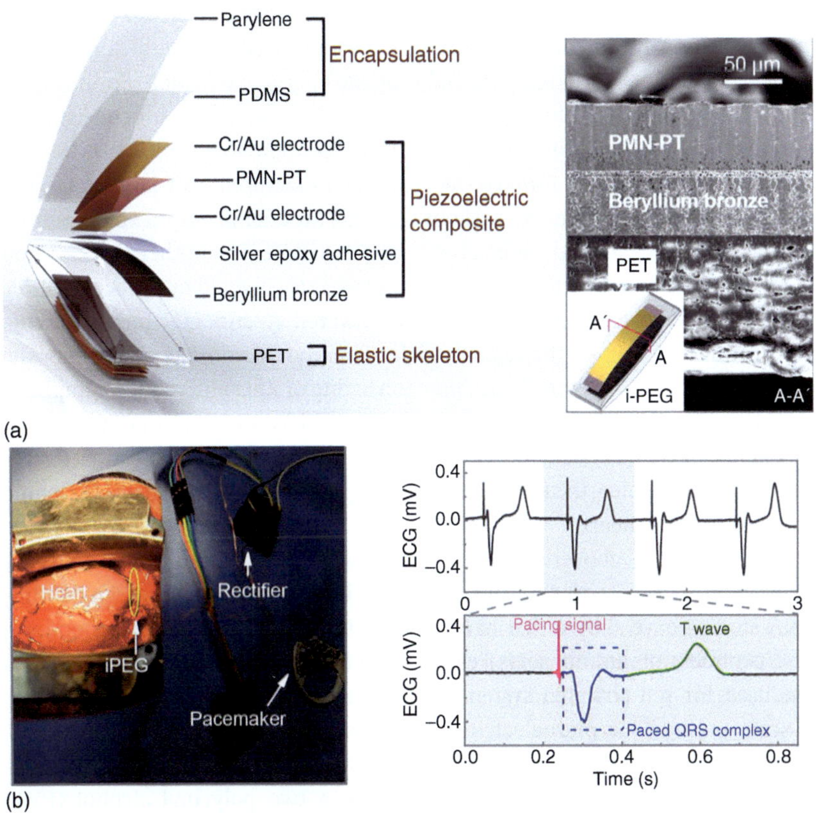

Figure 20.1 A two-layer PMN–PT implantable system for powering cardiac pacemaker: (a) schematic and structural characterization of the PMN–PT implantable system, (b) *in vivo* evaluation of the PMN–PT implantable system for powering pacemaker. Source: Li et al. [18]2019/Reproduced with permission of American Chemical Society.

There has been much research into piezoelectric polymer-based energy harvesters that harvest biomechanical energy from human activities, particularly those based on PVDF-related polymers. For instance, Cheng et al. [19] reported the feasibility and effectiveness of an implantable PVDF film-based nanogenerator for harvesting energy generated by blood circulation. The device was wrapped around the ascending porcine aorta, and the corresponding output voltage generated shows a similar trend in peak flow pressure. When blood pressure exceeds 140 mmHg, an LCD can be switched on to show the systolic blood pressure, validating how such a device can be used to predict the risk of hypertension. Recently, Nagayama et al. [20] reported a piezoelectric sensor array for multipoint detection of structural deformation in a pulsating 3D heart model using P(VDF-TrFE) films sealed with biocompatible parylene C. These piezoelectric signals can be translated into compressive stresses exerted on the flexible sensors by the pulsating motion of the 3D heart model, indicating that the contraction force generated by the myocardium (i.e. the heart muscle) can be directly assessed, showing potential in flexible implantable self-powered systems.

The preparation of piezoelectric composites is a compelling strategy that combines the advantages of both inorganic materials (e.g. high voltage electricity) and organic polymers (e.g. flexibility). This strategy has been widely used to manufacture flexible self-powered systems, particularly for implantable energy harvesting. For instance, Zhang et al. [21] reported a high-performance flexible nanogenerator consisting of a three-dimensional honeycomb structure of PZT ceramic foam-filled PDMS. This three-dimensional laminate film can be easily manufactured over large areas and is easily stretched/bent by fingers without damage, producing output voltages from 22 to 65 V as the tensile strain increases from 5% to 15%. In another work, Azimi et al. [22] demonstrated a battery-free pacemaker that is powered by a biocompatible and flexible piezoelectric polymer-based self-powered system, generating electricity from cardiac motion in the left ventricle. The system consists of a composite nanofiber of PVDF and a hybrid nanofiller consisting of ZnO and reduced graphene oxide. The composite nanofibers are optimized to achieve a high power output. The optimized self-powered system implanted in the body can successfully extract 0.487 J of energy from each heartbeat, which is much greater than the pacing threshold energy of the human heart.

Compared to the abovementioned material systems, natural piezoelectric materials tend to be convincing due to their high uniformity and degradability. Previous studies have shown that natural materials with (such as bones, tendons, viruses, peptides, or amino acids) can exhibit piezoelectric properties and can also be used for self-powered systems. For example, Yang et al. [23] reported a wafer-scale method for piezoelectric biomaterial films based on γ-glycine crystals. The film has a sandwich structure in which a crystalline glycine layer is self-assembled and automatically aligned between two polyvinyl alcohol (PVA) films. The piezoelectric coefficient of the heterostructured glycine–PVA films is 5.3 pC/N or 157.5×10^{-3} Vm/N. Glycine–PVA films have natural compatibility and biodegradability in a physiological environment, which provides the possibility for the development of transient implantable electromechanical devices.

In the past 10 years, piezoelectric powering systems have been extensively studied and deeply concerned and have been widely developed and applied in a range of applications besides implantation, such as wearable medical electronics. In conclusion, with the development of piezoelectric materials and device design, such self-power systems offer unlimited possibilities for the next generation of intelligent medical systems.

20.2.2 Ultrasonic Wireless-Powered Systems

Compared with harvesting energy from internal organ motions, wireless external sources can provide tunable and long-term sustainable transmission power, regardless of organ shape, implantation location, and body size. Meanwhile, a great deal of internal resistance (tens of megaohms) is generally obtained in traditional piezoelectric devices due to low-frequency body or organ movements, resulting in limited power output. Additionally, wireless energy harvesting enables such implants to operate safely and seamlessly while eliminating the requirement of transcutaneous or percutaneous wires that are cumbersome and prone to infection, especially during long-term therapy.

As an emerging wireless transfer technology, ultrasonic energy transfer (UET), which uses traveling ultrasonic waves to transmit energy, has been exploited and incorporated into various implantable biomedical systems for wireless power supply, medical monitoring, and information communication [24]. Ultrasound technique has long been used in medical therapeutics and diagnosis and is safe in most medical applications. In this wireless transfer scheme, ultrasound waves emitted by a transducer can be converted to electrical power by piezoelectric materials. Back in 2007, a ZnO nanowire array-based nanogenerator that can be driven by ultrasonic waves to yield continuous electrical output for implantable applications was proposed by Wang and coworkers [25] When the piezoelectric structure is deformed by ultrasonic waves, the charge separation process generates a piezoelectric potential that induces a current to drive electronics. In contrast to traditional wireless power transfer represented by electromagnetic coupling, UET has several major advantages; for example, ultrasound offers a higher FDA limit for power flux ($\approx$0.72 W/cm^2), lower attenuation ($\approx$0.5–1 dB cm/MHz) through tissue, much shorter wavelengths ($\approx$1.5 mm at 1 MHz) for more efficient coupling to miniaturized implants, and higher security in cyber communications [26]. As proof of cases, miniaturized implantable ultrasonic backscattering sensors composed of a piezocrystals, an application-specific integrated circuit (ASIC), and a pair of electrodes have been reported for electrical stimuli and recording of peripheral nerves, deep tissue oximetry monitoring, and real-time temperature monitoring, all of which use ultrasound power and control for downlink communication and readout [26–28]. Compared to conventional implanted electronics requiring multiple chips, packaging, and wires, these ultrasonic wireless-powered systems are highly volume efficient, with device sizes in millimeters or sub-millimeters, minimizing the burden on the organization.

Additionally, ultrasonic energy transfer, as an emerging technology has been integrated into a variety of biomedical systems in flexible, multichannel, and other forms for diagnosis, monitoring, and therapy functions. For example, Jin et al. [29] demonstrated an implantable acoustic energy transfer and communication device (AECD) for *in vivo* physiological monitoring and cardiac pacing. Furthermore, to improve the acoustic intensity of energy transfer, a convenient, effective, and easy-to-operate ultrasonic focusing method was developed through a flexible ultrasonic array that not only easily accomplishes rapid ultrasonic focusing but also controls the ultrasonic focusing position. An animal experiment was conducted involving the powering of the system by ultrasound and the creation of a cardiac arrest model. In this AECD device, high-performance electronic components (PZT transducers, temperature sensors, and energy management units) are assembled on soft, flexible substrates through the use of flexible electronics consisting of emerging design strategies and manufacturing technologies. The inner components are fully sealed by PDMS, which has high biocompatibility. The AECD's stretchability can be up to 20%. Another recent example of research is the flexible piezoelectric ultrasound energy-harvesting (PUEH) array for deep brain stimulation and analgesia applications [30]. The novel PUEH array is an arrangement of 6×6 and based on Sm-doped $Pb(Mg_{1/3}Nb_{2/3})O_3$–$PbTiO_3$ (Sm–PMN–PT) single crystal, whose piezoelectric coefficient (d_{33}), electromechanical coupling coefficient (k_{33}), and relative permittivity (ε_r) reach up to 4000 pC/N, 95%, and 13 000. The output power density of this device can reach up to 1.1 W/cm^2 *in vitro*, which is 18 times higher than the previous record (60 mW/cm^2). After being implanted in the rat brain, the as-developed array can produce an instantaneous effective output power of 280 μW under 1-MHz ultrasound with a safe intensity of 212 mW/cm^2, which can immediately activate the periaqueductal gray brain for analgesia applications.

One of the disadvantages of this ultrasonic transfer method is that the ultrasound waves propagate in a directional manner. As a result, slight misalignment and misdirection between the external transducer and the implanted receiver can lead to a reduction in coupling efficiency. In addition, the transducer must contact the skin to eliminate the large impedance mismatch at the tissue–air interface. To address this issue, Jiang et al. [31] developed an alternative method of generating ultrasound waves through the photoacoustic energy that was generated by pulsed light irradiating a flexible graphene nanoplatelet-loaded polydimethylsiloxane (GNPs-PDMS) patch. The photoacoustic signal has better spatial and temporal resolution and directionality than conventional piezoelectric ultrasound. In addition, the ultrasound harvester exploits a 1×13 linear array that can ensure 3D twining on the organ surface and achieve ultrasound capture from multiple angles without operational difficulty (Figure 20.2). Specifically, as the core components of the energy receiver, the ultrasonic system used KNN 1–3 composites, which possess environmental friendliness and biocompatibility. KNN-based lead-free piezoceramics are currently quite comparable to PZT in terms of the d_{33} piezoelectric coefficient, enabling them to be used in a range of high-performance lead-free ultrasound transfer systems, including a KNN composite-based piezoelectric

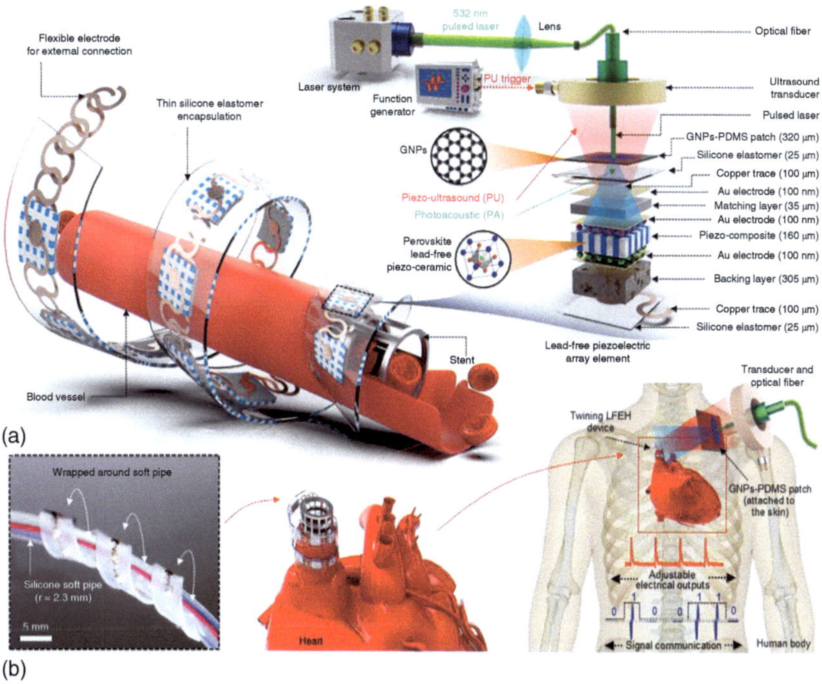

Figure 20.2 Photoacoustic and piezo-ultrasound hybrid-induced lead-free energy harvester for 3D twining wireless multifunctional implants: (a) schematic diagrams and the design of the hybrid-induced 3D twining lead-free device, where a focused ring piezo-transducer and a miniaturized fiber-photoacoustic converter as the hybrid acoustic source and a KNN-based lead-free linear piezo-array as the harvester, (b) photo and schematic diagrams showing the 3D twining of the device for wireless implants. Source: Reproduced with permission from Ref. [31]. Copyright 2021, RSC.

helix for transcutaneous energy harvesting [32], a flexible lead-free piezoelectric array for high-efficiency wireless power transfer and communication [33], and a millimeter-scale flexible ultrasound patch for potential retinal electrical stimulation application [16].

These developed ultrasonic wireless-powered systems are acoustically tunable and can provide sufficient power even though the devices are placed in several centimeters thick into *in vivo* tissues. Some representative systems are summarized in Table 20.1. For piezoelectric ultrasound power supply and communication technology, a number of common bioelectronic requirements have been met, and these systems will help realize the broader potential of the near-future biomedical systems that require high accuracies and multicontrol capabilities, such as *in vivo* diagnosis and treatment, real-time monitoring, and chronic disease monitoring and feedback.

Table 20.1 Representative ultrasonic wireless-powered systems.

Devices	Materials	Structure	Output signal	Applications	Refs.
AECD	PZT ceramic	Stretchable	75 mW	Cardiac pacing; temperature monitoring	[29]
Neural Dust	PZT ceramic	Miniaturized	120 μW	ENG and EMG recording	[34]
Stim Dust	PZT ceramic	Miniaturized	89 μW	Neural stimulation	[28]
O_2-sensing system	PZT ceramic	Miniaturized	150 μW	Monitoring deep-tissue oxygenation	[26]
Sub-0.1-mm^3 mote	PZT ceramic	Miniaturized	1.85 mW/mm^2	Temperature sensing	[27]
F-URSP	PMN–PT 1–3 composite	Flexible	22.6 mW/cm^2	Biomimetic visual prostheses to restore vision	[35]
Sm-PUEH	Sm-PMN–PT single crystal	Flexible	280 μW	Deep brain stimulation; analgesia applications	[30]
LFEH	KNN 1–3 composite	Flexible; multifunctional	7.5 mW/cm^2 at 1 MHz; 103.1 mW/cm^2 at 3.3 MHz	Transcutaneous wireless powering; signal communication	[31]
Piezo-helix	KNN 1–3 composite	Flexible	22.97 mW/cm^2	Transcutaneous wireless powering	[32]
f-LFPA	KNN 1–3 composite	Flexible	22.4 V; 0.145 W/cm^2	Transcutaneous wireless powering; signal communication	[33]

20.3 Ultrasound Imaging Transducers

Medical ultrasonic transducers that come in a variety of forms and sizes, ranging from single-element transducers for mechanical scanning, linear arrays, to multidimensional arrays for electronic scanning have been widely used for medical imaging, diagnosis, and therapy because of their noninvasiveness, high accuracy, great sensitivity, and strong penetration capabilities [36–38]. An ultrasonic transducer is basically composed of multilayer elements (Figure 20.3), mainly including a backing substrate that absorbs the acoustic emissions, an active layer made of piezoelectrics, and a matching layer compensating acoustic impedance mismatch between the piezoelectrics and the biological tissues [36]. The most crucial component of an ultrasonic transducer is the piezoelectric active layer that converts electrical energy into acoustic energy and vice versa through the piezoelectric effect. Thus, the properties of the piezoelectrics have a significant effect on the performance of the transducers. Due to their excellent electromechanical properties, lead-based piezoelectric materials (e.g. PZT ceramics and PMN–PT crystal) have been widely used in commercial ultrasound probes for clinical imaging applications. Nevertheless, there is an urgent need to develop environmentally friendly, nontoxic alternative materials due to increasing safety concerns, especially in clinical areas where there is direct contact with the human body. This section, therefore, focuses on the recent use of lead-free materials in ultrasound imaging transducers.

Over the past few decades, there have been numerous research studies on developing lead-free piezoelectrics for ultrasound transducers that are very promising as lead-based alternatives for next-generation medical devices that require environment-friendly features [39–41]. For example, Zhu et al. [42] developed a 52 MHz high-frequency intravascular ultrasound (IVUS) imaging transducer that is fabricated from a 35 μm free-standing (100)-textured Li-doped KNN (KNLN) thick film. The textured KNLN thick film was prepared by tape-casting technology using $NaNbO_3$ as a template. The developed transducer possesses a side-looking miniature architecture with an aperture of 0.4 mm × 0.4 mm and a bandwidth of

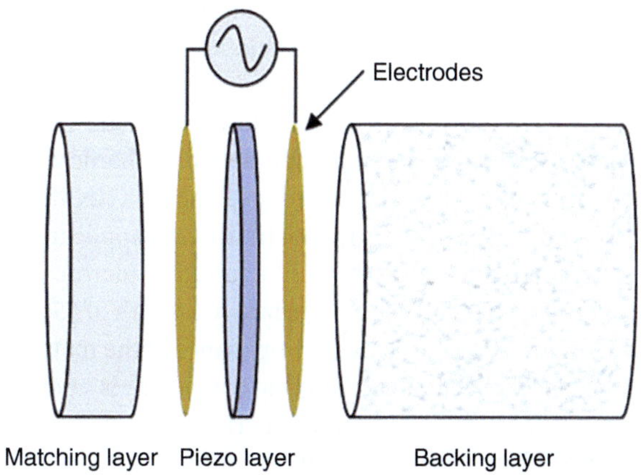

Figure 20.3 Schematic diagram of a single-element transducer.

61.5% at -6 dB. Its rabbit aorta image has a higher noise-to-signal ratio and better resolution than those of images acquired from a 40MHz PMN–PT single crystal transducer, indicating that the texture KNLN thick film is a promising material for high-frequency (>50 MHz) IVUS imaging. In comparison with pure ultrasound imaging, photoacoustic imaging, a hybrid imaging process that integrates a tiny single-element ultrasound transducer with a laser fiber, will meet the needs of a more precise diagnosis and an optimal treatment [43, 44]. In this context, Zhu et al. [45] further developed a tiny intravascular photoacoustic (IVPA) transducer for photoacoustic angiography of the atherosclerotic lesion. A $(K_{0.45}Na_{0.55})_{0.96}Li_{0.04}NbO_3$ (KNLN) single crystal with <100>-orientation synthesized by a seed-free solid-state crystal growth method was exploited as the core piezoelectric layer. The 38 MHz tiny IVPA probe possesses a 1 mm outside diameter and is combined with a 200 μm core multimode fiber. The axial resolution and lateral resolution of the IVPA probe are 60 and 220 μm, respectively. High-resolution time and frequency domain IVPA images of the human artery were achieved, indicating the feasibility of KNLN lead-free single crystals for angiography of atherosclerotic lesions.

It is generally accepted that the piezoelectricity and electromechanical coupling coefficient in the single crystal are higher than in other types of materials [36]. However, the single crystal growth process is complicated and time-consuming, which is not favorable to industrial production. On the contrary, piezoelectric ceramics are favored because of their simple preparation process and low cost. The performance of transducers based on lead-free ceramics has also achieved breakthroughs. For example, Chen et al. developed highly sensitive transducers based on a $(K,Na)NbO_3$–$KTiNbO_5$–$BaZrO_3$–Fe_2O_3–MgO (KNN–NTK–FM) lead-free piezoelectric ceramic for high-frequency ultrasonic imaging [46]. The fabricated transducer exhibits great performance with a high center frequency (52.6 MHz), a bandwidth (64.4% at −6 dB), and a very low two-way insertion loss (10.1 dB). Recently, a new KNN-based transducer with an ultra-broad bandwidth of 97% has also been reported [47]. The imaging ability of the new KNN-based transducer was evaluated by imaging the biological tissue structures of an excised porcine eyeball. The anatomical detail of the anterior portion, including the iris (I), the surface of the lens (L1), conjunctiva (C2), limbus (L2), the cornea (C1), and sclera (S), are clearly presented in the ultrasound bio-microscopy (UBM) images. Piezoelectric 1–3 composites have been developed to improve device performance because they provide many advantages over conventional monolithic piezoelectric ceramics and single crystals, including lower acoustic impedance (Z_a), lower mechanical and electric loss, and a higher electromechanical coupling coefficient (k_t) [48, 49]. Therefore, Jiang et al. further developed a KNN-based lead-free 1–3 piezocomposite for high-sensitivity ultrasonic transducer applications [40]. A modified dice-and-fill technique was used to manufacture the microscale piezocomposite, by which the ceramic pillars were miniaturized to a width of 55 μm with a kerf of 15 μm, as shown in Figure 20.4. The acoustic impedance of the material is lowered to 12.1 MRayl, and the electromechanical coupling coefficient is augmented to 0.59. The as-fabricated transducers exhibit enhanced performance with a high center frequency (16 MHz), a broad bandwidth (83%), and an ultra-low two-way insertion loss (9.8 dB), which is

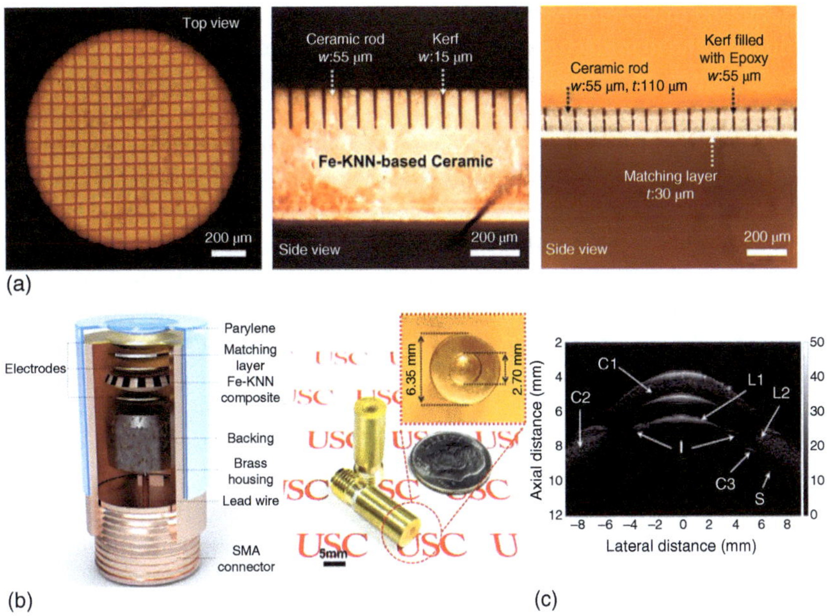

Figure 20.4 KNN 1–3 composites for ultrasound imaging transducers: (a) preparation of the KNN 1–3 composites, (b) schematic diagram and photographs of the KNN 1–3 composite transducer, (c) KNN 1–3 composite transducer for ultrasound imaging of the porcine eyeball. Source: Jiang et al. [40]/Reproduced with permission of AIP Publishing LLC.

the lowest as far as we know compared to previously reported lead-free transducers. If the sensitivity is improved, the organism can be interrogated under lower acoustic energy, and the ultrasonic biological effects can be minimized. Zheng et al. [41] also reported a lead-free composite based on a high d_{33} (~500 pC/N) KNN-based ceramic for ultrasonic imaging transducer with a center frequency of 5 MHz and a bandwidth of 81% at −6 dB, which is comparable to or much better than those of PZT-based transducers with the same center frequency.

High-frequency array transducers provide higher imaging resolution than traditional single-element transducers, thus producing finer images and resolving smaller features [50]. A 64-element 18.3 MHz lead-free linear array ultrasonic transducer based on $(K_{0.44}Na_{0.52}Li_{0.04})(Nb_{0.86}Ta_{0.1}Sb_{0.04})O_3$ (KNLNTS) piezoceramic was developed by Zhang et al. [51] for ultrasound medical imaging. Array elements with a pitch of 75 μm are interconnected by a customized flexible circuit. The active area of each array element is $3 \times 4.785\,mm^2$. The measured results demonstrated that the developed array transducer exhibited an average bandwidth of 42% at −6 dB, crosstalk between adjacent elements of less than −53 dB, an average insertion loss of 41.8 dB, and axial and lateral resolutions of 105 and 166 μm, respectively, enabling a viable alternative for medical ultrasound imaging applications. Moreover, conventional manufacturing is a significant bottleneck in the development processes of next-generation electronic products that use complex piezoelectric structures because each design adaptation requires production steps with a long lead time. Additive manufacturing (AM) or 3D printing technology has revolutionized the manufacturing world via its freedom of design, mass customization, waste minimization, ability to manufacture complex structures, and rapid prototyping [52]. Thus, Chen et al. [53] proposed mask-image-projection-based stereolithography (MIP-SLA) method by integrating a tape casting process for 3D printing focused piezoelectric elements for ultrasonic transducers to realize the energy focusing and ultrasonic sensing. A piezoelectric-composite slurry consisting of photosensitive resin and $BaTiO_3$ nanoparticles (100 nm) was used as raw materials. The printed ceramic samples exhibit a density of 5.64 g/cm^3 and were used for a 6.28 MHz ultrasonic imaging transducer to successfully visualize the tissue structure of a porcine eyeball. The same group further developed a MIP–SLA 3D printed piezoelectric array for improving spatial resolution and acoustic field in medical ultrasonic imaging [54]. The annular array consists of 5 concentric elements, and the thickness of the array is 400 μm. The area and center frequency of each working element were designed as 13.5 mm^2 and 6 MHz. Compared to the 3D-printed single-element transducer mentioned above, the printed array transducer significantly improves spatial resolution and successfully modifies the acoustic beam, laying the foundation for next-generation medical devices by putting forward a novel manufacturing concept.

20.4 Bone/Tissue Regeneration

Tissue engineering attempts to repair or replace damaged, nonfunctional organs and tissues through regeneration [55]. This technology is particularly useful for the rehabilitation of body parts damaged by trauma, congenital defects, or carcinogenic diseases, while traditional therapies are no longer effective. For tissue regeneration, scaffolds based on piezoelectric materials are an attractive strategy because they provide electrical stimulation to cells [56]. When cells are electrically stimulated, they undergo various phenotypic and genetic changes, thereby accelerating the process of tissue regeneration.

A growing number of studies have shown that endogenous electronegative potentials appear spontaneously at the site of bone defects and mediate healing. Therefore, external electric fields and applied currents can induce osteogenesis, promote consolidation of fracture calluses, and accelerate bone attachment. Among those, piezoelectric materials that can generate electricity in response to forces, especially environmentally friendly perovskite lead-free piezoelectric materials, have received increasing attention to exogenous electrical stimulation to repair bone defects [57]. As a successful example, Liu et al. [58] established a built-in electric field between electropositive ferroelectric BiFeO3 (BFO) nanofilms and electronegative bone defect walls to trigger osseointegration and bio-healing of the implant, as shown in Figure 20.5. First, the organization of crystalline panels at the atomic scale using epitaxial growth techniques has enabled the ferroelectric polarization of BFO nanofilms to match the amplitude and direction of the endogenous potential in the wall of the bone defect. Implants with BFO nanofilms with downward polarization

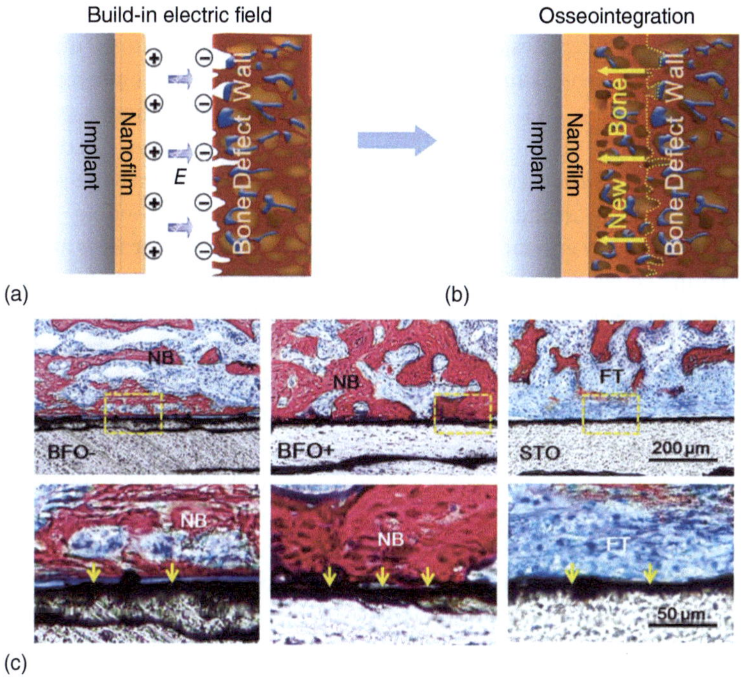

Figure 20.5 BFO nanofilms-induced enhancement of osseointegration: (a,b) illustration of built-in electric fields promoting implant osseointegration, (c) histological analysis at 2 weeks post-implantation demonstrating better osseointegration on BFO+ nanofilm implants than on BFO− and STO implants. Partially magnified images are shown below. Yellow arrows indicate the osseointegration interfaces. NB, nascent bone; FT, fibrous tissue. Source: Liu et al. [58]/Reproduced with permission of John Wiley & Sons.

(BFO+) showed rapid and superior osseointegration in rat femurs in the presence of a built-in electric field. BFO+ promoted protein adsorption and mesenchymal stem cell (MSC) attachment, spreading, and osteogenic differentiation, providing a new strategy for the development of implant surface modification techniques.

Additionally, the combination of bone grafting and barrier membranes is the classic approach to guided bone regeneration (GBR) therapy. To overcome the insufficient osteoconductivity of currently available barrier membranes, Bai et al. [59] exploited the piezoelectric properties of biomaterials to improve the osteoconductivity of barrier membranes. The method utilizes a flexible nanocomposite membrane that mimics the piezoelectric properties of natural bone as a barrier membrane. The nanocomposite membranes were made with uniformly distributed piezoelectric $BaTiO_3$ nanoparticles (BTO NPs) embedded in a poly(vinylidene fluoride) (P(VDF–TrFE)) matrix. Its efficacy in combination with allogenic grafts of deproteinized bovine bone (DBB) was investigated for the repair of critical-sized rabbit mandibular defects. Upon implantation, significantly earlier neovascularization and fully mature bone structure formation were induced within the critical size defect in the rabbit mandible, suggesting that the piezoelectric nanocomposite membranes may be suitable for improving the clinical efficacy of GBR.

In the past few decades, their applicability has expanded from bone to the regeneration of other soft tissues such as cartilage, nerves, skin, tendons, muscles, and ligaments [60–63]. For example, Wang et al. [63] reported a biocompatible self-powered piezoelectric PVDF-based hydrogel for diabetic wound repair, as shown in Figure 20.6. In this study, a self-powered piezoelectric PVA/PVDF composite hydrogel was constructed by establishing a unique freezing/thawing-solvent substitution–annealing–thawing preparation process. Hydrogen bonding in the hydrogel was significantly enhanced during the annealing–melting process and the hydrogen bonding between PVA/PVDF molecules was stronger than that between PVA molecules, facilitating the conversion of α-phase PVDF to electroactive β-phase PVDF and the formation of a crystalline structure with high cross-link density. As a result, the hydrogel achieves a significant piezoelectric response with a high piezoelectric coefficient and electrical signal output, superior stability and sensitivity, and good mechanical strength and tensile properties. In a diabetic rat wound model, the piezoelectric hydrogel not only rapidly attracts wound exudate and maintains a moist environment on the wound surface but also converts the mechanical energy generated by the rat's physical activity into electrical energy, thus providing uniform and symmetrical local piezoelectric stimulation of the wound surface in real time. Such an effect clearly promotes wound re-epithelialization and collagen deposition

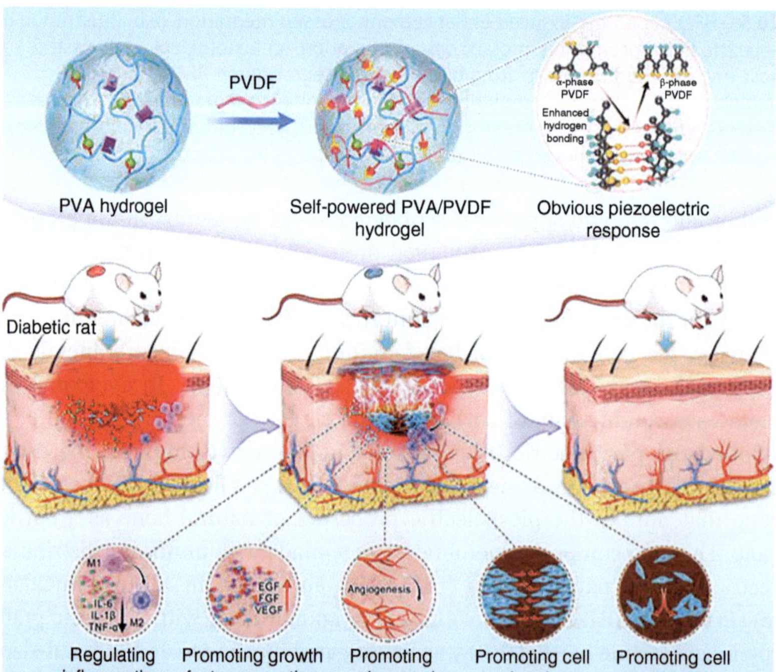

Figure 20.6 Illustration of a biocompatible self-powered piezoelectric PVA-based hydrogel for diabetic wound repair. Source: Reproduced with permission from Ref. [63]. Copyright 2017, ACS.

increases angiogenesis and growth factor secretion in wound tissue, providing a new therapeutic strategy for chronic wound healing.

The development of high-performance piezoelectric biomaterials that act as both biodegradable tissue scaffolds and controlled electrical stimulators remains a great challenge. Therefore, in another application case, Chen et al. [64] constructed a biodegradable, high-performance 3D piezoelectric scaffold with ultrasound-driven wireless electrical-stimulation capability and demonstrated its successful application in a rat model for repairing spinal cord injury. The 3D multichannel piezoelectric scaffold was prepared by combining electrospun polylactic acid (PLA) nanofibers with biodegradable high-performance piezoelectric potassium sodium niobate ($K_{0.5}Na_{0.5}NbO_3$, KNN) nanowires. Through programmed ultrasound irradiation as remote mechanical stimulation, the 3D piezoelectric scaffold can deliver electrical stimulation *in vivo* on demand with an adjustable timeline, duration, and intensity. Under appropriate ultrasound stimulation, 3D tissue scaffolds made from piezoelectric composite nanofibers can accelerate recovery of motor function and enhance the repair of spinal cord injury. Immunofluorescence investigations demonstrated that the 3D piezoelectric scaffold combined with ultrasound-driven *in vivo* electrical stimulation promoted neural stem cell differentiation and endogenous angiogenesis at the lesion site, highlighting the potential application of biodegradable, high-performance piezoelectric scaffolds to provide ultrasound-driven on-demand electrical cues in regenerative medicine.

The variety of piezoelectric materials for bone and soft tissue engineering demonstrates the importance of piezoelectric smart materials in regenerative medicine. Piezoelectric smart materials are therefore the best possible future materials for regenerative medicine.

20.5 Neural Stimulation

Direct electrical stimulations, as an artificial stimulus/modulation of the excitable nervous systems, have been widely applied in clinical practice and laboratory studies for a wide range of pathological conditions [65, 66]. Additionally, precise electrical nerve stimulation is a means of promoting neuronal regeneration and is a promising solution for patients with neurotrauma and neurodegenerative diseases. Traditional invasive methods based on electrodes and wires are often limited by the restricted accessibility of inner sites, and several innovative techniques are being explored in order to address this challenge. The possibility to achieve direct electrical stimulation by means of piezoelectric materials as novel interfaces with biological systems is of great interest to all biomedical research and has emerged in the latest decade as a most promising tool in many remote electrical stimulation/modulation bio-systems [67, 68], such as the ultrasound-induced wireless energy systems that use propagating ultrasound waves to carry the available energy and magneto-piezoelectric coupling [16, 69, 70].

Barium titanate nanoparticles (BTNPs) as ultrasound-sensitive nano-transducers for the wireless neuronal stimulation have been proposed by Marino et al. [71]

to enhance the SH–SY5Y-derived neurite outgrowth and neural differentiation. The first direct biological proof of cell activation was provided in response to ultrasound-induced piezoelectric stimulation. BTNPs were stabilized with gum Arabic and were found to be strongly associated with neuron membranes after a 24-h treatment, both at the level of cell bodies and their neurites. The effects of acute US-induced piezoelectric stimulation on ion fluxes were then investigated, illustrating that the synergic ultrasound + BTNPs stimulation was able to evoke high-amplitude Ca^{2+} and Na^{+} peaks in SH–SY5Y-derived neurons, resulting in a significant effect in promoting the maturation of neural networks, particularly by regulating the growth of neurites. The ion transient amplitudes were modulated by tuning the power intensity of the ultrasound and were sensitive to both tetrodotoxin (TTX) and Cd^{2+} blockers, thus indicating the involvement of voltage-gated ion channels in mediating the response to piezoelectric stimulation. Interestingly, ultrasound-induced stimulation with non-piezoelectric BTNPs did not elicit any significant cell response in terms of ion fluxes, highlighting the role of piezoelectrics in converting the pressure of wireless ultrasound into electrical stimulation.

In another application of stimulus, Liu et al. [69] proposed a wireless manipulation of magnetic/piezoelectric micromotors for precise neural stem-like cell stimulation, as shown in Figure 20.7. In this study, wirelessly controllable targeted motion and precise stimulation at the single-cell level using helical-shaped magnetic powered micromotors are successfully demonstrated for the first time. Such a micromotor was prepared by sequentially coating the *Spirulina platensis* with Fe_3O_4 and $BaTiO_3$ (*S.platensis*@Fe_3O_4@t$BaTiO_3$) nanoparticles through electrostatic adsorption. The micromotor system can be navigated in a highly controllable manner at low

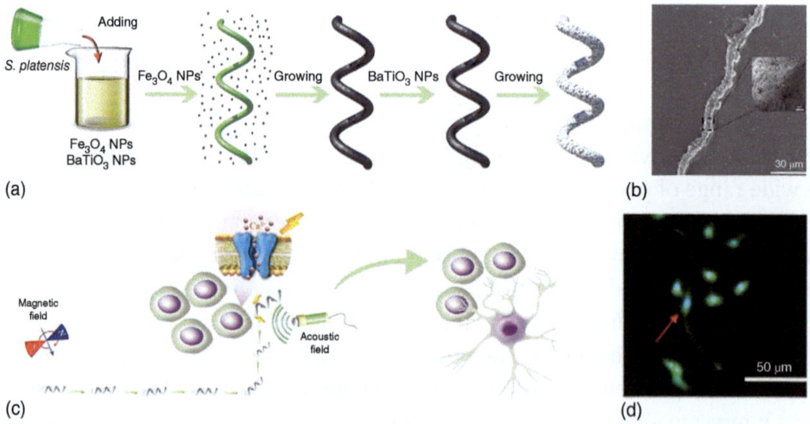

Figure 20.7 Wireless manipulation of magnetic/piezoelectric micromotors for precise neural stem-like cell stimulation: (a) schematic illustration of the preparation of *S. platensis*@Fe_3O_4@t$BaTiO_3$ micromotor, (b) SEM image of *S. platensis*@Fe_3O_4@t$BaTiO_3$, (c) schematic illustration of a highly controllable micromotor to induce the differentiation of the targeted neural stem-like cell, (d) fluorescent image of differentiated PC12 cells stimulated by *S. platensis*@Fe_3O_4@t$BaTiO_3$ with ultrasound. Source: Liu et al. [69]/Reproduced with permission of John Wiley & Sons.

intensities of rotating magnetic fields. In particular, the developed system enables single-cell directed motion. The ultrasonic energy is then converted into an *in situ* electrical signal by means of a piezoelectric effect to induce precise differentiation of the target neural stem-like cells. Given the high biocompatibility and controllability of this promising platform, this approach could lead to exciting opportunities in a wide range of precision neural regeneration applications.

Recently, a growing number of studies have shown that it is technically feasible to replace photoreceptor function with electronic visual prostheses [72]. The phenomenon of electrically excited phosphorescence has been reliably demonstrated in early studies of a prototype retinal implant. A typical retinal prosthesis system primarily consists of an implantable stimulator and an external device, which are connected to each other by a wired or wireless connection. A camera in the external unit captures the image and converts it into digital data. The command signal is appropriately encoded and transmitted to the implantable device, which receives

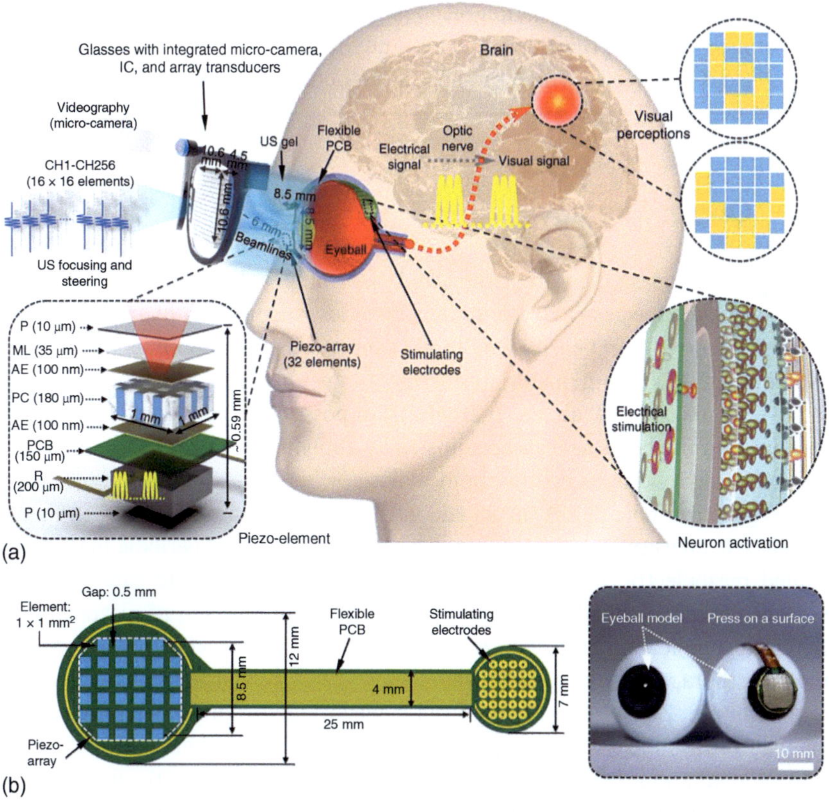

Figure 20.8 Flexible ultrasound-induced retinal stimulating piezo-arrays for biomimetic visual prostheses: (a) schematic diagram showing the design and working principle the F-URSP for biomimetic visual prostheses, (b) schematic diagram and photograph showing the structure and dimension of the F-URSP. Source: Jiang et al. [35]/Reproduced from Springer Nature/CC BY 4.0.

the signal, recovers the power and data from the signal, and sets the parameters of the stimulator. The stimulation pattern is then applied to the retina via an array of electrodes that dock directly to the retinal surface. In the context of this retinal prosthesis, a power supply and data transmission platform capable of providing sufficient power for neuronal responses within safe limits while ensuring the portability, longevity, and cytocompatibility of the device is required. Therefore, Jiang et al. [35] proposed a complex flexible ultrasound-induced retinal stimulation piezoelectric array (F-URSP) that can provide prosthetic eyes for people with acquired blindness, as shown in Figure 20.8. The device integrates a 2D piezoelectric array and a stimulating electrode array with 32 pixels on a flexible printed circuit board (PCB). An ultrasound field with spatial acoustic power distribution can be appropriately programmed and transmitted wirelessly via an external transducer (a 2D array). The emitted ultrasound power will be received by the piezoelectric array and then converted into electrical energy to initiate a neural response in the cell populations of the retina. Spatially complex stimulation patterns can be applied to the retina by stimulating the electrode arrays. Continuous acoustic field-induced pattern reconstruction and electrical stimulation responses in mouse retinal tissue *in vitro* were demonstrated by using the F-URSP. This study lays the foundation for the piezoelectric visual prosthesis, which provides a method for creating artificial vision in the blind.

20.6 Cancer Therapy

Cancers are increasing in frequency and becoming a leading cause of death. Reactive oxygen species (ROS), such as peroxides, superoxide, hydroxyl radicals, and singlet oxygen, can be exploited to kill cancer cells [73], and thus the selective generation of ROS from material surfaces is an emerging direction in cancer therapy. As mentioned above, piezoelectrics possess spontaneous polarization potential, and the potential can be adjusted by applying stress to the materials due to piezoelectricity. Analogous to photocatalysis, where the photo-induced electric charges (electron-hole pairs) participate in the catalytic redox reactions, in theory, piezoelectric charges (positive and negative ones) induced by periodic mechanical vibrated piezoelectric particles could also be applied to activate catalytic redox reactions, which is called piezo-catalysis. [74, 75] The research of piezo-catalysis extends the application field of piezoelectric materials and provides a novel strategy for cancer therapy [76].

For example, Zhu et al. [77] reported a piezocatalytic tumor therapy schema that uses ultrasound as a microscopic pressure resource to generate ROS under catalytic mediation by piezoelectric tetragonal $BaTiO_3$ (T-BTO), as shown in Figure 20.9. Under ultrasonic vibrations, electrons, and holes are unpaired and they are piezoelectrically separated, leading to the creation of a built-in electric field that subsequently catalyzes the *in situ* generation of ROS, such as toxic hydroxyl ($^{\bullet}OH$) and superoxide radicals ($^{\bullet}O_2^-$), to eradicate tumors. This approach shows intriguing advantages, such as a more stable sensitizer and dynamic

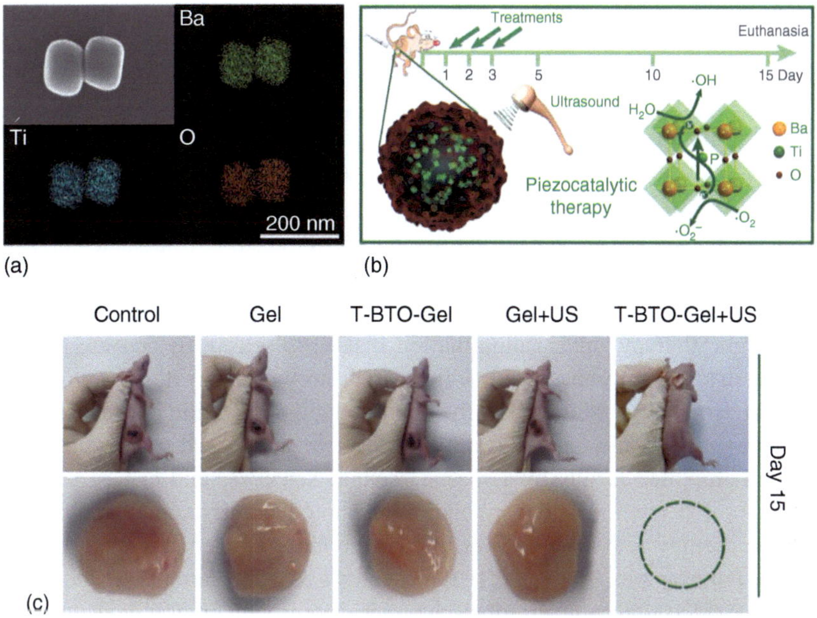

Figure 20.9 Piezocatalytic tumor therapy by ultrasound-triggered and BaTiO3-mediated piezoelectricity: (a) STEM image and the corresponding elemental mapping images of T-BTO nanoparticles, (b) schematic illustration of piezocatalytic therapy *in vivo*, (c) digital photos of 4T1-tumor-bearing mice and their tumor regions after various treatments. Source: Zhu et al. [77]/Reproduced with permission of John Wiley & Sons.

control of the outcome of the redox reaction. Furthermore, detailed *in vitro* cellular-level and *in vivo* tumor xenograft evaluations have shown that injectable T-BTO-nanoparticle-embedded thermosensitive hydrogels will substantially induce ultrasound irradiation-triggered cytotoxicity and pressure-sensitive catalytic tumor eradication, accompanied by high therapeutic biosafety *in vivo*. Meanwhile, Zhao et al. [78] fabricated $Cu_{2-x}O$–$BaTiO_3$ piezoelectric heterostructures as acoustic sensitizers and chemo dynamic agents for both improved ROS generation and cancer treatment efficacy. Under ultrasonic irradiation, $Cu_{2-x}O$–$BaTiO_3$ heterojunctions with piezoelectric effects exhibit high-performance singlet oxygen ($^1O_2$) and hydroxyl radical ($^{\cdot}OH$) production to sonodynamic therapy.

In another study, Wang et al. [79] reported a synergistic chemotherapy–piezoelectric therapy via defectively engineered lead-free piezoelectric (K,Na)NbO_3 (KNN), which generates ROS to effectively circumvent osteosarcoma. Osteosarcoma is a lethal malignancy of the bone, jeopardizing the life of an enormous population worldwide. In this work, tumor growth was almost halted after treatment with KNN due to oxygen vacancy-driven free radical catalysis (i.e. chemokinetic therapy), and these effects were enhanced by the introduction of ultrasound-triggered piezoelectric catalysis (piezoelectric therapy) to make the ultrasound vibrate. In addition, KNN was found to induce apoptosis and autophagy in osteosarcoma cells and showed good histocompatibility on evaluation in mouse models, with no killing of normal cells and no toxic effects on organs *in vivo*. Both *in vitro* cellular level assessment and *in vivo* osteosarcoma xenograft assessment showed that injectable KNN nanocrystals induce cytotoxicity and tumor eradication through a synergistic chemotherapy-compression catalytic effect. These present piezoelectric treatment modalities integrate piezoelectricity, ultrasound, and tumor therapy and provide a novel and promising strategy for noninvasive flanking tumor treatment in nanocatalytic medicine based on the designed piezoelectric nanoplatform.

Additionally, a number of new therapies have attracted considerable interest in the treatment of many cancer indications due to their safe modalities, such as the combination of photodynamic therapy and acoustodynamic therapy. In photodynamic therapy, optical illumination is used to selectively activate photosensitizing drugs (photosensitizers), which can destroy tumors without significant side effects. However, it is limited by the depth of light penetration in the body and tissues. Acoustodynamic therapy is a noninvasive therapy based on low-intensity ultrasound amplified locally by a special acoustic sensitizer and can be used to treat deep-seated tumors. The combination of photodynamic therapy and acoustodynamic therapy provides a powerful tool for more efficient and safe cancer treatment. For example, Guan et al. [80] presented a new implantable self-powered therapeutic pellet for wireless photodynamic–acoustodynamic hybrid therapy to inhibit cancer recurrence and allow tumors to regress. The pellet contains an integrated piezoelectric self-powered device, LED illuminator, and control circuitry. Its miniaturization allows it to be implanted directly into the target site, and the piezoelectric self-powered device captures ultrasound energy through a piezoelectric effect. Thus, a therapeutic dose of light can be delivered to the tissue as photodynamic therapy together with direct acoustodynamic therapy. This photodynamic–acoustodynamic hybrid therapy has shown powerful results in preclinical mouse models for the treatment of breast cancer (superior to single photodynamic therapy or acoustodynamic therapy) and could be further developed for clinical applications in humans. Furthermore, further studies of this implantable self-powered therapeutic particle for wireless photodynamic–acoustodynamic hybrid therapy *in vitro* and *in vivo* could elucidate the key interactions between existing clinical cancer therapies and future precision therapies.

20.7 Summary, Challenges, and Outlook

This chapter focuses on some of the applications of piezoelectric materials in biomedicine. These technologies have been used in a range of diagnostic, monitoring, and therapeutic applications, including wireless power for bio-implantable devices, ultrasound imaging transducers, direct cellular/neural electrical stimulation, tissue regeneration, and tumor therapy. Piezoelectricity is an area of interest to researchers around the world as a safe and effective medical strategy. Advances in nanotechnology, microelectronics, and biomedical systems are revolutionizing this technology. Their synergies can be greatly enhanced and can play an important role in a range of emerging applications. The future will continue to evolve from the perspectives of materials strategies, manufacturing technologies, design considerations, and functional applications.

First, new materials with high electromechanical coupling capability are developed. Currently, piezoelectric materials are limited to the traditional ZnO, PZT, $BaTiO_3$, and PVDF. In order to improve the conversion efficiency and extend the lifetime of the devices, there is a demand for new materials with enhanced properties and new structures, such as piezoelectric metamaterials [81]. In addition, with the development of bio-implantable devices, new materials with biocompatibility and even biodegradability, such as high-performance lead-free piezoelectric materials and bio-piezoelectric materials [5, 39], are also needed for the next generation of piezoelectric devices.

Second, in order to develop new types of piezoelectric devices, new manufacturing techniques, and structural designs will be required for specific applications. For example, 3D printing technology has been widely used to customize designs to create complex structures and has facilitated the fabrication of miniaturized and complex piezoelectric ceramic structures to achieve desired performance [82]. Furthermore, in terms of specific structural designs, applications on curved surfaces (human organs, tissues, etc.) require the development of devices with high tensile properties and long lifetimes. Currently, stretchable devices made from piezoelectric polymers have the disadvantage of low performance, while ceramic-based stretchable devices have the disadvantage of low stretchability. Therefore, new flexible and stretchable devices need to be designed to meet the needs of wearable and implantable devices. In addition, some bio-implantable devices need to change shape after implantation. For example, origami-based devices may be suitable for such applications, being small and easy to implant and changing their geometry after implantation to achieve high performance. Some bio-inspired designs would also be interesting and might guide the design of piezoelectric devices to achieve high performance.

Third, the translation of current piezoelectric technology into clinical medicine needs to be accelerated in order to facilitate the development of medical technology. The maturity of the piezoelectric technologies described in this section has increased considerably over the last few years to the extent that many therapeutic systems are now commercially available as research tools. Other systems are at an advanced stage of commercial development and are available not only for use in animal models and cellular assays but also for applications in human health monitoring and therapy applications.

In summary, we hope that this chapter will provide a review of current developments in piezoelectricity for biological applications for further research and inspire better design of systems for piezoelectricity-related applications. High-performance, autonomous, and even biodegradable electronics and nano/micro bio-robotics combined with piezoelectricity are the next generation of devices that are rapidly evolving. A suitable system that is structurally and functionally compatible with these systems is extremely important. The integration of these functions can broaden their practical functionality. In addition, evaluation criteria for medically relevant standards are necessary in order to qualitatively and quantitatively assess the applications of piezoelectricity and their potential for practical applications and commercialization.

References

1 Deng, W., Zhou, Y., Libanori, A. et al. (2022). Piezoelectric nanogenerators for personalized healthcare. *Chem. Soc. Rev.* 51 (9): 3380–3435.

2 Won, S.M., Cai, L., Gutruf, P., and Rogers, J.A. (2021). Wireless and battery-free technologies for neuroengineering. *Nat. Biomed. Eng.* 7: 405–413.

3 Sezer, N. and Koc, M. (2021). A comprehensive review on the state-of-the-art of piezoelectric energy harvesting. *Nano Energy* 80: 105567.

4 Lin, M., Hu, H., Zhou, S., and Xu, S. (2022). Soft wearable devices for deep-tissue sensing. *Nat. Rev. Mater.* 7 (11): 850–869.

5 Xu, Q., Gao, X., Zhao, S. et al. (2021). Construction of bio-piezoelectric platforms: from structures and synthesis to applications. *Adv. Mater.* 33 (27): 2008452.

6 Bazaka, K. and Jacob, M.V. (2013). Implantable devices: issues and challenges. *Electronics* 2 (1): 1–34.

7 Hannan, M.A., Mutashar, S., Samad, S.A., and Hussain, A. (2014). Energy harvesting for the implantable biomedical devices: issues and challenges. *Biomed. Eng. Online* 13 (1): 79.

8 Erturk, A. and Inman, D.J. (2011). *Piezoelectric Energy Harvesting*. John Wiley & Sons.

9 Toprak, A. and Tigli, O. (2014). Piezoelectric energy harvesting: state-of-the-art and challenges. *Appl. Phys. Rev.* 1 (3): 031104.

10 Yang, Y., Hu, H., Chen, Z. et al. (2020). Stretchable nanolayered thermoelectric energy harvester on complex and dynamic surfaces. *Nano Lett.* 20: 4445–4453.

11 Wang, Z.L. and Song, J. (2006). Piezoelectric nanogenerators based on zinc oxide nanowire arrays. *Science* 312 (5771): 242–246.

12 Yang, Z., Zhou, S., Zu, J., and Inman, D. (2018). High-performance piezoelectric energy harvesters and their applications. *Joule* 2 (4): 642–697.

13 Hehn, T. and Manoli, Y. (2015). *CMOS Circuits for Piezoelectric Energy Harvesters. Springer Series in Advanced Microelectronics*, vol. 38, 21–40. Springer.

14 Huan, Y., Zhang, X., Song, J. et al. (2018). High-performance piezoelectric composite nanogenerator based on Ag/(K, Na) NbO3 heterostructure. *Nano Energy* 50: 62–69.

15 Jeong, C.K., Han, J.H., Palneedi, H. et al. (2017). Comprehensive biocompatibility of nontoxic and high-output flexible energy harvester using lead-free piezoceramic thin film. *APL Mater.* 5 (7): 074102.

16 Jiang, L., Yang, Y., Chen, R. et al. (2019). Ultrasound-induced wireless energy harvesting for potential retinal electrical stimulation application. *Adv. Funct. Mater.* 29 (33): 1902522.

17 Park, K.-I., Xu, S., Liu, Y. et al. (2010). Piezoelectric BaTiO3 thin film nanogenerator on plastic substrates. *Nano Lett.* 10 (12): 4939–4943.

18 Li, N., Yi, Z., Ma, Y. et al. (2019). Direct powering a real cardiac pacemaker by natural energy of a heartbeat. *ACS Nano* 13 (3): 2822–2830.

19 Cheng, X., Xue, X., Ma, Y. et al. (2016). Implantable and self-powered blood pressure monitoring based on a piezoelectric thinfilm: simulated, in vitro and in vivo studies. *Nano Energy* 22: 453–460.

20 Nagayama, Y., Kondo, Y., Koshiba, Y. et al. (2022). Multipoint detection of structural deformation of pulsating 3D heart model using flexible organic piezoelectric-sensor array. *Jpn. J. Appl. Phys.* 61: https://doi.org/10.35848/1347-4065/ac5a04 (SE).

21 Zhang, G., Zhao, P., Zhang, X. et al. (2018). Flexible three-dimensional interconnected piezoelectric ceramic foam based composites for highly efficient concurrent mechanical and thermal energy harvesting. *Energy Environ. Sci.* 11 (8): 2046–2056.

22 Azimi, S., Golabchi, A., Nekookar, A. et al. (2021). Self-powered cardiac pacemaker by piezoelectric polymer nanogenerator implant. *Nano Energy* 83: 105781.

23 Yang, F., Li, J., Long, Y. et al. (2021). Wafer-scale heterostructured piezoelectric bio-organic thin films. *Science* 373 (6552): 337–342.

24 Jiang, L., Yang, Y., Chen, Y., and Zhou, Q. (2020). Ultrasound-induced wireless energy harvesting: from materials strategies to functional applications. *Nano Energy* 77: 105131.

25 Wang, X., Song, J., Liu, J., and Wang, Z.L. (2007). Direct-current nanogenerator driven by ultrasonic waves. *Science* 316 (5821): 102–105.

26 Sonmezoglu, S., Fineman, J.R., Maltepe, E., and Maharbiz, M.M. (2021). Monitoring deep-tissue oxygenation with a millimeter-scale ultrasonic implant. *Nat. Biotechnol.* 39 (7): 855–864.

27 Shi, C., Andino-Pavlovsky, V., Lee, S.A. et al. (2021). Application of a sub-0.1-mm^3 implantable mote for in vivo real-time wireless temperature sensing. *Sci. Adv.* 7 (19): eabf6312.

28 Piech, D.K., Johnson, B.C., Shen, K. et al. (2020). A wireless millimetre-scale implantable neural stimulator with ultrasonically powered bidirectional communication. *Nat. Biomed. Eng.* 4 (2): 207–222.

29 Jin, P., Fu, J., Wang, F. et al. (2021). A flexible, stretchable system for simultaneous acoustic energy transfer and communication. *Sci. Adv.* 7 (40): eabg2507.

30 Zhang, T., Liang, H., Wang, Z. et al. (2022). Piezoelectric ultrasound energy-harvesting device for deep brain stimulation and analgesia applications. *Sci. Adv.* 8 (15): https://doi.org/10.1126/sciadv.abk0159.

31 Jiang, L.M., Lu, G.X., Yang, Y. et al. (2021). Photoacoustic and piezo-ultrasound hybrid-induced energy transfer for 3D twining wireless multifunctional implants. *Energy Environ. Sci.* 14 (3): 1490–1505.

32 Jiang, L., Xue, H., Li, R., and Wu, J. (2022). A KNN composite-based piezoelectric helix for ultrasonic transcutaneous energy harvesting. *Appl. Phys. Lett.* 120 (23): 233504.

33 Jiang, L., Wu, B., Wei, X. et al. (2022). Flexible lead-free piezoelectric arrays for high-efficiency wireless ultrasonic energy transfer and communication. *Mater. Horiz.* 9: 2180–2190.

34 Seo, D., Neely, R.M., Shen, K. et al. (2016). Wireless recording in the peripheral nervous system with ultrasonic neural dust. *Neuron* 91 (3): 529–539.

35 Jiang, L., Lu, G., Zeng, Y. et al. (2022). Flexible ultrasound-induced retinal stimulating piezo-arrays for biomimetic visual prostheses. *Nat. Commun.* 13: 3853.

36 Zhou, Q., Lam, K.H., Zheng, H. et al. (2014). Piezoelectric single crystal ultrasonic transducers for biomedical applications. *Prog. Mater. Sci.* 66: 87–111.

37 Hu, H., Zhu, X., Wang, C. et al. (2018). Stretchable ultrasonic transducer arrays for three-dimensional imaging on complex surfaces. *Sci. Adv.* 4 (3): eaar3979.

38 Garroni, S., Senes, N., Iacomini, A. et al. (2018). Advanced synthesis on lead-free kxNa (1−x) NbO_3 piezoceramics for medical imaging applications. *Phys. Status Solidi A* 215 (16): 1700896.

39 Zheng, T., Wu, J., Xiao, D., and Zhu, J. (2018). Recent development in lead-free perovskite piezoelectric bulk materials. *Prog. Mater. Sci.* 98: 552–624.

40 Jiang, L., Chen, R., Xing, J. et al. (2019). Fabrication of a (K, Na) NbO3-based lead-free 1-3 piezocomposite for high-sensitivity ultrasonic transducers application. *J. Appl. Phys.* 125 (21): 214501.

41 Zheng, T., Zhang, Y., Ke, Q. et al. (2020). High-performance potassium sodium niobate piezoceramics for ultrasonic transducer. *Nano Energy* 70: 104559.

42 Zhu, B., Zhang, Z., Ma, T. et al. (2015). (100)-Textured KNN-based thick film with enhanced piezoelectric property for intravascular ultrasound imaging. *Appl. Phys. Lett.* 106 (17): 173504.

43 Xu, M. and Wang, L.V. (2006). Photoacoustic imaging in biomedicine. *Rev. Sci. Instrum.* 77 (4): 041101.

44 Wang, L.V. (2017). *Photoacoustic Imaging and Spectroscopy*. CRC Press.

45 Zhu, B., Zhu, Y., Yang, J. et al. (2016). New potassium sodium niobate single crystal with thickness-independent high-performance for photoacoustic angiography of atherosclerotic lesion. *Sci. Rep.* 6: 39679.

46 Chen, R., Jiang, L., Zhang, T. et al. (2018). Eco-friendly highly sensitive transducers based on a new KNN–NTK–FM lead-free piezoelectric ceramic for high-frequency biomedical ultrasonic imaging applications. *IEEE Trans. Biomed. Eng.* 66 (6): 1580–1587.

47 Xing, J., Jiang, L., Zhao, C. et al. (2020). Potassium sodium niobate based lead-free ceramic for high-frequency ultrasound transducer applications. *J. Materiomics* 6 (3): 513–522.

48 Jiang, L., Yang, Y., Chen, R. et al. (2019). Flexible piezoelectric ultrasonic energy harvester array for bio-implantable wireless generator. *Nano Energy* 56: 216–224.

49 Shen, Z.-Y., Xu, Y., and Li, J.-F. (2009). Fabrication and electromechanical properties of microscale 1–3-type piezoelectric composites using (Na, K) NbO_3-based Pb-free piezoceramics. *J. Appl. Phys.* 105 (10): 104103.

50 Drinkwater, B.W. and Wilcox, P.D. (2006). Ultrasonic arrays for non-destructive evaluation: a review. *NDT Int.* 39 (7): 525–541.

51 Zhang, Z., Chen, R., Wang, B. et al. (2018). Development of a KNN ceramic-based lead-free linear array ultrasonic transducer. *IEEE Trans. Ultrason. Ferroelectr. Freq. Control* 65 (11): 2113–2120.

52 Ngo, T.D., Kashani, A., Imbalzano, G. et al. (2018). Additive manufacturing (3D printing): a review of materials, methods, applications and challenges. *Compos. Part B* 143: 172–196.

53 Chen, Z., Song, X., Lei, L. et al. (2016). 3D printing of piezoelectric element for energy focusing and ultrasonic sensing. *Nano Energy* 27: 78–86.

54 Chen, Z., Qian, X., Song, X. et al. (2019). Three-dimensional printed piezoelectric array for improving acoustic field and spatial resolution in medical ultrasonic imaging. *Micromachines* 10 (3): 170.

55 Lanza, R., Langer, R., Vacanti, J.P., and Atala, A. (2020). *Principles of Tissue Engineering*. Academic Press.

56 Hollister, S.J. (2005). Porous scaffold design for tissue engineering. *Nat. Mater.* 4 (7): 518–524.

57 Tandon, B., Blaker, J.J., and Cartmell, S.H. (2018). Piezoelectric materials as stimulatory biomedical materials and scaffolds for bone repair. *Acta Biomater.* 73: 1–20.

58 Liu, Y., Zhang, X., Cao, C. et al. (2017). Built-in electric fields dramatically induce enhancement of osseointegration. *Adv. Funct. Mater.* 27 (47): 1703771.

59 Bai, Y., Dai, X., Yin, Y. et al. (2019). Biomimetic piezoelectric nanocomposite membranes synergistically enhance osteogenesis of deproteinized bovine bone grafts. *Int. J. Nanomedicine* 14: 3015–3026.

60 Kapat, K., Shubhra, Q.T., Zhou, M., and Leeuwenburgh, S. (2020). Piezoelectric nano-biomaterials for biomedicine and tissue regeneration. *Adv. Funct. Mater.* 30: 1909045.

61 Rajabi, A.H., Jaffe, M., and Arinzeh, T.L. (2015). Piezoelectric materials for tissue regeneration: a review. *Acta Biomater.* 24: 12–23.

62 More, N. and Kapusetti, G. (2017). Piezoelectric material–a promising approach for bone and cartilage regeneration. *Med. Hypotheses* 108: 10–16.

63 Wang, L., Yu, Y., Zhao, X. et al. (2022). A biocompatible self-powered piezoelectric poly(vinyl alcohol)-based hydrogel for diabetic wound repair. *ACS Appl. Mater. Interfaces* 14: –41.

64 Chen, P., Xu, C., Wu, P. et al. (2022). Wirelessly powered electrical-stimulation based on biodegradable 3D piezoelectric scaffolds promotes the spinal cord injury. *ACS Nano* 16: 16513–16528.

65 Borchers, S., Himmelbach, M., Logothetis, N., and Karnath, H.-O. (2012). Direct electrical stimulation of human cortex—the gold standard for mapping brain functions? *Nat. Rev. Neurosci.* 13 (1): 63–70.

66 Jacobs, J., Miller, J., Lee, S.A. et al. (2016). Direct electrical stimulation of the human entorhinal region and hippocampus impairs memory. *Neuron* 92 (5): 983–990.

67 Marino, A., Genchi, G.G., Sinibaldi, E., and Ciofani, G. (2017). Piezoelectric effects of materials on bio-interfaces. *ACS Appl. Mater. Interfaces* 9 (21): 17663–17680.

68 Genchi, G.G., Ceseracciu, L., Marino, A. et al. (2016). P (VDF-TrFE)/$BaTiO_3$ nanoparticle composite films mediate piezoelectric stimulation and promote differentiation of SH-SY5Y neuroblastoma cells. *Adv. Healthc. Mater.* 5 (14): 1808–1820.

69 Liu, L., Chen, B., Liu, K. et al. (2020). Wireless manipulation of magnetic/piezoelectric micromotors for precise neural stem-like cell stimulation. *Adv. Funct. Mater.* 30 (11): 1910108.

70 Yue, K., Guduru, R., Hong, J. et al. (2012). Magneto-electric nano-particles for non-invasive brain stimulation. *PLoS One* 7 (9): e44040.

71 Marino, A., Arai, S., Hou, Y. et al. (2015). Piezoelectric nanoparticle-assisted wireless neuronal stimulation. *ACS Nano* 9 (7): 7678–7689.

72 Suaning, G.J., Lovell, N.H., Schindhelm, K., and Coroneo, M.T. (1998). The bionic eye (electronic visual prosthesis): a review. *Aust. N. Z. J. Ophthalmol.* 26 (3): 195–202.

73 Sena, L.A. and Chandel, N.S. (2012). Physiological roles of mitochondrial reactive oxygen species. *Mol. Cell* 48 (2): 158–167.

74 Mushtaq, F., Chen, X., Hoop, M. et al. (2018). Piezoelectrically enhanced photocatalysis with $BiFeO_3$ nanostructures for efficient water remediation. *iScience* 4: 236–246.

75 You, H., Wu, Z., Zhang, L. et al. (2019). Harvesting the vibration energy of $BiFeO_3$ nanosheets for hydrogen evolution. *Angew. Chem. Int. Ed.* 58: 11779–11784.

76 Wang, Y., Wen, X., Jia, Y. et al. (2020). Piezo-catalysis for nondestructive tooth whitening. *Nat. Commun.* 11 (1): 1–11.

77 Zhu, P., Chen, Y., and Shi, J. (2020). Piezocatalytic tumor therapy by ultrasound-triggered and $BaTiO_3$-mediated piezoelectricity. *Adv. Mater.* 32 (29): 2001976.

78 Zhao, Y., Wang, S., Ding, Y. et al. (2022). Piezotronic effect-augmented $Cu_{2-x}O$–$BaTiO_3$ sonosensitizers for multifunctional cancer dynamic therapy. *ACS Nano* 16 (6): 9304–9316.

79 Wang, S., Chen, C., Wang, J. et al. (2022). Synergetic chemo-piezodynamic therapy of osteosarcoma enabled by defect-driven lead-free piezoelectrics. *Adv. Funct. Mater.* 32 (44): 2208128.

80 Guan, H., Zou, P., Lin, R. et al. (2023). Implantable self-powered therapeutic pellet for wireless photodynamic/sonodynamic hybrid therapy of cancer recurrence inhibition and tumor regression. *Nano Energy* 105: 108002.

81 Ji, G. and Huber, J. (2022). Recent progress in acoustic metamaterials and active piezoelectric acoustic metamaterials—a review. *Appl. Mater. Today* 26: 101260.

82 Zeng, Y., Jiang, L., He, Q. et al. (2022). Recent progress in 3D printing piezoelectric materials for biomedical applications. *J. Phys. D. Appl. Phys.* 55 (1): 013002.

Index

Piezoelectric Materials: From Fundamentals to Emerging Applications, First Edition. Edited by Jiagang Wu.

m

t

u

V

W

Z